AN ILLUSTRATED LABORATORY TEXT IN

ZOOLOGY

THIRD EDITION

AN ILLUSTRATED LABORATORY TEXT IN

ZOOLOGY

THIRD EDITION

RICHARD A. BOOLOOTIAN

Science Software Systems, Inc., formerly of University of California, Los Angeles

DONALD HEYNEMAN

Hooper Foundation, University of California, San Francisco

HOLT, RINEHART AND WINSTON
New York   Chicago   San Francisco   Atlanta   Dallas   Montreal   Toronto   London   Sydney

PREFACE

To the Student

Students in Introductory Zoology often complain that the lecture and the lab are two different courses. Actually, this is seldom true. The integration of seemingly diverse topics, however, such as the cellular details of a jellyfish and the evolutionary importance of the phylum it belongs to, is often difficult. Nor is it easy to relate *function*, as discussed in lecture, with *structure*, as seen in the laboratory. Yet a degree of understanding of the pickled frog on your dissection tray is basic to an understanding of the interrelationships of all living things, and it is your responsibility to build a conceptual bridge between the two. To do so, you should realize that however presented, the lectures and laboratory are two expressions of one living system.

This *laboratory textbook* is designed to help you to erect such a bridge. It will introduce much new information to you, but more important, we hope it will help you develop the ability to relate the specifics in this text and your lab with the concepts they illustrate. We want to assist you to formulate the connection between general lectures and specific dissections and exercises. Each set of laboratory instructions is preceded by an outline and an introductory discussion of the general animal group under study. This is intended to enable you to recognize the whole and not become lost in the welter of parts. After a bit, the facts and details will begin to make sense, enabling you to begin to appreciate the biological principles that underlie and govern the integrated complex of all living things in our environment. This text will be successful if it can help you to go beyond the initial stage of memory, to one of exploration, and, finally, to the development of a sense of correlation and understanding.

Not all the material pertinent to a course in introductory zoology can be laboratory-oriented or presented in a book of this kind. Your general textbook will cover in greater detail those topics that fall more logically into theoretical aspects of zoology. But often direct observation and introductory experiments such as are detailed here can yield information basic to the larger concepts.

The proper study of an animal begins with observing it in its natural habitat. These observations may be tested and varied, followed by analytic studies, physiological, biochemical, and biophysical investigations of increasing sophistication under the more controlled conditions of the research laboratory. Resynthesis of the whole complex is then possible, buttressed and tested by more disciplined field observations. The study of animal behavior is such a return to the natural laboratory. You will follow a similar sequence—from gross observations to a more detailed study of structure, preliminary experimental studies of function, and, as much as possible, to an integration of the two, followed by more knowledgeable observation and comparison with other animals, leading, we hope, to a greater appreciation of the biological continuum.

If your dissection shows one location or appearance of a structure and this book states another, stick with your dissection! Continue to expand your observations, studying other examples to see if your dissection is normal or is unusual or atypical. If possible, find out why the difference exists and what may prove to be a new finding— or a normal range of variability. Retain a questioning or doubting attitude, rather than a passive or accepting one. Many keen minds have worked over many years to supply the information recorded here, but some of these observations or conclusions doubtless are incomplete, misleading, or simply wrong. And you may be the one to find this out. Such hazards—and challenges—abound in the process of learning biology, as in all human endeavor. As experienced biologists appreciate, what we do *not* know greatly exceeds what we *do* know. Each breakthrough into a new realm of knowledge is not a set accomplishment, but, rather, a new vantage from which to ask other questions and probe to reach more basic levels of understanding. This recognition of our ignorance is an important part of knowledge. Humility in science, as elsewhere, grows in direct proportion to the extent of one's knowledge.

We would like you to think of this book, therefore, as a guide to the organization of animal types and their structural similarities and distinctions (relationships), but one that is only as reliable as changing knowledge permits. It is intended to outline, organize, illustrate, and guide you through a survey of the major groups of animals, in such a way as to clarify and make sense of the great association and groupings of animals and their enormous range of forms and adaptations. It is then your task to integrate this information into the more basic framework of biological processes and relationships. We also hope that it may serve as a challenge to those who want more information or wish to ask other, possibly more meaningful, questions.

To the Instructor

Courses in zoology can be taught in many ways. Most of them fall into one of two categories: a course of principles, dealing with a succession of key topics, employing examples freely through the phyletic spectrum; or a course with a phylogenetic orientation, dealing with the evolutionary succession of animal groups, with biological principles being developed and illustrated where convenient by appropriate animal types. Both procedures have advocates and advantages, varying with the degree of interest and the orientation of the student. Nonetheless, both routes still require a firm base in the animal itself. This laboratory is therefore geared to a group-by-group or phyletic approach, followed by a consideration of the major disciplines applicable to all living things, such as cell and tissue structure and function, embryology, genetics, and ecology. Evolution, the great synthesis of biology, is implicit and explicit throughout the book. Either a principles approach or a phylogenetic one can be tied to a laboratory presentation such as is employed here. The sequence can, of course, be altered or followed selectively to suit your particular requirements. We have designed this text chiefly to help anchor the student's grasp of the major animal groups, first, in terms of their structural body-plan, then, via more detailed consideration of specific examples, usually those generally available in introductory courses. In our experience, a physiological or purely conceptual approach, without a phylogenetic base developed concomitantly, leaves many students unable to make the essential tieup between electron transport, for example, or DNA decoding mechanisms, and the real world. They often show insufficient appreciation of the extent and interrelationships of the principles studied (often *in vacuo* so far as the student is concerned) with the living complex in which it all operates. We would like to help you strengthen this integration of the universal with the particular. We feel that your effort to build up the student's understanding of levels of biological organization can most easily be met by erecting the phyletic framework in the laboratory first, and then moving from that toward the more fundamental building blocks and functional units. The final tieup, we believe, is an overall integration at the population and community levels, which is far more meaningfully appreciated by students with a thorough foundation in the phyla and in the animals themselves.

This book has enlarged its scope and has changed over the years, but it retains the basic organization developed through the authors' teaching experience in an introductory course in zoology at UCLA. In turn, it also reflects the teachings and writings of those instructors who were influential in the authors' training, to whom we owe thanks for a lifetime of pleasure and intellectual excitement.

The basic format of the first two editions and the responsibilities of the authors as outlined there are retained in this edition. More experimental exercises have been added and the later chapters substantially rewritten. A new section on dissection of the fetal pig has been added to Chapter 14 (Mammalia), which should be useful to the many instructors employing this excellent instructional organism. A section on early chick development has been added to Chapter 17 (Embryology); an Introductory experiment on population genetics using *Drosophila* has been added to Chapter 18 (Genetics), as well as a detailed consideration of the Hardy-Weinberg principle. The animal behavior experiments in Chapter 19 are more fully presented in this edition, rather than simply outlined as student projects. The final chapter, Ecology, is completely new, with experiments described for the study of biological succession, interspecific competition, tide-pool ecology, small-mammal ecology (laid out as group projects), and an extended section on human ecology and environmental decay included for individual or group long-term student projects. A number of new illustrations have also been added and others replaced.

Without repeating the authors' specific thanks to colleagues and contributors named in the previous editions, we wish to extend to them and to many others gratitude for expressing their views and for giving us the benefit of their teaching experience. To this latter group in particular, the instructors—and to their students as well—we wish to extend our special thanks. We share with you a concern that the excitement of the living world and the rapidly enlarging concepts to account for it be presented in the most meaningful and significant way. The contents of this work—errors and omissions included—are our responsibility. But the real responsibility is *yours*: the task and privilege of stimulating, presenting, interpreting—of *teaching*.

Our gratitude is extended to Miss Elaine Fingerette, Mr. John R. Edwards, and Mr. Steven M. Sachs for their assistance in the preparation of this book.

R.A.B.
D.H.

Los Angeles, California
San Francisco, California
February 1975

CONTENTS

AN ILLUSTRATED LABORATORY TEXT IN

ZOOLOGY

THIRD EDITION

INTRODUCTION

This laboratory text is designed to be an integral part of your introduction to structure, relationships, and life styles among animals. It can supplement, but never replace, your examination of the living or preserved animal itself. In fact, nothing can ever substitute for the most important research tools you have—your eyes and hands to see and sense the object, your mind to interpret and explain. Direct experience with the animal itself is the only real way to *know*. "Study nature, not books," said Louis Agassiz, as he threw a fish to his students at Harvard in the 1860s. That was the basic laboratory, and for the receptive students it was enough.

A laboratory textbook should help you become familiar with the different animal types, and direct your studies to an understanding of biological principles. This text is both a map to the anatomy and classification of representative animals, and a guide to the laws that govern living material. Organization of the text emphasizes these two phases of your work: first, *analysis*—the study of structure by dissection and careful review of key parts of selected animals; second, *synthesis*—the integration of this information into a larger scheme to help you appreciate relationships and biological principles within the animal world. However, if you approach the dissections in a cookbook fashion, you will prob-

ably find them both useless and dull. The degree of your effort to relate specific organisms to their general group—for example, earthworm to the ANNELIDA, frog to the CHORDATA—will measure your appreciation of the purpose of the laboratory. Recognition of this aim will increase your interest and pleasure in the work, giving you a clear orientation, and an understanding of what you are trying to do and why. Since clarification of animal structure and relationships also means a new concept of yourself, your attention and perception will be sharpened to an even greater degree.

Give your natural curiosity reign, and this often neglected resource will help you through the exercises with increased interest and understanding, and, hopefully, with the zest and appreciation for zoology that a connoisseur has for a '61 Châteaux Lafite-Rothschild.

Chapter Organization

The chapter sequence is a phylogenetic or evolutionary one, from protists to vertebrates. Most of the major animal groups are covered insofar as laboratory time and availability of specimens permit. A more complete listing of animal groups, including many not discussed in the text, will be found at the end of Chapter 2. Included in Chapters 2–14 are the following:

1. General introduction to the phylum; statement of the basic characteristics and points of interest of the animals included.
2. Systematic arrangement of the group as a whole and its major subdivisions.
3. Laboratory instructions for dissection or observation of the animal selected to represent the entire group or particular division of the group.
4. Additional experiments, or special projects, to emphasize field or experimental possibilities available to students with more time or specialized interests, and to emphasize interesting phenomena characteristic of the group.

Text[1]

The textual material is laboratory oriented. It is intended to serve as a link between generalizations considered chiefly in lectures, and organisms seen only in the laboratory. Students often find it difficult to relate these two aspects of zoology, even to the point of treating them as different subjects. The more quickly you can dispel this view, the more easily you can grasp the details and understand the principles underlying zoological concepts. You may be able to derive some of these principles on your own.

This laboratory textbook will, therefore, introduce basic material and review biological principles best demonstrated by each phylum; it will then guide you through the various laboratory procedures to be performed. Each topic should be studied *in advance* of the scheduled laboratory. You will be asked to compare, correlate, and integrate the new material with the knowledge you have already acquired. Emphasis on relationship and comparative structure builds a conceptual framework to which the details of this book can be tied, enabling you to look at nature as an integrated matrix. With-

out it, zoology is reduced to bewildering detail and terminology. Many former students recall their zoology laboratory exercises as a series of slimy snails, smelly frogs, and horrid dead cats. We hope that you will see these exercises as doorways to the understanding of living structure and processes in every animal—including yourself.

The most important of the integrating ideas and unifying principles to be considered is that of *evolution*. From a knowledge of evolution, we can gain insights into the bewildering variety of living creatures that surround us, and from this determine the degree of relationship among similar forms (the probability of common descent of organisms with similar patterns of structure and development). Our notion of relationship is built from combining traits shared by different species. For this reason we emphasize basic characteristics common to large numbers of organisms. Thus, we can conceive of a generalized animal, built from these traits, which we call a *basic body type* for each of the major groups of animals. This imaginary creature is a useful simplification, an "average" to represent a particular type, an archetype for the phylum. It may incidentally resemble the primitive species, thought to be ancestral to all of the present-day specialized examples.

Groups of animals with common traits are arranged by biologists into *taxa* (singular: *taxon*)—species, genera, families, and larger categories. The student learns by inductive methods to group animals in various categories, just as biologists have learned to do over the past 200 years. Each category into which animals are placed can, in turn, be grouped systematically into a succession of more inclusive, and therefore more generalized, categories. This is the *taxonomic scheme*, and *taxonomy* is defined as an organization of living things visualized as a hierarchy of groups showing varying degrees of similarity. These groupings lead to our ideas of relationship, and to our picture of evolution.

The best way to approach the huge assortment of animal types and forms is to appreciate the convenience and value of the taxonomic filing system, and to adopt it as a means to efficiently organize whatever body of facts you wish to encompass. As long as the relatively few basic body types are clearly kept in mind, additional information about the group will remain well ordered. In other words, it will make sense, and most students will find that their interest and understanding increase appreciably when they realize for themselves that

---

[1] Some instructors prefer to start the study of zoology with the frog or some other vertebrate type, so that the student feels that he is starting on familiar ground. He soon realizes, however, that the ground is far from familiar. Since the vertebrate and invertebrate portions fall into two rather distinct groups, often used as semester divisions, the arrangement can follow the particular lecture sequence preferred by the instructor. The phylogenetic succession, however, is the preference of the authors.

zoology *does* make sense, that it is an integrated science rather than a conglomeration of facts.

Examine carefully the section of classification (Chapter 2). The terms and groups discussed will be referred to again and again, to emphasize general relationships between the specific form you are studying and the successively larger taxa to which that form belongs.

Photographs

A properly selected photograph is often second only to the animal itself in conveying the message of overall organization and specific structure. Within limits of depth of field, lighting, and other technical factors, a photograph is an exact illustration of a specific organism, a particular dissection, or a microscopic image. Of course, the individual organism being viewed by you may well appear somewhat different. Such discrepancies may be due to normal variability within the species, the age of the preparation, the type of stain used in a microscopic preparation, or the dye to make your specimen clearer, a particular angle of view being photographed, the layout of the organs, or partial dissections. Obviously, a photograph, while an exact rendition of a particular specimen, is to be used by you as a *general* representation of the subject. Part of the design of these laboratory exercises is to develop your ability to differentiate between characteristics that are variable, therefore less significant, and those that are actually representative of the population as a whole.

The photographs and labels used in this text should help you to distinguish and to name parts quickly, without need of a greatly detailed description. With your own specimen and the labeled photograph for orientation, you should then become familiar with its particular characteristics, and from them be able to make the important transition from specific to general. To determine by induction the construction, organization, or structural pattern within each group of animals from your specific example, or from a number of such examples, is a basic method in science. Your understanding of this method is one of the primary goals of this course.

Variation in structure among individuals of the same species (and variation in dissection assignments) precludes your using these illustrations to replace your own specimens. You will discover that it is preferable to look at your specimen, compare it with a photograph or sketch of a similar but still distinctive example, and then return to your *own*

material for final study and drawing. In doing so, you will enjoy one of the greatest pleasures of science: independent observation and discovery.

Artwork

A drawing is an abstraction, usually a composite of many examples. It is often a highly simplified or diagrammatic view of extremely complex structures (such as the nervous or circulatory systems), or a simple outline of selected morphological features. Both photographs and drawings have a definite place in zoology; each performs a useful, necessary function. The advantage of a drawing is that it can *select*, simplify, emphasize, and thereby clarify by dropping unnecessary detail.

Student Drawings

Most students are convinced that they cannot draw, never could, and never will. Much of this fear is simply that they do not know what is expected of them and assume that nothing short of a perfect, detailed copy will do. Drawings in this course have *one* purpose: to force you to observe. You must observe so carefully that the outline of what you see can be transferred to paper. Looking at a slide or specimen well enough to recognize the parts and get a general idea of shape and arrangement is one thing; but looking closely enough to *reproduce* these parts, even in barest outline, on a piece of paper is quite another. For example, try looking in a store window just carefully enough to recognize the types of items represented and their relationships to one another. Then look again, knowing that this time you will be asked to name from memory or sketch on paper the articles and their precise locations. There is quite a difference in the way you look the second time! This difference separates casual from critical observation.

The mechanics of drawing vary, but the rudiments are simple.

1. Decide precisely what you are going to show. Select a *typical* part or animal showing the required aspect to the fullest degree.
2. Select the appropriate scale or magnification. *Make your drawing large enough to show full detail.*
3. Divide your paper into quadrants with light pencil lines (#4), then with the aid of these divisions sketch lightly the *outline* of your subject in order to produce the proper symmetry. For most students this is the hardest part, but with practice and the use of

guide lines or squares it can be mastered readily.

4. Replace your sketch lines with firm unbroken outlines. Your drawings need not be suitable for framing! Artistic effects such as shading or feathery border lines are unnecessary (in fact, they may obscure biologically important details). Your goal, remember, is to produce an accurate outline that shows correct proportion. You can then block in important structural details. You need not be artistic or photographic, but *you must be selective* since your drawing includes only those features you consider essential for the designated purpose.

5. Fill in structural detail, but no more than necessary. If your specimen is bilaterally symmetrical, details on one side are usually enough. If it is radially symmetrical, cellular detail can be confined to a small pie-shaped wedge.

6. *Labels* are extremely important. They should be accurate and neat without interfering with the general appearance of the subject. Confine them to the margins, *print* clearly, and connect them by light, solid lines to the object or structure named. Arrange the labels so that their lines do not cross one another.

7. Complete the drawing assignment with a full title and magnification scale (*not* simply the scale of the microscopic image). State the name or identification of the subject, view or angle illustrated, sex, and other pertinent details. It is often helpful to place the full classification in an upper corner of your page.

Emphasis in these drawings is on accuracy of observation, not on artistic talent. Confine the details and labels to what *your own specimen shows*. Other information may be added if desired, but the source should be stated. If you believe that your specimen is atypical, ask your laboratory instructor for interpretation or suggestions.

Questions

Many questions are found in the text. They should cause you to think back over ground covered earlier or to retrace what you have just done, often from a different vantage point. Sometimes questions are introduced to anticipate relationships or concepts to follow. Answering these questions will help you to integrate laboratory and theoretical aspects of the course.

Experiments

Emphasis during introductory laboratory periods must necessarily be on structure rather than function. Before you can fully comprehend *physiology*, and functioning of an organism, you should first familiarize yourself with its structural form or *anatomy*. With further study you will begin to realize that physiology and anatomy are essentially inseparable.

Once the structural basis is learned, you can look more closely at physiological processes. The laboratory exercises in this book will stress physiological and experimental approaches in the study of heart, muscle, and nerve, and more general physiological processes such as regeneration in planaria, embryology of the sea star, genetics of the fruitfly. Other experiments, involving application of structural information and dissection techniques, are suggested where appropriate.

Since most course instructors will select particular laboratory exercises or examples within an exercise, some repetition of dissections or of experiments from which to choose have been included to ensure coverage of important topics.

Succeeding zoology courses that will become available to you will stress more functional, cellular, and biochemical aspects of animals. Still other studies will emphasize group interrelationships and behavioral activities of individuals and communities of animals in nature. These courses will give you a chance to study living processes at work in the normal environment and to appreciate functioning of the intact, beautifully adapted *living* animal. Your ability to pursue and understand these subjects will be based on your familiarity with the concepts and animal groups to be studied now.

Textbooks

Suggested readings at the end of each chapter will guide you to more specialized or detailed references and aspects that can only be briefly covered in the text. We also include, where appropriate, general textbooks or illustrated guides and monographs that we feel would be valuable to you. In addition, each chapter is keyed out to specific pages in a number of the more popular current textbooks in zoology. This cross-referenced guide is found on the inside front and back covers of this book.

CHAPTER 1

THE MICROSCOPE

1. Introduction

No tool is more characteristic of biologists than the microscope. It is an instrument uniquely able to carry them into the world of the infinitely small. To the new observer, the microscope opens a world of unexpected complexity and beauty. To scientists it is a great deal more. It is the extension of their most critical sense—vision—that has made fundamental biological advances possible. Without the microscope, our notion of life would have remained archaic. For you, the microscope will become a tool enabling you to see for yourself the structure of muscles, cells, protists, eggs—many of the basic units of life. It is an instrument that must be cared for, but *used*, and used properly. By the end of the course you should have the skill to wander freely in a world 10 to 1000 times beyond your ordinary range of vision, into the land of the micron or mu (μ), a metric measure equal to one millionth of a meter.

Other optical instruments (see Section 4 at the end of this chapter) can utilize nonvisible portions of the electromagnetic spectrum (ultraviolet light, x-rays) and electron beams, considerably extending the microscope's range into smaller realms. Employment of these shorter wavelengths has vastly increased both optical resolving power and magnification. The most spectacular development in recent years has been the *electron microscope*, employing beams of electrons that can resolve objects a thousandth of a micron apart. For example, viruses, the tiniest things we know possessing the qualities of life, have become visible for the first time through electron microscope studies. These instruments are tools for the research specialist. We will utilize two basic types of light microscopes, each better optically than those used to make the pioneering advances in histology and microanatomy. These are the principal optical instruments employed by biologists today.

The first of these is the *stereoscopic dissecting microscope*, with a single-lens system (Fig. 1.1). The second is the *compound microscope*, either *monocular* or *binocular*, with a compound-lens system (Fig. 1.2). The dissecting microscope offers several advantages: (1) direct magnification without reversal of the image; (2) stereoscopic vision; (3) a large working distance between the object and the lens, allowing dissection or manipulation of the object being observed; (4) use of either transmitted or reflected light; and (5) observation of objects too large for the field of a compound microscope. The degree of magnification, from 5 to

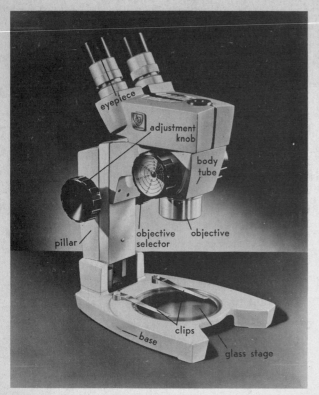

Figure 1.1. Stereoscopic (dissecting) microscope. (Photograph courtesy of the American Optical Company.)

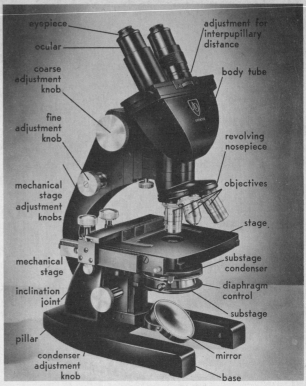

Figure 1.2. Parts of a compound microscope. (Photograph courtesy of the American Optical Company.)

50 diameters, is, however, much less than that of the compound microscope, though sufficient for relatively noncritical study or for work requiring manual manipulation (a large *working distance* between the objective lens and the object).

Microscopic studies of small, translucent animals, tissues, or cells utilize the compound microscope. The objective lens of this instrument creates a *primary image*, which is reversed and inverted in position (right to left and top to bottom) and then magnified again by the ocular lens to produce a *virtual image* seen by the eye. With the compound microscope, magnifications ranging from 100 to 1000 times are obtained, equivalent to seeing an object approximately 1 km distant as if it were 1 m away.

2. Handling and Special Care

Optical instruments are expensive and fragile, but with proper care they should last for many years. Protection of lenses from shock and abrasion is especially important. You should observe the following precautions:

1. Carry the instrument by the base and pillar, cradled vertically in both hands. Tipping it backwards may have disastrous results, as the student whose oculars are sent crashing to the floor will quickly appreciate.
2. Put the instrument down carefully to cushion any shock from the hard surface of your laboratory bench.
3. Keep all exposed lens surfaces clean, but use fresh *lens paper* only. (Your handkerchief carries lint and highly abrasive dust and dirt particles that will scratch the polished, coated lens surfaces.) Never touch a lens with your fingers—the oil attracts dust, obscures vision, and is difficult to remove.
4. Check cleanliness of oculars by rotating them while looking through them to see whether dust particles move. Make sure the mirror is clean. Be particularly cautious with the *objectives*, the most valuable part of the microscope.
5. Clean your instrument before each exercise. *Return it clean* to its case or shelf, with the *lowest power* objective in position over the stage and the coarse adjustment racked

down. (Why?) Always double-check to be certain that no microscope slide remains on the stage and that the latter is carefully cleaned before returning the microscope to its case. These precautions will help prolong the life of the microscope for your own use, and for that of your fellow students.

6. Do not remove the oculars from their tubes except to clean them, as removal permits dust to enter the instrument.

When you first inspect the microscope, report to your instructor any scratches, oil, or other apparent misuse or damage to any part. This attention will ensure prompt repair or adjustment.

3. Parts of the Instrument

I. Stereoscopic Dissecting Microscope

Locate, name, and review the function of each major part:

1. *Base*, with glass *stage* and slide-holding *clips*.
2. *Substage*, with mirror (may be absent from your microscope).
3. *Pillar*, supporting stand for the microscope *body*, housing of the lens system.
4. *Objective lenses*, contained in a *rotating* or *movable drum* or *pod*.

■ How many lenses are there and of what magnifying power?

5. *Ocular lenses*, recessed in *ocular tubes* in upper portion of body.

■ What is their magnifying power?
■ How is this magnification expressed?

II. Compound Microscope (may be *monocular* or *binocular*, with *vertical* or *inclined* oculars)

1. *Base*, or supporting stand, generally horseshoe-shaped.
2. *Arm*, attached to base by *inclination joint*, supports *body tube*, and serves as the handle for carrying instrument.
3. *Body tube*, the optical housing for the lenses. In newer binocular instruments, the body tube and oculars are inclined for more convenient viewing.
4. *Coarse* and *fine-adjustment knobs.* Coarse for low-power work and initial focusing, fine for final adjustment and variation of *plane of focus* for viewing an object at different depths. These are your "steering wheels," the controls that bring the object into proper focus so that it not only becomes sharp and clear, but can also be viewed three-dimensionally as well by continuous movement of the fine-adjustment knob.
5. *Substage condenser* with *iris diaphragm,* controlled by a projecting handle for management of light intensity and focus, a key control device for proper microscopy.
6. *Condenser adjustment knob* for control of condenser level, hence to control the plane-of-light focus and concentration of light reflected from the mirror onto the microscope stage and on or through the object viewed. In instruments without a condenser, this function is performed by the concave mirror.
7. *Reflecting mirror* with *concave* (*parabolic*) and *flat* surfaces.

■ Why are two types of mirror needed?

8. *Stage* with *microscope slide clips.* Research instruments usually have a *graduated mechanical stage*, a special slide-holding device marked by graduations, with two *mechanical stage adjustment knobs* for accurate movement of the slide.

■ How are the graduated markings used?

Most student microscopes are equipped with a nongraduated mechanical stage controlled by adjustment knobs (Fig. 1.3).
9. *Ocular lens* or *lenses.* You may have two sets for different magnifications ($6\times$ and $10\times$, or $10\times$ and $15\times$). Usually, better results are obtained with a lower-power ocular and high-power objective rather than a high-power ocular and lower-power objective, even though the actual magnification may be the same. This is because correcting lenses are built into the objective, the most optically critical part of the microscope.
10. *Objective lenses* are screwed into a ro-

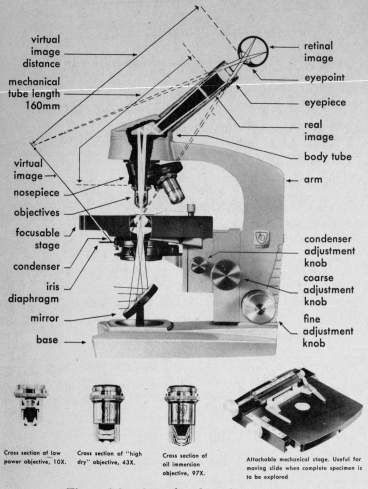

virtual image distance

mechanical tube length 160mm

virtual image →

nosepiece

objectives

focusable stage

condenser

iris diaphragm

mirror

base

retinal image

eyepoint

eyepiece

real image

body tube

arm

condenser adjustment knob

coarse adjustment knob

fine adjustment knob

Cross section of low power objective, 10X.

Cross section of "high dry" objective, 43X.

Cross section of oil immersion objective, 97X.

Attachable mechanical stage. Useful for moving slide when complete specimen is to be explored

Figure 1.3. Optical and mechanical features of a microscope, showing the path of light. (Photograph courtesy of the American Optical Company.)

tating head, the *revolving nosepiece*. Observe that markings on these lenses state their magnifying power (10×, 43×, or 97×). The 10× is called the *low-power* objective (or *low-dry*), the 43× is the *high-power* (or *high-dry*) lens, and the 97× is the *oil-immersion* lens. The latter is always used with a drop of oil to form a liquid bridge between itself and the slide surface examined. For still more critical work, an oil bridge is also placed between the *lower* surface of the slide and the top lens of the condenser.

■ Can you state the optical advantage of having light pass from the object viewed through oil rather than air before entering the lens and passing to your eye?

Use of the oil immersion lens will be restricted to special demonstrations or to work demanding the optical limits of your instrument.

■ Review with your instructor the function and basic optics of the lens system; be able to name and describe each part of the instrument. Observe differences as well as similarities between your microscope and those illustrated here. (Figs. 1.1 and 1.2). You should know both definition and significance of each of the following terms:

resolving power   transmitted light
numerical aperture   reflected light
working distance   focusing
image reversal   parfocal
primary and virtual image   depth of field
focal plane   magnification

Be able to trace the path of light reflected by the mirror from its source, through condenser and object on the slide, into objective lens, optical tube, ocular lens (or lenses), to your eye, noting carefully the primary and virtual images produced (Fig. 1.3).

■ What is the effect on the light path of a change in position of the condenser? of the use of the parabolic in place of the flat mirror?

Remember that the main purpose of your microscope is to enhance *detail*, not simply to give magnification. You can project an image on a huge screen, but see nothing in greater clarity. The key is *resolving power*, the ability of your lens system to separate two points. The closer the points, the greater the resolving power needed to see them distinctly as two points, rather than as one blur. Resolving power as a measure of quality of your instrument varies *inversely* with the wavelength of light used (the shorter the wavelength, the greater the resolving power), and *directly* with the ability of the condenser and objective lens to gather light, a measurement called the *numerical aperture*. Additional optical information may be found in a booklet prepared by the American Optical Company, *The Effective Use and Proper Care of the Microscope*; or by the Bausch and Lomb Company, *The Use and Care of the Microscope*. (Copies are available on request.)

4. Procedure for Proper Use of the Microscope

1. Position the instrument carefully at your desk, with the arm facing you. Incline the

instrument to a comfortable position (unless your microscope has an inclined head). Do *not* incline the arm if fresh material is to be used, or you will find the water and sample dripping into your lap.

2. Adjust the microscope lamp at least 6 inches behind the instrument, shielded so that light shines directly on the mirror, not on the stage or into your eyes.

3. Turn the *lowest* power objective into vertical position over the opening in the stage.

4. Adjust the mirror so that the light beam is reflected directly up through the center of the stage opening. (Check this by looking from the side.) There should be a light bulb image in the center of your field of vision. Use the curved parabolic surface of the mirror *only* if there is no condenser in your microscope; with the condenser always use the *flat* surface of the mirror.

5. Bring the substage condenser up to its highest position, then close down the iris diaphragm until the light is diffused evenly across the entire field of vision. It may be necessary to rack the condenser down to a lower position to reduce the intensity of light for lower-power work. Remember to raise it again for high-power objectives when greater light intensity is required.

NOTE

Always start with the lowest *power objective and the* coarse adjustment *since this offers a much larger field and allows more rapid orientation and preliminary focusing. The fine adjustment, in fact, need be used only with the high-dry lens and* must *be used with the oil-immersion lens.* Avoid glare; *reduce light intensity with the iris diaphragm or condenser, or preferably by closing down the diaphragm on your lamp if such a control is present, or by moving the light source farther back and readjusting the mirror. Do not* simply change the mirror angle; this misdirects the light. You will discover that reduced light intensity adds considerable clarity and provides both a greater degree of visual control and a more three-dimensional image.*

Practice proper microscopic procedures (focusing, proper alignment of light and control of its intensity) by examining a millimeter rule through the various objectives. Then observe a piece of lens paper placed on a slide in a drop of water, covered by a coverslip. *Place the coverslip in position slowly to prevent entrapment of air bubbles. Focus on individual fibers under low and high power, using various intensities of illumination. Continue experimenting until you can confidently position, illuminate properly, and focus on a selected object or optical plane.*

6. After the above steps and practice procedures are completed, place a prepared microscope slide on the stage. This will probably be a thin tissue slide, specially stained and cleared for microscopic study. It was mounted on the slide in an optically transparent fluid medium such as Canada balsam, then covered by a coverslip under which the medium hardened. You may also be given a mounted intact animal ("whole mount") that is suitably stained and prepared for microscopic examination. If you have *living* material, such as a culture of protozoa, place a drop or two on the center of a slide. This should be enough to fill the space under an 18-mm diameter coverslip. Place a coverslip *slowly* over the drop, as previously practiced, to avoid trapping bubbles or forcing the organisms out from under the glass. Move the slide about on the stage until the area to be viewed lies directly in the beam of light coming up through the center of the stage opening.

7. With the *low-power* objective in position (16-mm working distance), and while watching from one side, turn the coarse-adjustment knob in order to rack the objective down to within 6 mm of the coverslip. Now look through the ocular, *keeping both eyes open*, and turn the coarse adjustment toward you (rack up) until the object becomes fairly clear. Move the slide slowly with the stage-adjustment knobs or by hand until you find a field of interest.

8. Switch to the *high-power* objective. It has only a 4-mm working distance, so be certain that your microscope is parfocal. Focus with careful movement of the fine-adjustment knob. Remember too that the limited working distance of the high-dry objective lens will not permit you to rack down more than a few millimeters. Always remember to start with the objective slightly *below* the 4-mm distance so that you bring the

object in view by racking *up* with the fine-adjustment knob. Racking down may force the objective into the coverslip, producing a smashed slide, damaged objective, and shattered student morale. Such careless focusing is the beginning microscopist's prime hazard, a costly and inescapable declaration of faulty technique.

More light will be needed with the high-dry objective than with the low-power objective. (*Why?*) Control the light as before by judicious use of the iris diaphragm, position of the condenser, or distance of the light source. Switch to low power and back to high-dry. *Observe* that the *field* is sharply reduced, *light intensity* is reduced, and *working distance* reduced when magnification is *increased.* (*Why?*) For this reason always orient yourself first under low power.

5. Laboratory Exercises

1. **Letter e slide** With the compound microscope study a slide with the letter *e* marked on it. Note the position of the letter. *Draw* it at low and high magnification, stressing its size relative to that of the field. State magnification employed. Do this by drawing a circle to represent the field and showing the *e* or that part of it that is visible within the circle. Now examine the same slide under each magnification of your *dissecting microscope.* Why is the working distance greater than with the compound instrument? Which instrument and which lens combination would you use for each of the following:

a.   dissection of a mosquito
b.   examination of frog blood
c.   examination of a living tadpole
d.   examination of the surface of your index finger
e.   examination of a hair
f.   examination of a living amoeba

2. **Specially prepared stained slides** Tissue, insect, or protist slides are all excellent for this practice. Practice proper focusing procedure with each specimen and make drawings suggested by your instructor.

If you keep one hand on the fine-adjustment knob and constantly work it up and down, you will learn to see in depth and to appreciate various levels of focus until you

can picture the specimen in a fully three-dimensional view. Remember that microscopic proficiency is essential throughout the entire course.

· Whenever an illustration is requested, make it carefully, drawing an accurate outline with proper proportion and symmetry. Review the discussion of drawing procedure in the Introduction. You will soon see the difference between simply *looking* and *observing* well enough to reproduce what you see.

3. **Slide preparation of scrapings from teeth or gums** Obtain material with a toothpick and place it in a drop of normal saline solution on a slide and add a coverslip. Neutral red or methylene blue will stain the cells, coloring the nucleus, cytoplasm, and various organelles to different intensities, allowing you to see these parts more clearly. Add a drop of dilute stain under one corner of your coverslip, then draw it under the coverslip by using a bit of filter paper as a wick at the opposite corner of the coverslip. With some practice, this technique will become most useful.

4. **Freshly prepared and stained sample of frog or human blood** You can make your own stained preparation in the following manner: Obtain a blood smear (a droplet of blood evenly spread across the slide with the end of another slide, as demonstrated by your instructor, is then air-dried and "fixed" by brief immersion in methyl alcohol). Place Wright's stain drop by drop onto the slide until the stain covers the smear. Leave it for about 2 min, then add about 2 ml of distilled water. Leave the mixture for 4 or 5 min, then pour it off and wash the slide quickly but carefully in distilled water. Add a coverslip and examine.

5. **Preparation of frog skin** Compare with a bit of your own skin. Mount the skin in saline and cover with a coverslip. Add a drop or two of stain, using the procedure already learned.

6. **Living ciliated cells** Place a bit of tissue from the roof of the mouth of a freshly killed frog in a drop of saline on a slide. Cover without trapping air bubbles and examine.

7. **Living protists from a mixed culture** Place a drop of culture medium on your slide, cover, and examine. How many differ-

ent types of organisms can you find? Detailed study of these will be delayed until the next laboratory period, but avoid spending 30 min staring at an air bubble!

8. **Special exercise: measurement and computation of magnification in your drawings** Simple magnification is computed by taking the product of the objective and ocular lenses. (For example, 10× ocular and 43× high-dry gives a visual magnification of 430 times what the naked eye can see. It is written 430 or 430× diameters.) It does not, however, give a direct reading in units of size nor does it accurately tell the magnification on your drawing. A simple way to determine this magnification is to lay a metric ruler alongside the right side of the stage of your instrument, 25 cm from the ocular lens and in the line of vision of your right eye, while you look through the eyepiece with your left eye. (Use the *right* eyepiece on a binocular microscope.) For left-handed students, the opposite arrangement is followed.

Practice this until you can train your eyes to juxtapose the two images and actually *read the diameter of the entire microscope field or the object you are observing through the microscope, measured against the ruler.* To determine the actual magnification of your drawing as opposed to that of the lens system (the virtual image), substitute in the following simple ratio:

$$\frac{\text{lens magnification (ocular times objective)}}{\text{size of virtual image}}$$

$$= \frac{\text{magnification of drawing}}{\text{size of drawing}}$$

Solving for the third element of this proportion will then give the final magnification.

Calibration of your microscope for *direct measurement* of object size in known units can be done with an *ocular micrometer* and a *graduated stage* (actually a microruler) that enables you to measure accurately the total field of your microscope, or the size of any object viewed, at each power of the objective lens. The ocular micrometer fits into the ocular tube, the graduated stage is placed on the microscope stage. Divisions seen through the ocular micrometer are then measured off against the divisions (in tenths and hundredths of a millimeter) on the graduated

stage. With each lens combination, a specific reading is obtained (that is, one ocular division might equal 11.5 stage divisions). Ocular units can be determined in terms of the known units of the stage micrometer. This gives you a rule in the ocular which can then be aligned with any object being studied to give a direct reading in *ocular units*, quickly converted to microns by use of the predetermined calibration. Review these steps carefully with your instructor and calibrate your microscope for several combinations of lenses or observe a demonstration of the process.

Measurement of cell (egg) volume by this technique is a valuable exercise.

■ **Can you think of any other applications of direct measurement of microscopic structures that might be made using the calibrated ocular micrometer?**

6. Other Microscopes

Although introductory courses use the microscopes described below only for demonstration, you should be acquainted with their use. These instruments allow better contrast in observing living, unstained structures (paragraphs 1 to 3), or a vastly greater resolving power (paragraphs 4 and 5), revealing a new level of organization—the *subcellular* or *macromolecular* size range.

1. The *phase contrast microscope* has wide application in biological research. Structures in living cells may be seen when a greater contrast is produced between them and their surroundings. This is accomplished by equipping an ordinary light microscope with a special *diaphragm* in the condenser and a transparent disc (called a *phase plate*) in the objective, thereby modifying a portion of the light that passes through the microscope. In doing so, the refractive index of the specimen is exaggerated. It then becomes possible to distinguish structural details that vary only slightly in thickness or in refractive index and are not discernible without intensification of contrast.

2. The *interference microscope* provides an additional means of studying transparent objects such as living cells. This instrument combines a double beam *interferometer*

with a *polarizing* microscope. The polarization and wave interference produce greater contrast than is available with phase contrast, and produce images in brilliant color as well. This microscope also permits determination of dry mass in a particular portion of a cell. Direct measurement of dry mass is based on the fact that the greater the amount of dry material observed, the greater the *optical path difference* between it and the surrounding fluid. This difference can be measured and converted into units of mass.

3.   The *ultraviolet microscope* looks like a conventional light microscope, but its entire lens system is constructed of quartz because glass is opaque to the shorter ultraviolet wavelengths. It permits only a twofold increase of resolving power over that of the conventional light microscope, but since ultraviolet light is absorbed more by certain parts of a cell than by other parts, contrasts are created in otherwise indistinguishable structures. Direct observation by ultraviolet light is not possible, so the object must be studied by means of photographs or by the television camera.

4.   The *electron microscope* differs in many respects from the microscopes discussed previously in this chapter. As the name suggests, the electron microscope employs a beam of electrons, instead of a beam of light to produce a magnified image. The image is viewed when it impinges on a fluorescent screen or is photographed by use of a specially positioned camera. Electrons have wavelengths that are much shorter than those of light, and hence they can be used to distinguish between closer points. The resolving power is 50 to 100 times that of a light microscope. Electrons emitted from a filament in a strong vacuum move toward the target at close to the speed of light and are focused by powerful electromagnet "lenses."

■   **Why must a vacuum be maintained in the path of the electron beam?**

As the electron beam passes through the specimen, some electrons are scattered, and the remaining beam is "projected" (by more electromagnetic lenses) onto a fluorescent screen or a photographic plate. The projecting lenses can produce a magnification of over 200,000× (some reach 2,000,000×, equivalent to bringing an object 2000 km away to a distance of 1 m!).

As a result of its extraordinary magnification powers, the electron microscope has extended our range of observation into the realm of the viruses and intracellular macromolecular structures, previously known only by experimental inference. The electron microscope has had an important role in the revolutionary advances in biology in recent years. These extraordinary developments have opened new fields of research (molecular biology) and resulted in an epoch comparable to the great period of advance in theoretical physics a few decades ago. For the first time we can view the previously invisible, explore directly the innermost functional secrets of the cell and, ultimately, of life itself.

5.   The *scanning electron microscope* (SEM). The light microscope and standard electron microscope differ greatly in resolving power but work on the same principle of employing a beam of radiation to produce an image. While the light microscope yields very good information, the resolution obtained with the standard electron microscope is of a far higher order. However, the total amount of information carried by the electron beam is very small.

With the development of the *scanning electron microscope* (SEM) a great deal more information can be obtained about the specimen. In essence, the SEM regularly moves a narrow beam of electrons over the specimen. Simultaneously, another beam of electrons is passed across the face of a cathode ray tube. The narrow beam of electrons scanning the surface of the specimen stimulates the production of secondary radiation. A detector passes these converted signals from the specimen to modulate the beam crossing the cathode ray tube.

The unique advantage of the SEM is that the secondary electrons that modulate the cathode ray electron beam impart a three-dimensionality to a two-dimensional signal from the cathode ray tube. The result is an enormous increase in depth of field (about 500 times that of a light microscope), with high resolution (10× the light microscope) and a very wide range of magnification (50 to over 100,000×). Another important advantage is the speed gained in preparing specimens. There is no need for either ultrathin sections or the high vacuum required by the standard electron microscope.

To appreciate more fully the subcellular level and its extraordinary complexity, we first should view the *micro* and *macro* aspects of living organisms. Therefore, we shall return to the light microscope to observe the *cellular* and *tissue* levels of living organization. The knowledge obtained then can be employed in our view of the function and appearance of *organs* and *organ systems*. This in turn will lead us into the macroscopic world, in which all of the subordinated levels interact to produce the living, whole animal—the *organismal* level. Our view must encompass still larger groupings. As we are so keenly aware, organisms interact in *communities*, and communities interact within the environmental complex—an *ecological framework* when viewed in a given time interval, an *evolutionary framework* when viewed over vast spans of time.

Suggested References

Bradbury, S., 1967. *The Evolution of the Microscope*. New York: Pergamon Press.

Gray, P., 1967. *The Use of the Microscope*. New York: McGraw-Hill.

Martin, L. C., and B. K. Johnson, 1958. *Practical Microscopy*, 3rd ed. London: Blackie & Son.

Needham, G. H., 1958. *The Practical Use of the Microscope, Including Photomicrography*. Springfield, Ill.: Charles C. Thomas.

Wyckoff, R. W. G., 1958. *The World of the Electron Microscope*. New Haven, Conn.: Yale University Press.

CHAPTER 2

CLASSIFICATION AND PHYLOGENY

1. Introduction

Classification, or *taxonomy*, has a dual purpose—to identify an organism and to arrange various organisms into a manageable classification system. This means that each organism must have a name that indicates its position in the system. Degrees of morphological, physiological, cytological, genetic, and ecological similarity indicate degrees of *relationship*. But to a remarkable degree, morphological evidence parallels the other indicators and serves as the most generally employed and most readily studied line of evidence. Hence, by using these criteria to arrange the most similar animals into particular groups, and these groups into larger and larger groups, we construct a formal classification scheme consisting of a series of categories, or *taxa*. Ordinarily, seven of these taxa are used. First and least inclusive is the *species* (plural: *species*[1]), composed of organisms that can interbreed (such as the varieties of house cats). Then, distinct but similar species (such as the domestic house cat, mountain lion, African lion, and tiger)

are placed together into a more inclusive category, called the *genus* (plural: *genera*), in this case, the genus *Felis*. Similar genera comprise a *family*, as Felidae, the cat family. (Note that the ending *idae* is used for all animal family names.) Thousands of different families have been described, each consisting of one to many genera that show some measure of affinity. Families that have some affinities are grouped into a yet more inclusive category, the *order* (for example, the CARNIVORA, or meat eaters, the ANTHROPOIDEA, or man-monkey-ape group). Orders showing some degree of similarity are collected into a *class*. Thus, CARNIVORA, ANTHROPOIDEA, and all other animals that have hair and suckle their young are combined into the class MAMMALIA. (Other classes combine all birds, or all reptiles, all frogs and salamanders, all insects, and so on.) The classes that show an even more general degree of similarity are combined into an enormous assemblage, the *phylum* (plural: *phyla*). The phylum **CHORDATA** includes fishes, amphibians, reptiles, birds, and mammals, as well as some "primitive" organisms that lack vertebrae. (The latter are usually little known to beginning students but very important from an evolutionary, or *phylogenetic*, standpoint.) Finally, all phyla of the most general or fundamental similarity are placed in one

---

[1] Observe that the term *species* is spelled the same in singular and plural; "specie" means gold or silver coin, not the name of an organism.

of the *kingdoms*, as ANAMALIA, the animal kingdom, or PLANTAE, the plant kingdom.

Although our primary concern in this course must be larger assemblages, phyla and classes we will utilize certain species as examples that serve to demonstrate characteristics of the entire group. When we look at a specimen it must be first and always a *species*, that is, it must have a specific identity, a name. Then it can be categorized as a member of larger, more inclusive groups.

■ With an arbitrary assemblage of 24 species (labeled 1 to 24), prepare a diagrammatic chart to show how these 24 species *could* be arranged into successive taxa to form a hierarchy employing all seven categories considered above. The arrangement you select is artificial, of course, but it should demonstrate the simplest basic taxonomic principles.
■ What do we mean by this statement: "Taxonomy arranges organisms into a series of categories of successively greater generality and inclusiveness"?

Further refinements are used to break down the seven taxa into still more categories. For example, *subspecies*, *subgenera*, *suborders*, and *subclasses* denote subdivisions of species, genera, orders, and classes into additional systematic levels or taxa. *Superspecies*, *superfamilies*, and *superclasses* also can be used. Many such higher or lower groupings can be interpolated into the taxonomic scheme, a mark of its flexibility and usefulness, enabling it to classify, arrange, and catalog into serviceable units.

Taxonomy consists of more than naming and recording diverse organisms. Taxonomic arrangement is our chart of relationship, and hence of evolution. In terms of change throughout the long history of life, divergence from an ancestral group is evolution. We attempt to construct a *natural classification* to represent true evolutionary patterns, rather than an *artificial classification*, such as amassing all swimming animals in one group and all flying animals in another. Evolutionary orderliness is the *sense* of taxonomy, the reason we emphasize larger categories and major patterns of relationship. Morphological divergences are the bedrock upon which both classification and our conception of evolutionary pattern are based.

The broadest taxonomic divisions, those between *phylum* and *kingdom*, are for our purposes chiefly of theoretical importance. These are the so-called *supraphyletic* categories — *branch*, *grade*, *level*, and *superphylum*—which deal with the most fundamental of biological characteristics. Criteria that are used to arrange the phyla suggest distant hereditary relationships and indicate clues to the earliest branching points in the history of life. The significance of such features as type of body cavity (which determines the taxon *level*), basic body symmetry (the *grade*), or fundamental arrangement of cells and tissues (the *branch*) cannot be studied in detail here. These differences are of particular interest to scholars of evolution. The overall taxonomic system is one combining logic, convenience, and enormous usefulness.

Nomenclature, the mechanics of naming, remains to be considered. Very specific rules ensure reasonable uniformity and wide international and interdisciplinary acceptance of names given to newly described organisms. These rules allow for the creation of new names when undescribed species are discovered. They also allow a degree of flexibility toward errors or changes required when names are found to be duplicated or a single species twice named. Rigorous standards are now required for a new name to qualify. Rules of nomenclature still must allow for disagreement among specialists and provide a scheme of priority to credit the first name given. In fact, an International Commission on Zoological Nomenclature, empowered to decide disputed cases (of which there are quite a number), adjudicates a complex code adhered to by scientists around the world. In addition, congresses of specialists convene from time to time to reconsider and revise the rules. Naming an organism is a demanding and specialized science; formulating the guiding rules is an even more specialized (and sometimes tedious but nonetheless important) responsibility. *Systematics* (the study of the system as a whole), *nomenclature* (the process of naming), and *taxonomy* (the placement of organisms in taxa) underlie all biology. Each biological name becomes part of an integrated system, a code respected and recognized everywhere. Each name then is of theoretical significance and practical usefulness as it provides both a handle and a lead to the entire taxonomic arrangement, telling us the probable relationships of each animal.

Since there are two parts to each animal's name, genus and species, the system of nomenclature is often called *binomial*. A complete species name also includes the name of the person who first described and established it, and the date this designation was published, thus: *Felis domesticus*

Linnaeus, 1759. Note that the name of an organism such as *Amoeba proteus* or *Homo sapiens* always includes the *generic* name (with first letter capitalized) and the *specific* name (first letter in lowercase). Reference to a "species" as the *name* of an animal should always include its genus in addition to its species designation, because identical species names (such as *domesticus, familiaris, americanus,* and so on) may be applied to a number of animals in different genera. The only way to sort out a particular kind of organism is to use the genus in addition to the species.

The generic name may be abbreviated to the first letter in a list or paragraph when there is no possibility of confusion. Both terms are *italicized* or underscored. Latin or Greek endings and forms are used, giving greater permanence and universality to the system, whether the scientist be Japanese, German, Indian, or an American college student.

The scheme of classification now in use for all plants and animals, living and extinct, goes back to the Swedish botanist Carl Linnaeus who established it in the eighteenth century. In fact, 1759, the date of publication of the tenth edition of Linnaeus' great work, *Systema Naturae*, marks the formal beginning of modern taxonomy. No name established prior to 1759 is considered valid. To be universally accepted, any name given to an organism after that date must conform to certain internationally accepted requirements. Once these conditions are met, the name is permanently enrolled in scientific literature and will not be removed unless a reevaluation justifies the conclusion that the animal was indeed not distinct enough to warrant being named a new species.

The following charts summarize, for present reference and future review, the major supraphyletic categories of animals, and then the subdivisions within each phylum. You should add the names of the examples you study to the sections where they belong, as a means of helping you to review the system and to recognize the larger categories. Finally, on the last page of this chapter, the inferred or hypothetical stages of complexity through which animals have passed in the course of evolution (as suggested by a great variety of biological, geological, and other evidence) is charted in a highly diagrammatic and necessarily conjectural fashion. Much of your course will be devoted to studying examples of this sequence and the nature of the evidence justifying the arrangement.

2. Major or Supraphyletic Categories and Phyla

Kingdom ARCHETISTA—acellular packets of DNA or RNA: viruses, rickettsiae, pleuropneumonia-like organisms

Kingdom MONERA—unicellular organisms lacking a nucleus (procaryotes): blue-green algae, bacteria

Kingdom PROTISTA—unicellular organisms with nucleus (eucaryotes): protozoa,[2] algae (except blue-green and green algae)

Kingdom PLANTAE—multicellular plants: green algae, nonseed plants, vascular plants

Kingdom FUNGI—yeasts and multicellular fungi

Kingdom ANIMALIA—multicellular animals

Branch MESOZOA	Phylum **MESOZOA**
Branch PARAZOA	Phylum **PORIFERA**
Branch EUMETAZOA	
Grade RADIATA	Phylum **CNIDARIA (COELENTERATA)**
	Phylum **CTENOPHORA**
Grade BILATERIA	
Level ACOELOMATA	Phylum **PLATYHELMINTHES**
	Phylum **NEMERTINEA (RHYNCHOCOELA)**
Level PSEUDOCOELOMATA	Phylum **ACANTHOCEPHALA**
	Phylum[3] **ASCHELMINTHES**

---

[2] Arranged by many authorities in a subkingdom PROTOZOA in the kingdom ANIMALIA, with the remaining animals placed in a subkingdom METAZOA. For our purposes, we will consider the PROTOZOA a phylum in the kingdom PROTISTA.

[3] Classes listed here are treated by some zoologists as six separate phyla.

Class NEMATODA
Class NEMATOMORPHA (or GORDIACEA)
Class ROTIFERA
Class GASTROTRICHA
Class KINORHYNCHA (or ECHINODERA)
Class PRIAPULIDA
Class GNATHOSTOMULIDA
Phylum **ENTOPROCTA**

Level EUCOELOMATA
    Superphylum SCHIZOCOELA   Phylum **BRYOZOA (ECTOPROCTA)**
                        Phylum **SIPUNCULOIDEA**
      (PROTOSTOMIA)   Phylum **ECHIUROIDEA**
                        Phylum **ANNELIDA**
                        Phylum **ARTHROPODA**
                        Phylum **MOLLUSCA**

    Superphylum ENTEROCOELA   Phylum **POGONOPHORA (BRACHIATA)**
                        Phylum **PHORONIDA** (relationship uncertain,
                           probably with enterocoels)
      (DEUTEROSTOMIA)   Phylum **CHAETOGNATHA**
                        Phylum **ECHINODERMATA**
    Term "Protochordata" includes   Phylum **HEMICHORDATA**
    Hemichordates, Urochordates,   Phylum **CHORDATA**
    and Cephalochordates
    (see p. 23)

3. Primary Subdivisions of the Phyla

Underlined names should become thoroughly familiar to you. The others, somewhat less important for our purpose, should be reviewed, though you will not be required to know their distinguishing characteristics unless this is specified by your instructor. Additional taxa (*group, superclass, division, suborder*) are used when necessary and are designated in the preceding larger unit being separated (for example, "phylum CHORDATA [chordates]. Separated into the following 2 *groups*: 1 . . . , 2 . . .").

    Animal groupings marked by an asterisk (*) are extinct.

PHYLUM	SUBPHYLUM	CLASS	SUBCLASS	ORDER

PROTOZOA

    SARCOMASTIGOPHORA (ameba-flagellate complex). Separated into 3 superclasses:
    1. MASTIGOPHORA
                PHYTOMASTIGOPHOREA (plant-like flagellates; *Euglena*)
                      (10 orders)
                ZOOMASTIGOPHOREA (animal-like flagellates; *Trypanosoma*)
                      (9 orders)
    2. OPALINATA (*Opalina*)
    3. SARCODINA (pseudopodia; true amebae)
                RHIZOPODEA (naked-bodied and testate amebae; *Amoeba,*
                    Entamoeba, Arcella, Difflugia)
                      (5 subclasses)

PHYLUM	SUBPHYLUM	CLASS	SUBCLASS	ORDER

GRANULORETICULOSEA (slender, granular reticulopodia)
>> FORAMINIFERIDA (chambered test; "forams," *Globigerina*)

ACTINOPODEA (spherical, with axopodia, silica or strontium test)
>> RADIOLARIA (radiolarians; *Hexacontium*)
>> HELIOZOIA (heliozoans; *Actinosphaerium*)

APICOMPLEXA

SPOROZOEA
>> GREGARINIDA (gregorines; *Monocystis*)
>> COCCIDIA (coccidia; *Eimeria, Plasmodium, Toxoplasma*)

PIROPLASMEA (small blood parasites, *Babesia, Theileria*)

MICROSPORA (*Nosema*)
MYXOSPORA (*Myxidium*)
CILIOPHORA   CILIATEA HOLOTRICHIA (*Paramecium; Tetrahymena*)
>> (7 orders)
>> PERITRICHIA (*Vorticella*)
>> SUCTORIA (*Podophyra*)
>> SPIROTRICHIA (*Euplotes, Stentor, Spirostomum*)
>> (6 orders)

__MESOZOA__ (primarily wormlike endoparasitic organisms; *Dicyema, Rhopalura, Microcyema*)

__PORIFERA__ (sponges)

CALCAREA (or CALCISPONGIAE) (calcareous sponges; *Scypha, Leucosolenia*)

HEXACTINELLIDA (or HYALOSPONGIAE) (siliceous or glass sponges; *Euplectella, Hyalonema*)

DEMOSPONGIAE (horny or bath sponges; *Cliona, Spongia, Euspongia*)

__CNIDARIA__ (__COELENTERATA__) (hydroids, jellyfishes, corals, sea anemones, and others)

HYDROZOA (hydrozoans)
>> HYROIDA (hydroids; *Hydra, Hydractinia, Obelia, Bougainvillia*)
>> TRACHYLINA (*Gonionemus Liriope*)
>> SIPHONOPHORA (Portuguese man-of-war, *Physalia*)
>> (hydrocorals)
>>> CHONDROPHORA (*Velella*)
>>> MILLEPORINA (*Millepora*)
>>> STYLASTERINA (*Stylantheca*)

SCYPHOZOA (jellyfishes; *Aurelia, Cassiopeia, Haliclystus*)

ANTHOZOA (sea anemones, corals)
>> ALCYONARIA (soft and horny corals; *Alcyonaria, Tubipora, Gorgonia, Pennatula, Renilla*)
>> ZOANTHARIA (stony corals and sea anemones; *Corynactis, Cerianthus, Metridium*)

__CTENOPHORA__ (comb jellies; *Pleurobrachia, Cestum, Beröe*)

__PLATYHELMINTHES__ (flatworm, flukes, tapeworms)

TURBELLARIA (free-living flatworms)
>> ACOELA (primitive flatworms; *Convoluta*)
>> RHABDOCOELA (flatworms with saclike gut; *Stenostomum, Microstomum*)
>> ALLOEOCOELA (*Plagiostomum*)

PHYLUM	SUBPHYLUM	CLASS	SUBCLASS	ORDER

TRICLADIDA (planarians; *Dugesia, Bipalium, Bdelloura*)

POLYCLADIDA (marine flatworms; *Hoploplana, Planocera*)

TREMATODA (flukes)

MONOGENEA (flukes with simple life cycles; *Polystoma, Gyrodactylus*)

DIGENEA (flukes with complex life cycles; *Fasciola, Schistosoma, Opisthorchis*)

CESTODA (tapeworms)

CESTODARIA (tapeworms with no proglottids; *Amphilina, Gyrocotyle*)

EUCESTODA (few to many proglottids; *Taenia, Diphyllobothrium*)

NEMERTINEA (RHYNCHOCOELA) (ribbon worms; *Protostoma, Tubulanus*)

ACANTHOCEPHALA (spiny-headed worms; *Echinorhynchus, Macracanthorhynchus*)

ASCHELMINTHES (pseudocoels)

NEMATODA (nematodes, roundworms; *Ascaris, Rhabditis, Trichinella, Enterobius*)

NEMATOMORPHA (GORDIACEA) (horsehair worms; *Paragordius, Gordius*)

ROTIFERA (rotifers, *Hydatina, Asplanchna, Philodina*)

GASTROTRICHA (gastrotrichs; *Chaetonotus*)

KINORHYNCHA (ECHINODERA) (*Centroderes, Echinoderella*)

PRIAPULIDA (*Priapulus*)

GNATHOSTOMULIDA (complex jaws with teeth) *Holectypus, Echinoneus*)

ENTOPROCTA (nodding heads; *Pedicellina, Loxosoma*)

BRYOZOA (ECTOPROCTA) (moss animals; *Bugula, Pectinatella*)

SIPUNCULOIDEA (peanut worms; *Sipunculus, Phaseolosoma*)

ECHIUROIDEA (spoon worms; *Urechis, Echiurus*)

ANNELIDA

OLIGOCHAETA (earthworms; *Lumbricus, Tubifex, Eisenia*)

POLYCHAETA (bristle worms; *Neanthes, Arenicola, Amphitrite, Chaetopterus, Sabellaria*)

ARCHIANNELIDA (simple marine worms; *Polygordius, Dinophilus, Chaetogordius*)

HIRUDINEA (leeches, *Hirudo, Acanthobdella, Glossiphonia*)

ARTHROPODA (arthropods)

ONYCOPHORA (*Peripatus;* "missing-link"—Arthropod–annelid group containing characteristics of both)

MANDIBULATA

CRUSTACEA (crustaceans)

BRANCHIOPODA (primitive crustacea with many leaflike appendages)

ANOSTRACA (fairy shrimps, brine shrimps; *Artemia*)

NOTOSTRACA (branchiopods with shieldlike carapace; tadpole shrimps; *Triops*)

CONCHOSTRACA (clam shrips; *Cyzicus*)

CLADOCERA (water fleas; *Daphnia*)

PHYLUM	SUBPHYLUM	CLASS	Subclass	order

Ostracoda (small, bean-shaped bivalved crustaceans; *Cypridina, Cypris*)

Branchiura (ectoparasite of fish; *Argulus*)

Copepoda (copepods; *Cyclops, Calanus*)

Cirripedia (barnacles; *Lepas, Balanus*)

Malacostraca (higher crustaceans). Separated into the following 5 *divisions* or *superorders*:

1. Phyllocarida (Leptostraca) (*Nebalia*)
2. Hoplocarida (Stomatopoda) (mantis shrimps; *Squilla*)
3. Syncarida (primitive fresh-water crustaceans; *Anaspides*)
4. Peracarida (crustaceans with a brood pouch)
   Mysidacea (abundant planktonic shrimps; *Mysis*)
   Cumacea (small marine crustaceans; *Diastylis*)
   Tanaidacea (*Tanais*)
   Isopoda (pill bugs, sow bugs; *Porcellio, Ligia, Oniscus*)
   Amphipoda (beach fleas; *Gammarus*)
5. Eucarida (crayfish, crabs, lobsters, shrimps, and allied forms)
   Euphausiacea (krill, planktonic shrimps; *Euphausia*)
   Decapoda (crayfish, lobsters, crabs, and shrimps). Separated into 3 *suborders*:
   1. Macrura (shrimps (*Cranson*); crayfish (*Astacus Cambarus*); lobsters *Homarus*)
   2. Anomura (hermit crabs; *Pagurus*)
   3. Brachyura (true crabs; *Cancer*)

MYRIAPODOUS ARTHROPODS

Chilopoda (centipedes; *Scutigera*)

Diplopoda (millipedes; *Spirobolus*)

Symphyla (centipede-like arthropods; *Scutigerella*)

Pauropoda (soft-bodied, blind, myriapods; *Pauropus*)

INSECTA

Apterygota (primitive wingless insects)
   Protura (small, soft protoinsects; *Acerentulus*)
   Collembola (springtails; *Entomobyra*)
   Thysanura (silverfish; *Lepisma*)
   Diplura (japygids; *Campodea*)

Pterygota (winged insects). Separated into the following 2 *divisions*:

1. Hemimetabola (development by nymphal stages)
   Orthoptera (grasshoppers, crickets; *Microcentrum*)
   Dictyoptera (roaches; *Periplaneta*)
   Dermaptera (earwigs; *Forficula*)
   Plecoptera (stoneflies; *Pteronarcys*)

PHYLUM   SUBPHYLUM   CLASS   SUBCLASS   ORDER

ISOPTERA (termites; *Kalotermes*)
ODONATA (dragonflies; *Gomphus*)
EMBIOPTERA (embiids; *Haploembia*)
EPHEMEROPTERA (mayflies; *Ephemera*)
MALLOPHAGA (biting lice; *Trichodectes*)
ANOPLURA (sucking lice; *Pediculus*)
CORRODENTIA (book lice; *Liposcelis*)
THYSANOPTERA (thrips; *Thrips*)
HEMIPTERA (true bugs; *Cimex*)
HOMOPTERA (aphids, scale insects; *Magicicada*)

2. HOLOMETABOLA (4-stage development with complete metamorphosis)

MECOPTERA (scorpionflies; *Panorpa*)
NEUROPTERA (lacewings; *Sialis*)
TRICHOPTERA (caddisflies; *Phryganea*)
LEPIDOPTERA (butterflies, moths; *Colias*)
DIPTERA (flies; *Musca, Tabanus*)
SIPHONAPTERA (fleas; *Pulex*)
COLEOPTERA (beetles; *Tenebrio*)
STREPSIPTERA (stylopids; *Stylops*)
HYMENOPTERA (ants, wasps, bees; *Bombus, Apis*)

CHELICERATA

*TRILOBITA (abundant in Palaeozoic seas; *Trianthrus*)
MEROSTOMATA
XIPHOSURA (horseshoe crabs; *Limulus*)
*EURYPTERIDA (giant water scorpions; *Eurypterus*)
ARACHNIDA (spiders, scorpions, and allies)

SCORPIONIDA (scorpions; *Hadrurus*)
PALPIGRADA (small, primitive arachnids; *Prokoenia*)
PEDIPALPI (whip scorpions; *Hypoctonus*)
PSEUDOSCORPIONIDA (false scorpions; *Chelifer*)
SOLPUGIDA (sun spiders; *Eremobates, Tarantula*)
PHALANGIDA (harvestmen or daddy long-legs; *Phalangium*)
ARANEIDA (spiders; *Latrodectus*)
ACARINA (mites, ticks; *Argas, Dermacentor*)

PYCNOGONIDA (sea spiders; *Nymphon*)
PENTASTOMIDA (*Porocephalus*) (tongue worms or linguatulids, parasites of uncertain affinities)
TARDIGRADA (water bears; *Macrobiotus*)
MOLLUSCA (snails, clams, octopuses, and allies)

MONOPLACOPHORA (recently discovered metameric molluscs thought to be extinct for 300 million years; *Neopilina*)
AMPHINEURA (chitons, coat-of-mail shells; *Chiton, Mopalia*)
GASTROPODA (snails and allies)

PROSOBRANCHIA (sea snails, limpets; *Haliotis, Tegula, Acmaea, Fissurella*)

PHYLUM	SUBPHYLUM	CLASS	Subclass	order

OPISTHOBRANCHIA (sea slugs; *Aplysia Aeolis, Clione, Doris*)

PULMONATA (most fresh-water and land snails; *Helix, Lymnaea, Physa, Planorbis, Limax*)

SCAPHOPODA (tooth shells; *Dentalium*)

PELECYPODA (BIVALVIA) (clams, oysters, mussels; *Anodonta, Mercenaria, Mya, Mytilus*)

CEPHALOPODA (nautilus, squid, octopus; *Nautilus, Loligo, Sepia, Octopus*)

POGONOPHORA (BRACHIATA) (beardworms; *Polybrachia*)

PHORONIDA (*Phoronis*)

BRACHIOPODA (lampshells; *Lingula*)

CHAETOGNATHA (arrowworms; *Sagitta*)

ECHINODERMATA (echinoderms)

PELMATAZOA

CRINOIDEA (sea lilies, also includes several related classes known only as fossils; *Antedon, Heliometra*)

ELEUTHEROZOA

ASTEROIDEA (sea stars, *Pisaster, Asterias, Patiria*)

OPHIUROIDEA (brittle stars; *Ophiura, Ophioplocus, Gorgonocephalus*)

ECHINOIDEA (sea urchins, sand dollars; *Arbacia, Echinus, Strongylocentrotus, Dendraster*)

HOLOTHUROIDEA (sea cucumbers; *Holothuria, Cucumaria, Stichopus, Thyone, Leptosynapta*)

HEMICHORDATA (acorn worms; *Balanoglossus, Saccoglossus, Cephalodiscus*)

CHORDATA (Chordates). Separated into the following 2 *groups*:

1. **ACRANIATA**

TUNICATA (or **UROCHORDATA**) (tunicates)

LARVACEA (planktonic, neotenic forms; *Appendicularia*)

ASIDIACEA (sea squirts; *Ciona, Ascidia*)

THALLIACEA (planktonic, chain tunicates; *Salpa*)

CEPHALOCHORDATA (amphioxus, lancelets; *Branchiostoma*)

2. **CRANIATA**

AGNATHA (vertebrates without jaws)

*OSTRACODERMI (primitive jawless armored fishes; *Pterolepis*)

CYCLOSTOMATA (lampreys and hagfish, living jawless fishes; *Petromyzon*)

GNATHOSTOMATA (vertebrates with jaws). Separated into the following 2 *superclasses*:

1. **PISCES** (fishes)

*PLACODERMI (early jawed fishes; *Climatius*)

CHONDRICHTHYES (cartilaginous fishes, sharks, rays, chimaeras)

SELACHII (sharks, rays; *Raja*)

HOLOCEPHALI (chimaeras; *Chimaera*)

OSTEICHTHYES (bony fishes)

PALAEOPTERYGII (sturgeons, spoonbills, and allies; *Acipenser*), 4 orders.

NEOPTERYGII (ganoid and teleost fishes; *Hippocampus*), 31 orders

CHOANICHTHYES (lungfish, lobe-finned fishes; *Neoceratodus*), 2 orders

PHYLUM   SUBPHYLUM   CLASS   Subclass   order

2. <u>TETRAPODA</u> (four-legged animals)

<u>AMPHIBIA</u> (amphibians)

Stegocephalia, 4 orders, including

*labyrinthodonti (ancient giant salamanders)

gymnophiona (or APODA) (caecilians; legless amphibians; *Ichthyophis*)

<u>Caudata</u> (salamanders, newts)

proteida (*Necturus*)

mutabilia (true salamanders; *Ambystoma*)

meantes (*Siren*)

<u>Salientia</u> (or <u>Anura</u>) (frogs and toads), 5 orders, including

procoela (true toads; *Bufo*)

diplasiocoela (true frogs; *Rana*)

<u>REPTILIA</u> (snakes, lizards, turtles, alligators)

<u>chelonia</u> (turtles; *Chelonia*)

rhynchocephalia (*Sphenodon*)

<u>squamata</u> (lizards, snakes; *Python*)

<u>crocodilia</u> (alligators, crocodiles; *Crocodylus*)

*cotylosaurs (primitive reptiles)

ichthyosaurs (short-necked fishlike marine reptiles)

*pleisiosaurs (long-necked fishlike marine reptiles)

*mosasaurs (lizardlike marine reptiles)

*thecodonts (early running reptiles)

*dinosaurs, 2 orders

*mammal-like reptiles, 2 orders

<u>AVES</u> (birds)

*Archaeornithyes (ancient birds; *Archaeopteryx*)

Neornithyes (true birds), 30 orders

Examples:

pelecaniformes (pelicans; *Pelecanus*)

anseriformes (ducks; *Anas*)

falconiformes (falcons and hawks; *Falco*)

galliformes (chickens, fowl; *Gallus*)

columbiformes (pigeons; *Columba*)

passeriformes (perching birds; *Troglodytes*)

<u>MAMMALIA</u> (mammals)

Prototheria (monotremes, platypus; *Ornithorhynchus*)

*Allotheria (multituberculates)

Theria (marsupials and placentals), separated into the following 3 *divisions*:

1. *Pantotheria

2. <u>Metatheria</u> (marsupials; kangaroos, wallabies; *Didelphis*)

3. <u>Eutheria</u> (placental mammals), 25 orders, many extinct.

PHYLUM	**SUBPHYLUM**	**CLASS**	SUBCLASS	ORDER
				INSECTIVORA (shrews, moles; *Sorex*)
				CHIROPTERA (bats; *Myotis*)
				DERMOPTERA (flying lemurs; *Galeopithecus*)
				PRIMATES (lemurs, monkeys, apes, man; *Macaca, Gorilla, Homo*)
				EDENTATA (sloths; *Bradypus*)
				PHOLIDOTA (scaly anteaters; *Manis*)
				TUBULIDENTATA (aardvarks; *Orycteropus*)
				LAGOMORPHA (hares, rabbits; *Lepus*)
				RODENTIA (rodents, *Marmota, Rattus*)
				CETACEA (whales, dolphins, porpoises; *Tursiops, Balaena*)
				CARNIVORA (cats, dogs, and similar carnivores; *Canis, Felis*)
				PROBOSCIDEA (elephants; *Loxodonta*)
				HYRACOIDEA (hyraxes; *Procavia*)
				SIRENIA (manatees, dugongs; *Trichechus*)
				PERISSODACTYLA (odd-toed hoofed herbivores: horses, asses, zebras; *Equus*)
				ARTIODACTYLA (even-toed hoofed herbivores: pigs, camels, deer, goats, sheep, cattle; *Sus, Ovis, Bison*)

4. Evolution Chart

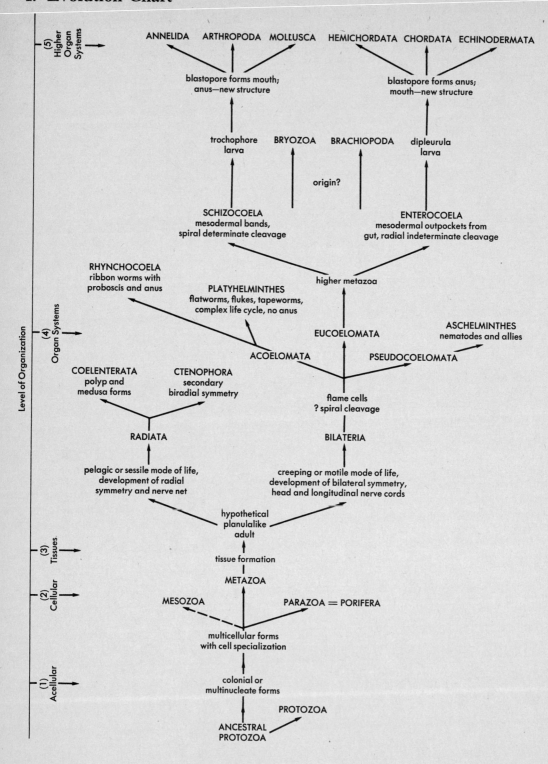

Suggested References

De Beer, G., 1954. "The evolution of the Metazoa," in *Evolution as a Process*, pp. 22–33, J. C. Huxley, A. L. Hardy, and E. B. Ford (eds.). London: G. Allen.

———, 1958. *Embryos and Ancestors*, 3rd ed. London: Oxford University Press.

Dodson, E. O., 1971. "The Kingdoms of Organisms." *Syst. Zool.*, 20:265.

Hadzi, J., 1963. *The Evolution of the Metazoa*. New York: Pergamon Press.

Hand, C., 1959. "On the origin and phylogeny of the Coelenterates," *Syst. Zool.*, 8:163.

Hanson, E., 1958. "On the origin of the Eumetazoa." *Syst. Zool.*, 7:16.

Hardy, A. C., 1953. "On the origin of the Metazoa." *Quart. J. Microsc. Sci.*, 94:441.

Honigberg, B. M., *et al.*, 1964. "A revised classification of the Phylum Protozoa." *J. Protozool.*, 11:7.

Huxley, J. (ed.), 1940. *The New Systematics* (A symposium on problems of systematics). New York: Oxford University Press.

Jagersten, G., 1950. "On the early phylogeny of the Metazoa." *Zool. Bidrag.*, 30:321.

Kerkut, G. A., 1960. *The Implications of Evolution*. New York: Pergamon Press.

Mayr, E., 1966. *Animal Species and Evolution*. Cambridge, Mass.: Harvard University Press.

———, E. G. Linsley, and R. L. Usinger, 1953. *Methods and Principles of Systematic Zoology*. New York: McGraw-Hill.

Oparin, A., 1945. *The Origin of Life*, 3rd ed. New York: Dover Publications.

Simpson, G. G., 1961. *Principles of Animal Taxonomy*. New York: Columbia University Press.

Stunkard, H. W., 1954. "The life-history and systematic relations of the Mesozoa." *Quart. Rev. Biol.*, 29:230.

Whittaker, R. H., 1959. "On the broad classification of organisms." *Quart. Rev. Biol.*, 34:210.

CHAPTER 3

PHYLUM PROTOZOA

1. Introduction

The first sight of *Paramecium* (Fig. 3.14) racing across the microscope field or of an *Amoeba* (Fig. 3.4) extending its granule-filled pseudopod is a sight few biologists ever forget. To realize the abundance, variety, and complexity of these minute creatures is to extend one's knowledge into a new dimension. Once we recognize the role that protozoa play in human life and in the total economy of nature, we can also appreciate their importance in contemporary biological research. We then look at these single-celled organisms with increased respect and wonder.

Protozoa are not confined to "simple" *Amoeba, Paramecium,* or other textbook examples. Members of this enormous group range from pathogenic organisms (such as those causing malaria or African sleeping sickness) to marine organisms of incredible beauty, complexity, and abundance. A vast range of forms occupy equally varied habitats. (See Figs. 3.1, 3.2, and 3.3.) The remains of marine protozoa cover thousands of square miles of ocean bottom, sometimes a mile or more deep. A few straws placed in a jar of water will yield a culture of myriads of protozoa whose species com-

position will change from day to day—a spectacular assemblage of enormous variety.

We will study the PROTISTA by considering the **PROTOZOA**, a large assemblage of protists with animal-like characteristics. The **PROTOZOA** are frequently considered a phylum in the kingdom ANIMALIA, but this classification scheme causes considerable confusion. Certain protozoans, chlorophyll-containing *Euglena*, for instance, *also* have plantlike characteristics that contradict their placement in the animal world (Fig. 3.8). The classification of protozoans as protists is a convenience that enables us to look at them as a phylum, organisms with very general similarities that may be placed together. By placing the **PROTOZOA** in the PROTISTA we avoid a confusing and often meaningless division of the groups into plant and animal kingdoms.

A protist is a complete organism within a single, undivided unit. It seems incongruous to some experts to call the organism a "cell," since the cell performs basically the same functions as do the many trillion cells of higher organisms. If a cell is, properly speaking, only a *part* of a living whole, then the **PROTOZOA** (as well as all protists) clearly are exceptions, for they carry on all the busi-

Figure 3.1. Selected examples of Protozoa (Protista). (a) Two gregarines in syzygy; (b) Isolated suctorian from amphipod appendage, ×200; (c) *Ceratium*; (d) *Opalina*; (e) *Vorticella*; (f) *Spirostomum*; (g) *Stentor*; (h) *Actinosphaerium*; (i) *Arcella*: (*top*) side view, (*bottom*) top view.

gregarines ←

ceratium

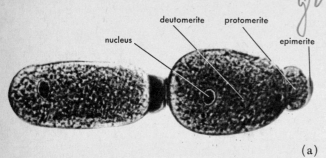

nucleus · deutomerite · protomerite · epimerite

(a)

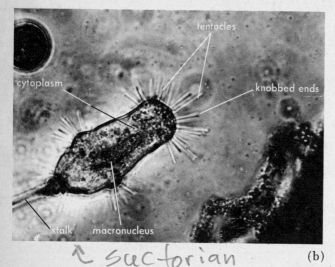

tentacles · knobbed ends · cytoplasm · stalk · macronucleus

← *suctorian*

(b)

ness of living in a single unit. Hence, many biologists prefer to call members of this group *acellular* rather than *unicellular*. All agree, nonetheless, that **PROTOZOA** are far from simple. These organisms carry out essential life processes in a small but incredibly complex unit—appreciated now more fully through electron microscopy and other modern research methods. Protozoans are also proving to be invaluable organisms for investigating such processes as reproduction and cell division (Figs. 3.10, 3.13, 3.18, 3.19, 3.20, and 3.21), cell movement, transfer of substances through cell membranes, and other physical and biochemical problems in physiology, genetics, ecology, and evolution. Results from protozoan research are often pertinent to understanding the biological processes and functions of higher plants and animals.

In this laboratory we can study only a few organisms. Hence, we emphasize that the **PRO-**

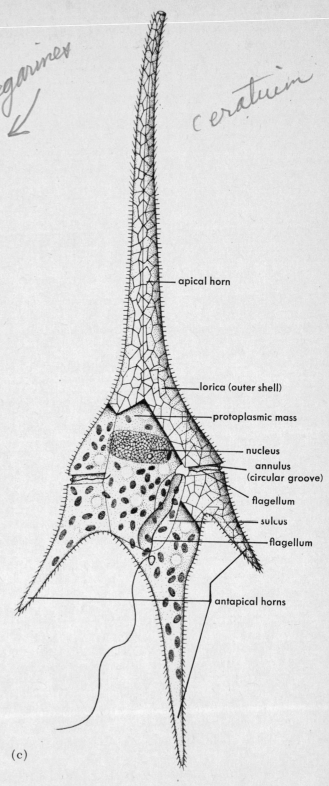

apical horn

lorica (outer shell)

protoplasmic mass

nucleus

annulus (circular groove)

flagellum

sulcus

flagellum

antapical horns

(c)

TOZOA are a vast and varied assemblage found wherever living things can survive, though only a few can be observed in our survey.

opalina

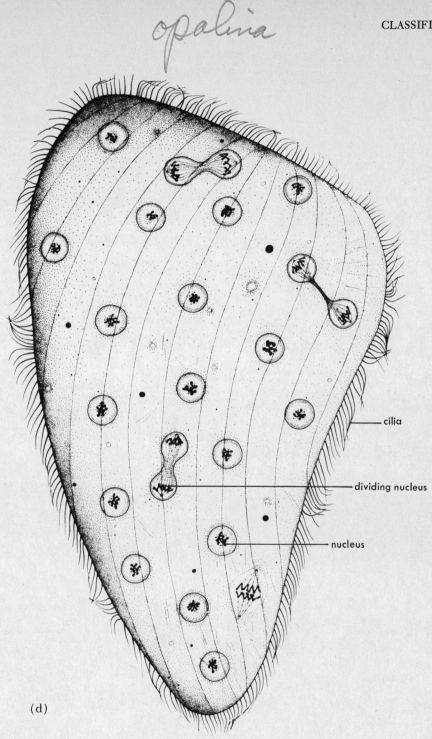

cilia

dividing nucleus

nucleus

(d)

■ **List some uses of protozoa in theoretical biology and research.**

■ **Why should research on protozoa be of value to understanding human processes and functions?**

■ **Compare the term unicellular with acellular. What does each imply?**

2. Classification

Since acellular organisms (including many nonprotozoans) include members whose affinities lie with the plant kingdom or the animal kingdom or perhaps with both, they are thought to form the evolutionary base of life.

Vorticella

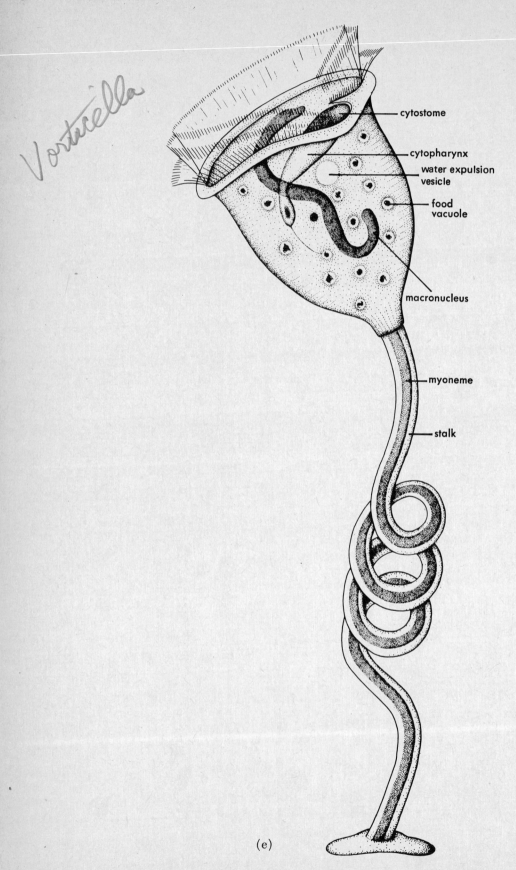

cytostome

cytopharynx

water expulsion
vesicle

food
vacuole

macronucleus

myoneme

stalk

(e)

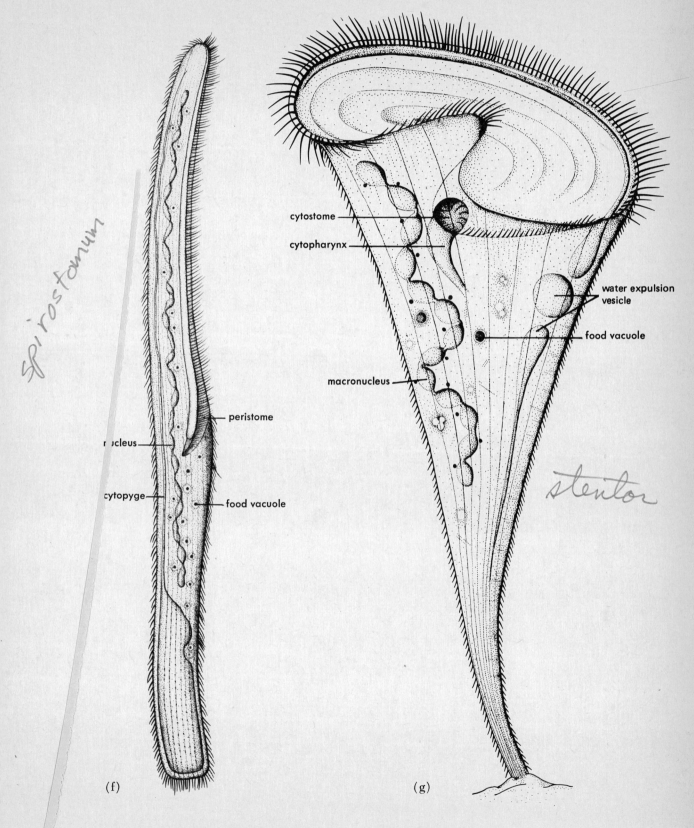

Spirostomum

cytostome

cytopharynx

water expulsion
vesicle

food vacuole

macronucleus

peristome

nucleus

cytopyge

food vacuole

stentor

(f)   (g)

We therefore have placed the protozoa within a kingdom apart from all multicellular animals and plants—the PROTISTA. The PROTISTA is divided into several very large groups, here considered phyla as a matter of convenience, but called subkingdoms by some authors. Do not be disturbed by such differing ideas. They reflect changing ideas and the dynamic state of the science. At the present

Actinosphaerium

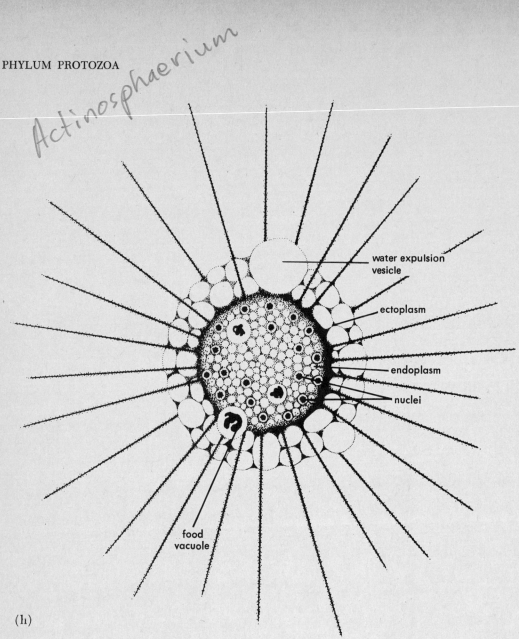

water expulsion vesicle

ectoplasm

endoplasm

nuclei

food vacuole

(h)

time, no expert and certainly no textbook can present more than an opinion or a consolidation of many viewpoints.

The following arrangement of the protozoa is a modification of the scheme used by many modern protozoologists:

Classification of the Phylum PROTOZOA. Adapted from Honigberg *et al.* (1964), Sherman and Sherman (1970), Hammond and Long (1973), and other sources (see references at end of chapter).

Phylum **PROTOZOA**

Subphylum I   **SARCOMASTIGOPHORA**
Having flagella, pseudopodia, or both; single type of nucleus.

Superclass I   MASTIGOPHORA
One or more flagella. Asexual reproduction by longitudinal binary fission.

Class 1   PHYTOMASTIGOPHOREA
Chromatophores, and one or two flagella most commonly.

Order 1   CHRYSOMONADIDA
One to three flagella, yellow-brown chromatophores. Store leucosin and lipid.

Order 2   CRYPTOMONADIDA
Two flagella, body compressed, typically two chromatophores. Store starch, e.g., *Chilomonas.*

Order 3   DINOFLAGELLIDA
Two flagella, one transverse and one trailing, body with trans-

Arcella

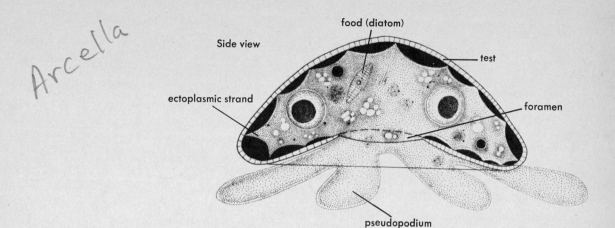

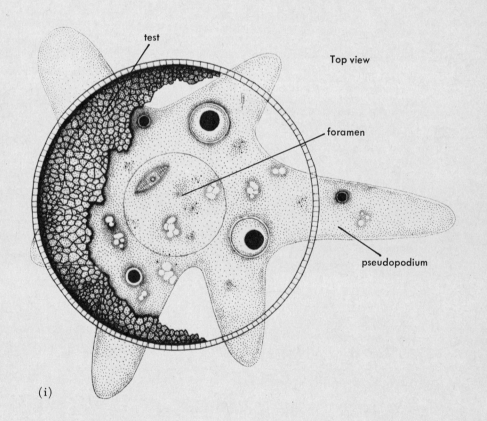

(i)

verse and longitudinal grooves.
Store starch and lipids, e.g.,
Ceratium.

Order 4   EUGLENIDA
One or two flagella, green chromatophores, body plastic. Store
paramylum, e.g., *Euglena, Asta-
sia*.

Order 5   CHLOROMONADIDA
Two flagella, one trailing, body
flattened. Store lipids and glycogen.

Order 6   VOLVOCIDA
Two to four flagella, green chro-
matophores. Store starch, e.g.,
Chlamydomonas, Volvox.

Class 2   ZOOMASTIGOPHOREA
Chromatophores absent; one to
many flagella.

Order 1   CHOANOFLAGELLIDA
Anterior flagellum surrounded
by a collar, e.g., *Codosiga, Pro-
terospongia*.

Figure 3.2. Assorted *Radiolaria*, ×80 (Courtesy CCM: General Biological, Inc., Chicago.)

Order 2   RHIZOMASTIGIDA
Pseudopodia and one to four flagella, e.g., *Naegleria*.

Order 3   KINETOPLASTIDA
One to four flagella; kinetoplast present, e.g., *Trypanosoma*.

Order 4   HYPERMASTIGIDA
Numerous flagella, e.g., *Trichonympha*.

Order 5   TRICHOMONADIDA
Four to six flagella, with one recurrent, e.g., *Trichomonas*.

Superclass II   OPALINATA
Numerous cilialike organelles, many similar nuclei, e.g., *Opalina*.

Superclass III   SARCODINA
Pseudopodia; flagella, if present, restricted to developmental stages.

Class 1   RHIZOPODEA
Locomotion by lobopodia, filopodia, or reticulopodia.

Subclass 1   LOBOSIA
Pseudopodia lobose, rarely filiform or anastomosing.

Order 1   AMOEBIDA
Naked, typically uninucleate, e.g., *Amoeba, Pelomyxa, Entamoeba*.

Subclass 2   GRANULORETICULOSIA
Fine, granular reticulopodia.

Order 1   FORAMINIFERIDA
Test with chambers; reticulopods protrude from aperture or perforations or both, e.g., *Globigerina, Dentalina, Cornuspira*.

Class 2   ACTINOPODEA
Spherical, axopodia, test of silica or strontium sulfate.

Subclass 1   RADIOLARIA
Siliceous skeleton with central capsule. Marine, e.g., *Hexacontium, Aulucantha*.

Subclass 2   HELIOZOA
No central capsule, skeleton siliceous. Marine or freshwater, e.g., *Actinosphaerium, Clathrulina*.

Subphylum II   **APICOMPLEXA**
All species parasitic; apical complex present, spores typically found.

Class 1   SPOROZOEA
Oocysts or spores found; locomotion by gliding or body flexion, cilia absent.

Subclass 1   GREGARINIDA
Extracellular parasites, chiefly of invertebrates, e.g., *Monocystis, Gregarina, Stylocephalus*.

Subclass 2   COCCIDIA
Typically intracellular parasites of vertebrates; alternation of sexual and asexual reproduction, e.g., *Eimeria, Isospora, Toxoplasma, Adelina, Plasmodium*.

Class 2   PIROPLASMEA
Small intracellular parasites, chiefly of vertebrate blood cells, apical complex reduced, spores absent, transmitted by ticks, e.g., *Babesia, Theileria*.

Subphylum III   **MICROSPORA**
All species parasitic; spores of unicellular origin, e.g., *Nosema, Glugea*.

Subphylum IV   **MYXOSPORA**
Chiefly fish parasites, spores of multicellular origin, e.g., *Ceratomyxa, Myxidium*.

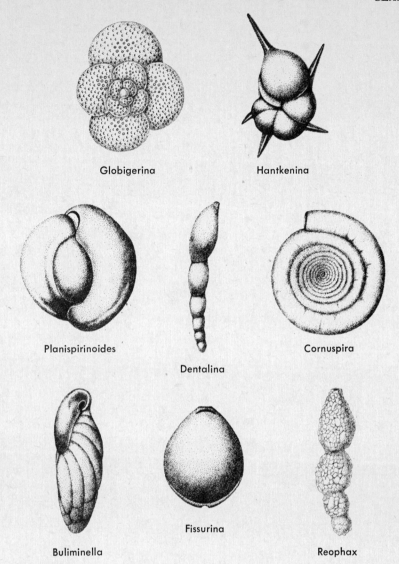

Globigerina

Hantkenina

Planispirinoides

Dentalina

Cornuspira

Buliminella

Fissurina

Reophax

Figure 3.3. Assorted *Foraminifera*.

Subphylum V	**CILIOPHORA** Cilia at some stage in the life cycle; two types of nuclei.	Subclass 4	SPIROTRICHIA Body cilia sparse, adoral zone with membranelles winding clockwise toward cytostome.
Subclass 1	HOLOTRICHIA Uniform ciliation on body, e.g., *Paramecium, Tetrahymena*.	Order 1	HETEROTRICHIDA Somatic cilia when present uniform, body large, often pigmented, e.g., *Stentor, Blepharisma, Spirostomum*.
Subclass 2	PERITRICHIA Body ciliation reduced or absent; oral cilia conspicuous, winding around apical pole counterclockwise toward cytostome. Body often attached, e.g., *Vorticella*.	Order 2	OLIGOTRICHIDA
		Order 3	TINTINNIDA Having lorica.
Subclass 3	SUCTORIA Adults lack cilia. Typically sessile with ingestion by sucking tentacles, e.g., *Tokophyra, Podophyra*.	Order 4	HYPOTRICHIDA Cirri on ventral surface, adoral zone of membranelles prominent. Organism flattened, e.g., *Euplotes*.

Try to place your laboratory examples within the nearest major category shown in this classification. After becoming more familiar with them, you may also wish to place them within their smaller taxonomic ranks, *order, family,* and *species.*

■ Why do classification systems undergo continual change and reevaluation?

■ What are the main objections to placing the PROTOZOA in the kingdom ANIMALIA?

3. Laboratory Instructions

I. *Amoeba proteus* (Figs. 3.4 and 3.5) (Which superclass and class does this organism represent?)

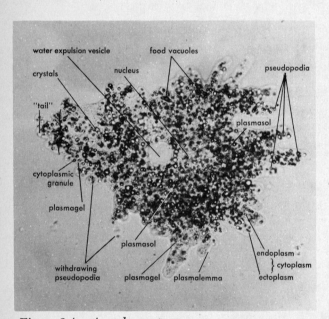

Figure 3.4.   *Amoeba proteus.*

1.   Structure   Examine closely a stained slide of *Amoeba.* Note the general outline and structures often difficult to distinguish in living amebae,[1] such as *ectoplasm, endo-*

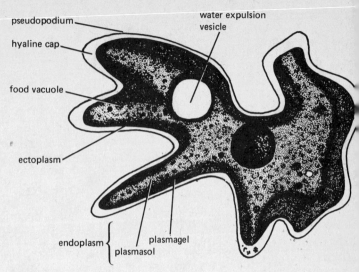

Figure 3.5.   A diagrammatic representation of *Amoeba proteus* showing main morphological features.

plasm, food vacuoles, nucleus, pseudopodium, and *water expulsion vesicle* (formerly called contractile vacuole).[2] See Figure 3.6.

Vary the light intensity with the iris diaphragm and condenser and observe the result. Learn to achieve a three-dimensional effect by constantly raising and lowering the fine

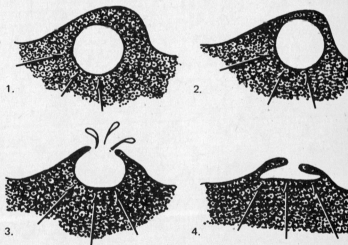

Figure 3.6.   Endoplasm forces the water vesicle against the cell membrane and both rupture. The water expulsion vesicle collapses under cytoplasmic pressure and its contents are forced out of the cell body.

---

[1] Often spelled in simplified form when term is used as a common name. Such changes, however, are not permitted for the proper binomial name under the rules governing systematic nomenclature.

[2] The more descriptive term is used since it is now known that this organelle has no muscular contraction. Cytoplasmic pressure collapses the vesicle.

focus, enabling you to see different depths in the animal. This will give you an idea of the organism's shape.

Now examine a drop of material containing living amebae. Amebae can be found in fresh and salt water, in the soil, and as parasites within animals (including man). Near the drop, place bits of filter paper, glass chips, or two short pieces of thread across the slide to support the coverslip and prevent the crushing of the specimens as the fluid evaporates. Be careful not to push the organisms out from under the coverslip when you place it over your preparation.

First search the field under low power to locate the specimens. The protozoa will appear gray or bluish and show fine granulations. Adjust the light and then examine your specimen under high power.

Watch carefully, look for granular movement, which indicates protoplasmic streaming. Observe pseudopod formation (temporary protoplasmic protrusions for movement or for enveloping food), retraction of a pseudopod and replacement by other pseudopodia.

■ How does the ameba form pseudopodia? (Even experts argue about this. See Figure 3.7.)

Draw the ameba in four or five outlines made 1 or 2 min apart. Show the *actual sequence* of movement of your specimen.

■ Can you trace protoplasmic streaming during formation of a pseudopod?

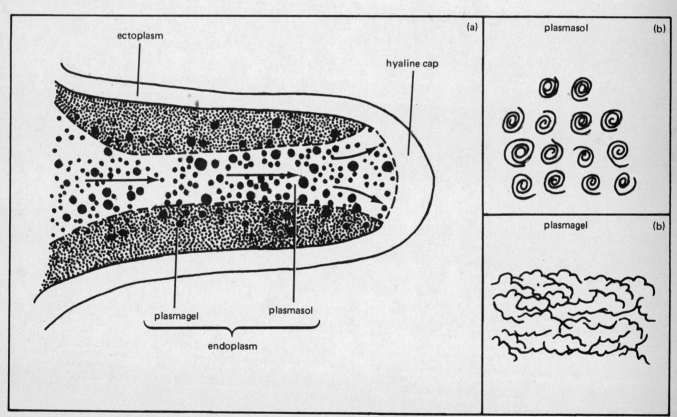

Figure 3.7. Locomotion in *Amoeba proteus*. (a) Posterior plasmagel becomes plasmasol and moves (possibly by contraction of posterior plasmagel, though this is argued) forward through the plasmagel tube into the pseudopodium, where it is turned outward and becomes plasmagel. The more rigid plasmagel tube is continually built up in the direction of movement, and broken down at the opposite end; (b) The transition from sol to gel is analogous to cooking an egg. In the sol state, molecules are folded up into small spheres that do not interfere with the movement of other molecules. In the gel state, these molecules have unfolded, offering interference to the movement of neighboring molecules. The gel to sol transition is simply the reverse of this process.

Study amebae from a starved culture to which paramecia have recently been added.

Watch for food capture or the presence of living paramecia or other organisms trapped and held in *food vacuoles* within the ameba. Later, egestion of undigestible portions of the prey will occur.

■ **Are the prey still moving about in the vacuole? What finally kills them?**

Look for the large, clear, bubblelike water expulsion vesicle usually seen in the trailing portion of the animal. Watch it fill and suddenly empty.

Draw a large specimen and include as many of the following structures as you can find (in your *specimen,* not the text): *nucleus, water expulsion vesicle, food vacuoles, granules.* Observe the *ectoplasm* with *plasmalemma* (outermost membrane enclosing a thin clear zone). The bulk of the ameba is endoplasm divided into a gelatinous *plasmagel* underlying the ectoplasm, and a central liquid *plasmasol,* marked by protoplasmic streaming.

■ **Can you see changes in the plasmasol and plasmagel?**
■ **How are they related to motility?**

2. Observations Watch the tip of a pseudopod and the posterior or trailing portion of endoplasm for changes in their consistency.

Record your observations by sketches and notes and try to develop some conception of the intricate structure, movements, feedings, and other habits of this allegedly simple animal.

■ **What is the function of the water expulsion vesicle?**
■ **What is implied by a high rate of pumping? a slow rate?**
■ **Why do many marine rhizopods lack water expulsion vesicles?**

Check the response by the ameba to a light tap on the coverslip and watch it change shape. Try to observe one floating or slowly sinking.

■ **What shape is it?**

■ **What happens to the shape after contact with glass?**
■ **What is the normal habitat of *Amoeba proteus*?**
■ **What rhizopods are found in the ocean?**
■ **Which ones inhabit the human intestine?**
■ **Which occur in the human mouth?**
■ **How are fossil amebae utilized in oil geology?**

II. *Euglena viridis* (What superclass and class are represented by this form?)

Examine microscopically a drop from a culture of living *Euglena* (Fig. 3.8). These green protists live in fresh-water ponds where they often coat the surface and may turn the pond green (an algal "bloom"). Observe your specimen carefully under reduced illumination, at both low and high power.

Notice the slow *directed* movement.

■ **Can you observe wormlike flexing as well as free swimming movements? Focus critically.**
■ **Can you see the undulipodium[3] whipping about near the narrow anterior end?**
■ **What technique might you use to observe *Euglena* in order to discover the way undulipodia propel this organism? (Figure 3.9)**
■ **Can you find the red *stigma* or eyespot?**

Notice the flexibility as some specimens coil about and appear to twist, extend, and contract. Continue observation under varying light, *continually changing the fine adjustment to develop your three-dimensional sense.* Observe the striated appearance of the outer *pellicle,* then locate internal structures. After 15–20 min of careful observation, aided by sketches and notes, carefully make a large drawing to show the normal or expanded profile.

_____

[3] Undulipodia is the term we will frequently use for the filamentous projections separately termed *cilium* (plural: *cilia*) and *flagellum,* (plural: *flagella*).

Undulipodium (plural: *-ia*): undul (L. *undulatus,* wavy, from *unda,* a wave) + i + podium (Gr. *pous, podos,* foot) . . . waving foot! This term can be used synonymously for cilia *or* flagella, which grossly differ (in length) but have essentially the same function and identical submicroscopic structure as seen in the electron microscope.

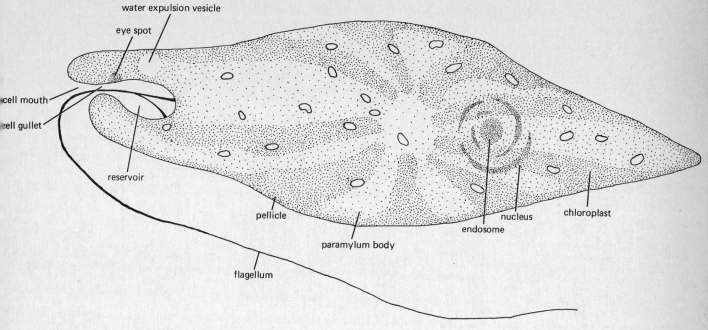

water expulsion vesicle

eye spot

cell mouth

cell gullet

reservoir

pellicle

paramylum body

flagellum

endosome

nucleus

chloroplast

Figure 3.8. *Euglena viridis.*

Illustrate the *gullet, stigma, chloroplasts, unduli-podium, pyrenoid granules, nucleus,* and *water expulsion vesicles* (grouped near the stigma). Add notes on motility or action you observed. Include with your sketch the *magnification* of the drawing (not just of the lens system) and the *classification* of the organism.

Increased differentiation of certain structures is possible with a light iodine stain (which colors special storage granules containing starch or paramylum). Vital dyes such as methyl violet or methylene blue may help reveal the undulipodium.

Study stained demonstration slides of the division process, if they are available (Fig. 3.10). As noted, *Euglena* is sometimes in such concentrations that it colors pond water green and forms a surface scum.

■ How does *Euglena* react to light? (Why is it concentrated at the surface?)

■ Is *Euglena* a plant? Explain.

■ What would you call a *Euglena* grown in the dark, lacking chloroplasts, which feeds on dissolved nutrients in the medium?

■ If *Peranema* (Fig. 3.11) are available, compare them with *Euglena.* What similarities and differences do you observe?

■ How do these mastigophores resist drying out or freezing over of ponds?

III. *Volvox*

Volvox is a colonial mastigophore (Fig. 3.12). Each unit is a biundulated green protist rather like an individual *Chlamydomonas.* In fact, *Volvox* is considered to be the end of a hypothetical series of evolutionary transistions among these primitive phytomastigophores. They lead from *Chlamydomonas*-like forms (one-celled, but four-celled at certain stages of its life cycle), through similar phytomastigophores consisting of flat plates of cells such as *Gonium* or of small spheres such as *Pandorina,* and finally to *Volvox,* which may consist of several thousand cells. If it were not for these intervening forms, it might be questioned whether *Volvox* is a colonial form or a true multicellular organism.

Volvox clearly represents a high level of protistan complexity, showing controlled movement, intercellular communicating fibers (protoplasmic strands), and functional specialization involving reproductive and vegetative (feeding and locomotory) cells.

Study living specimens if available, or stained preparations. Identify daughter colonies.

■ How are the undulipodia oriented?

Draw a typical *Volvox* showing a few cells with their interconnecting strands and outlining the

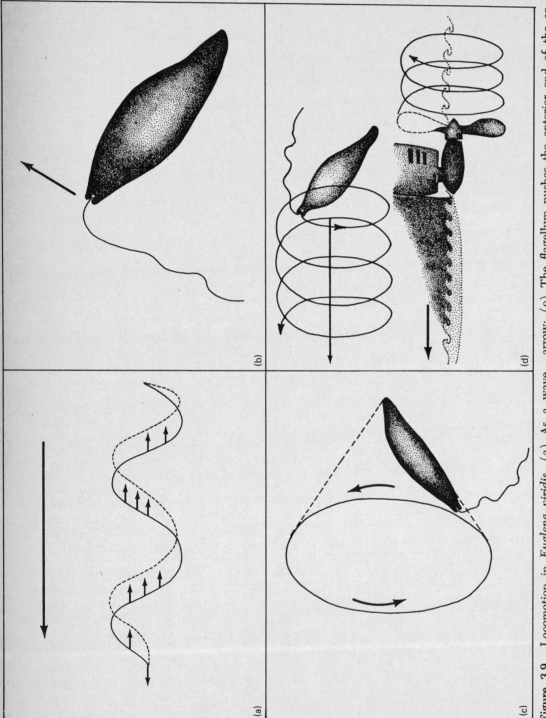

Figure 3.9. Locomotion in *Euglena viridis*. (a) As a wave travels down the flagellum, it moves from the solid line to the dotted line. Water is pushed in the same direction (small arrows). The flagellum as a whole moves in the opposite direction (large arrow); (b) The flagellum is oriented nearly perpendicular to the body. Thus, when the flagellum beats, the organism will be pushed in the direction indicated by the arrow; (c) The flagellum pushes the anterior end of the organism, the posterior end "drags." If we leave out the forward component of its motion, the *Euglena's* motion would describe a section of a cone; (d) The gyrating *Euglena* is analogous to a propeller blade (except that the propeller blade is driven from the center). Water is pushed behind the *Euglena*, the water offers resistance, and the organism moves forward.

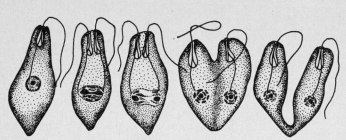

Figure 3.10. Stages in longitudinal fission in *Euglena viridis*.

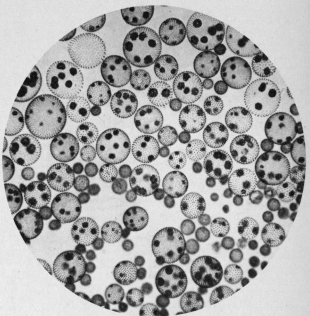

Figure 3.12. *Volvox*, ×48. (Courtesy CCM: General Biological, Inc., Chicago.)

others (Fig. 3.13). Show colonies and, if possible, germ cells. Identify and label the cellular detail you can actually find in your specimen.

■ **Does *Volvox* also show cell specialization?**

If material is available, compare *Volvox* with *Chlamydomonas, Gonium sociale* (four cells per plate colony), *G. pectorale* (16 cells), *Pandorina* (16 cells), *Eudorina* (colonies sperical and of distinct sexes), *Pleodorina* (32 or 128 cells with differentiation between germ or reproductive cells and somatic or vegetative cells).

IV. *Paramecium* (Figs. 3.14–3.20) (Representing which subphylum?)

Few organisms have been peered at by more inquisitive eyes than *Paramecium*. Not only are countless student hours spent examining this "slipper animalcule," but biologists of many disciplines and research interests look at *Paramecium* with equal interest.

1. **Observations** Obtain a drop from a healthy *Paramecium* culture. Examine it with the naked eye to locate the reflecting specks of fast-moving organisms. Look at them under your dissecting microscope at various

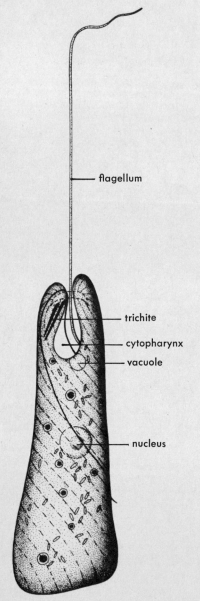

flagellum

trichite

cytopharynx

vacuole

nucleus

Figure 3.11. *Peranema*.

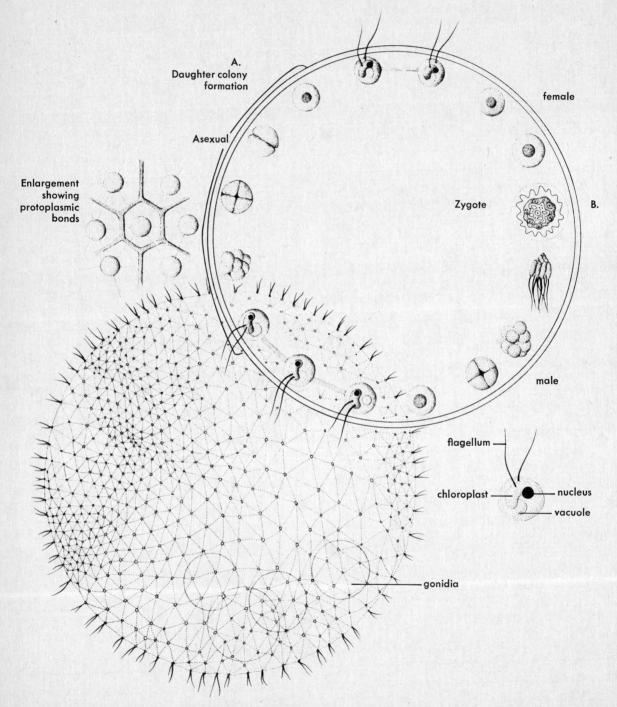

Figure 3.13. Structural details and life cycle of *Volvox*.

magnifications. Try to follow their general activity. Then cover and examine again under low power of your compound microscope. Observe motility, shape, and orientation. (Methyl cellulose,[4] a highly viscous fluid, can be used with a drop of culture fluid to show relatively motionless organisms.) In ordinary culture medium look near the edge of the fluid or find specimens crowded together with bits of debris.

■ **What kind of symmetry has *Paramecium*?**

■ **Which end is anterior? Does it always swim in one direction?**

Place a strong light near one end of the microscope stage.

■ **How do the paramecia orient themselves?**

■ **How do they react to tapping?**

■ **How do they react to a few salt grains placed at one end of the slide? What would you conclude from this reaction?**

Watch them move about among bits of debris on the slide.

■ **What restricts their change of shape as they squeeze into crevices?**

With the aid of methocel, slow down their movement; and find a quiet individual and switch to high-dry for a more critical morphological study.

2. **Structure** Carefully work out the structures visible in *Paramecium*; try to locate those shown in Fig. 3.15A.

Make a drawing at least 5 inches in length. Look at several paramecia to make certain the example you select is typical. Examine and draw the feeding apparatus, showing *oral groove, mouth,* and *gullet* (Figs. 3.14, 3.15, 3.16). The gullet ends in a small esophageal sac surrounded by fine fibers. The food vacuole formed here is re-

[4] To make 100 ml of solution, add 10 grams of methocel to 45 ml of boiling water, soak 20 min, then add 45 ml cold water, and cool until clear. Place a small ring of this fluid in the center of your slide; inside the ring add a drop of culture fluid containing the organisms. Add cover *slowly* and examine. Locate an organism trapped in the sticky material.

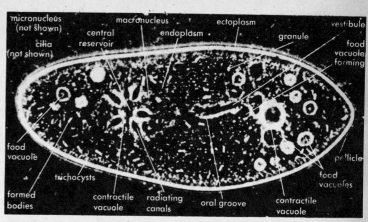

Figure 3.14. *Paramecium caudatum,* phase-contrast microphotograph ×470.

leased into the cytoplasm, and then proceeds along a spiral path, during which time the contents undergo digestion, induced by enzymes passed into the vacuoles by submicroscopic *lysosomes.* Find and draw the *cilia, pellicle, macronucleus, food vacuoles, granules,* and the stellate *water expulsion vesicles.*

Watch as *water expulsion vesicles* are filled by the *radial canals,* then empty into the *central reservoir* and quickly discharge externally through a permanent pore (Fig. 3.15B).

■ **How many discharges can you count per minute? Compare this with counts by other observers in your class.**

■ **How could a small amount of sea water affect the count? Test your hypothesis.**

3. **Special procedures** Certain structures show up well only when specific dyes are used. Add a droplet of one of the dyes suggested below, using a filter paper wick as described in Chapter 2. (Do not try to stain the methocel preparations, however.)

a. *Methyl green.* This dye will stain the macronucleus.

■ **Does this indicate a chemical difference between the macronucleus and cytoplasm?**

b. *Fountain pen ink.* Parker's ink is an excellent excitant to cause paramecia to eject rodlike or hairlike *trichocysts* (Fig. 3.17), said to be protective in function. Other stains such as iodine,

(a)

anterior end

undulipodium

trichocyst

food vacuole 3

water expulsion
vesicle

pellicle

ectoplasm

endoplasm

food vacuole 2

bacteria being ingested

oral groove

micronucleus

macronucleus

cell mouth

cell gullet

food vacuole forming

cell anus

radiating canal
(ampulla) of water
expulsion vesicle

food vacuole 1

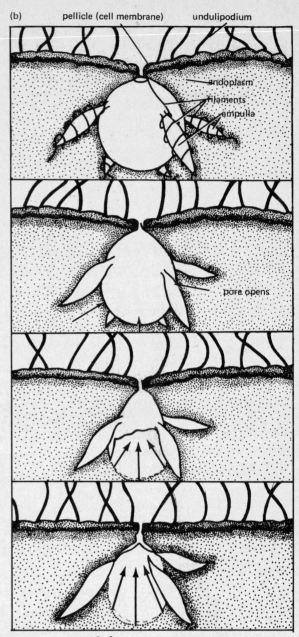

(b)   pellicle (cell membrane)   undulipodium

endoplasm

filaments

ampulla

pore opens

Figure 3.15. *Paramecium caudatum.* (a) Morphology
and structures; (b) Water expulsion vesicle expelling
water.

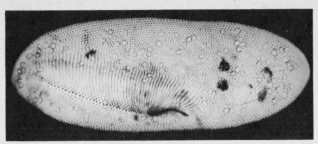

Figure 3.16. Ciliature pattern of *Paramecium*. (Courtesy CCM; General Biological, Inc., Chicago.)

methyl green, or dilute acetic acid also may be used to demonstrate trichocysts.

c. *Congo red.* Stain a suspension of heat-killed yeast cells with Congo red and feed a drop to paramecia. When the ciliates ingest the dyed yeast, digestive processes can be followed inside the food vacuole, as Congo red is an *indicator dye* that will gradually turn blue as the yeast particles become more acid. Watch carefully to observe swallowing and food vacuole formation. Then follow the winding course as vacuoles move through the endoplasm.

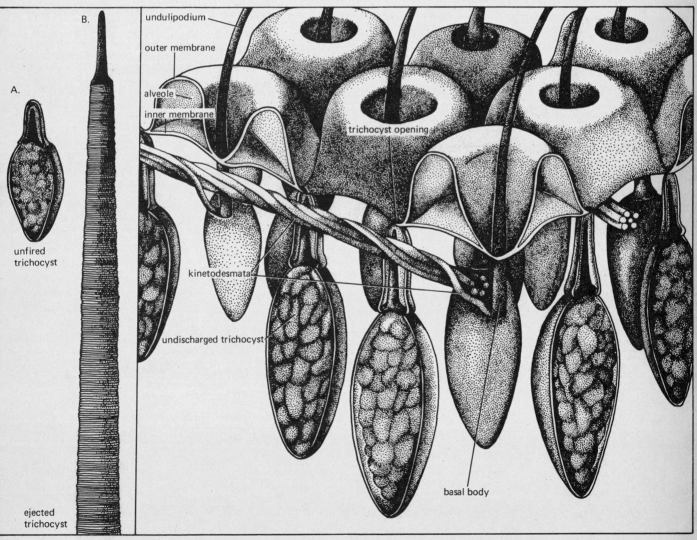

Figure 3.17. Trichocysts in *Paramecium*: (*left*) undischarged trichocysts and extremely magnified version of the structure of an ejected trichocyst; (*right*) arrangement of trichocysts beneath the pellicle.

■ **How is this color change related to digestion?**

■ **Is this process similar in paramecia and vertebrates?**

Outline a *Paramecium* and trace the path of a food vacuole within it, noting color changes en route.

4. **Asexual reproduction** If permanent stained slides are available, try to locate several *division stages* and nuclear changes during division. (Refer to lecture or text discussion of the stages of nuclear and cell division in *Paramecium*.)

■ **What is the plane of division?**

■ **Why is this type of division considered an asexual process?**

■ **Describe the precise steps before and during transverse binary fission.**

■ **What happens to the macronucleus? to the micronuclei?**

Sketch three or four division stages, including elongation and breakup of the macronucleus, dividing micronuclei, and gradual separation of daughter cells.

5. **Sexual reproduction or conjugation** (Figs. 3.18, 3.19, and 3.20) (Refer to discussion of this phenomenon in lecture or text.) Conjugation can be studied in fixed, stained preparations or in living material, but the latter gives a far better concept of this remarkable process. If appropriate *mat-*

ing types are available, obtain a drop of the freshly mixed *clones* and watch the stages of *clumping, pairing* (Fig. 3.18), and finally *exconjugant formation* (Figs. 3.19 and 3.20), followed in several hours by *asexual fission* of each exconjugant. Search for characteristic nuclear changes in the stained preparations. Review the nuclear steps in conjugation, the genetic significance of which will be considered later in the course.

■ **How can the process of conjugation be compared with metazoan sexual processes?**

V. *Monocystis* (Represents which subphylum, class, and subclass?)

Many earthworms are infected in the seminal vesicles with the sporozoan *Monocystis* or other members of this group. The complete life cycle of this parasite can be seen on a single slide. Smear some seminal vesicle material from a living worm into saline solution (0.5 percent), cover, and examine microscopically.

■ **How many different stages can you identify?** (See Fig. 3.21.)

Encysted stages will usually be the most common. Look for large *cysts* containing *spores* at various stages of development, particularly the infective stage when they are spindle-shaped bodies with eight *sporozoites* each. Active stages include large *trophozoites* that may appear undulipodiated (ciliated) due to the presence of degenerating worm spermatozoa around them. Sketch the forms you can find on your slide, then study the sequence of steps in the life cycle of *Monocystis*, referring to Fig. 3.21 and one of the appropriate references given at the end of this chapter.

VI. *Blepharisma*

Blepharisma is a ciliophore (which subclass and order?) that commonly lives at the bottom of fresh-water ponds. This protist is easily grown in an infusion of powdered cereal grains. Cultures should be kept out of bright sunlight, however, since the red pigment (giving *Blepharismas* their characteristic pink color) is highly photosensitive. Bright sunlight (even passing through glass) will kill them.

When its aqueous environment either dries

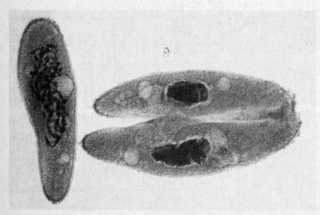

Figure 3.18. *Paramecium* conjugation (pairing), ×420. (Courtesy CCM: General Biological, Inc., Chicago.)

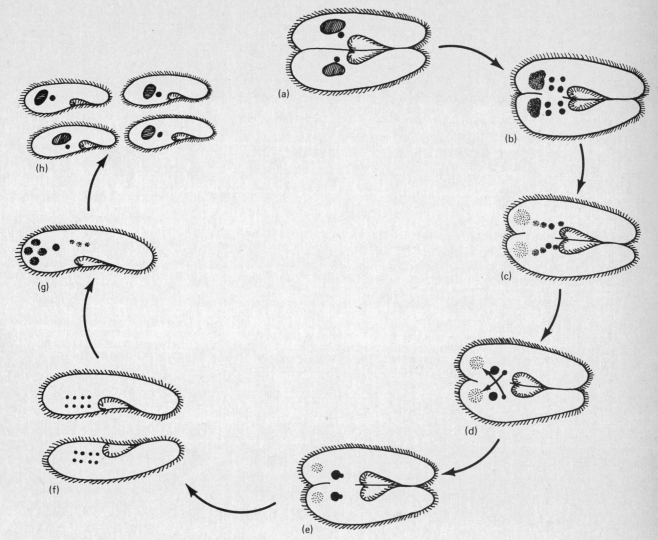

Figure 3.19. *Paramecium* conjugation, beginning with pairing and ending with exconjugants. You should be able to describe the activities of the nuclei shown in this diagram.

up or freezes, *Blepharisma* encysts until conditions improve. The ability to form cysts in response to harsh environmental conditions appears also in many multicellular animals, such as among members of the phylum Arthropoda. As we shall see,

Blepharisma also has remarkable powers of regeneration. See Giese, 1973 (references) for details.

Studying a living specimen under the microscope, refer to Fig. 3.22 and locate the following structures: macronucleus, food vacuole, water expulsion vesicle, peristome (mouthparts), and the numerous undulipodia. (The micronuclei are located near the macronucleus but can be seen only when stained.)

■ What is the orientation of the undulipodia?

■ How does the movement of *Blepharisma* compare with that of *Paramecium*?

■ What structural details help you to differentiate between *Paramecia* and *Blepharisma*?

Figure 3.20. *Paramecium exconjugant,* ×630. (Courtesy CCM: General Biological, Inc., Chicago.)

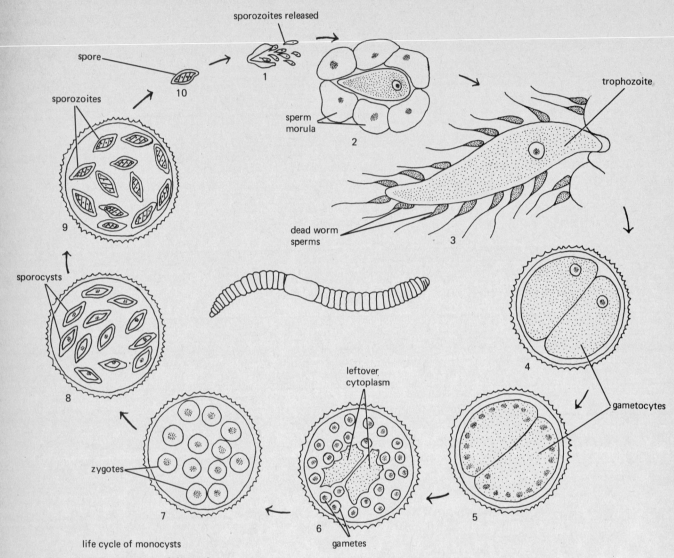

Figure 3.21.   *Monocystis* life cycle.

Regeneration For the study of regeneration in *Blepharisma* you will need the following equipment:

dissection scope
cool lamp
dishes with concavity or depression slides
petri dish (to hold the depression slides)
glass rods (4) 7 cm long and 4–5 cm in diameter
Bunsen burner
wood rectangles 4 cm by 8 cm by 2.5 cm
corrugated paper (glue it to the surface of a wooden block as a holding rack for the glass needles)
pipettes (4), tips 200 μm in diameter
1 meter of rubber tubing (to be used in making transfer pipettes)
medicine dropper
test tube, 2.5 cm diameter, 20 cm long (for pipette sterilization)
nonabsorbent cotton
petroleum jelly (vaseline is suitable)
glass slides and coverslips.

For cutting *Blepharisma* you will need to prepare glass needles from the glass rods. Using the Bunsen burner, simultaneously heat the tips of two rods, the first slightly more than the second. Then bring the two tips into contact and quickly pull the first rod (heated more than the other) away from the second rod. This should produce a strong, flexible cutting needle with a fine tip.

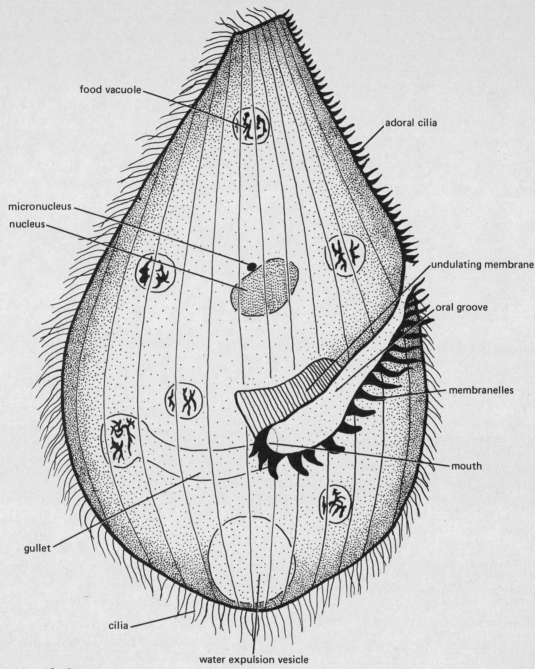

food vacuole

adoral cilia

micronucleus

nucleus

undulating membrane

oral groove

membranelles

mouth

gullet

cilia

water expulsion vesicle

Figure 3.22. *Blepharisma.*

Transfer the needles prepared in this manner to the holder you have constructed by gluing corrugated paper on the top of a wooden block.

The culture pipettes and transfer pipette should be placed narrow-end down on a cotton pad at the bottom of a test tube, the large ends of the pipettes plugged with cotton. They can then be sterilized. The transfer pipette is made by placing

the rubber tubing on the wide end of a sterilized glass pipette. On the other end of the tubing insert the narrow end of the medicine dropper. The wide end of the dropper forms a suitable mouthpiece to enable you to draw up or blow out samples of the *Blepharisma* culture.

Before cutting the test organisms, you will want to prepare a slide on which to transfer the pieces of *Blepharisma*. Melt some petroleum jelly in a beaker. Warm a pipette over the flame of your Bunsen burner; draw up the melted jelly into the warmed pipette and heat the tip again gently. Blow a thin layer of the jelly on a clean slide forming a ring just smaller than the coverslip to be placed over it.

Now you are ready to transfer part of the *Blepharisma* culture into the depression slide which should be resting in a petri dish on the microscope stage. Using a magnification of 15–20, place your cutting needle in the medium on the depression slide. When an individual swims under the needle, cut the *Blepharisma* into two parts by carefully moving the needle back and forth. In order to get an anterior and a posterior segment of the organism (important for your regeneration studies), make the cut directly below the mouthparts (peristome). Now, draw up the cut pieces of *Blepharisma* from the depression slide using the transfer pipette and place these in the center of the jelly ring on your slide. Carefully place the coverslip over the slide and press cautiously with a pencil, creating an air-tight moist chamber in which the split portions can live. (Instead, you may wish to place the cut pieces onto the coverslip and then invert the slide with petroleum jelly over it.)

Observe under higher magnification, and *draw* the segments.

Observe six hours later or the next day.

■ **How much regeneration has occurred? Make a careful drawing and compare it with those made the previous day.**

VII. Additional Studies

1. Culturing techniques Your work with these protists no doubt has convinced you of the vast diversity inherent in unicellular life. If proper facilities are available, you may want to do additional experiments in genetics, ecology, and physiology of the protists. These organisms are now the basic research tool for the study of many essential processes and phenomena characteristic of the higher or multicellular organisms.

In your further studies, culturing protists in the lab may be necessary. There are numerous methods for cultivating the strains commercially available and collected in the field. It is best to consult research articles that describe culture techniques for the par-

ticular organism you are working with. Here, we offer some general suggestions for culturing protists. In the subsequent section we describe methods for culturing several specific protists.

a. Maintain cultures at constant temperature, between 18° and 21°C(64°–70°F).
b. Maintain a neutral pH (7).
c. Do not expose cultures to fumes of concentrated acids or bases.
d. Protect cultures from sudden drafts that can transport contaminants.
e. Use moderate light. *Direct* sunlight may be used for culturing certain photosynthetic forms (in *Euglena* one hour of direct sunlight per day is maximum). For other nonphotosynthetic forms it can be harmful, as it raises the temperature of the medium.
f. To clean glassware, use low-suds detergent and a *series* of washes in distilled water for the final rinsing.

2. Hay infusion One method, the hay infusion, is a useful culture technique, one you can use in culturing amebae. According to Halsey, boil a mixture of eight Timothy hay stalks (1-inch lengths) in 100 ml of spring water for 10 min. Let this stand for 24 hours. Add *Colpidium* or *Chilomonas* in large quantity. After allowing the culture to stand for 2 or 3 days, introduce the amebae. As the population grows and available food is depleted, divide the culture medium in half. To each half add an equal volume of freshly prepared hay infusion already injected with *Colpidium* or *Chilomonas*. For each 50 ml of culture medium added, add two grains of uncooked rice or boiled wheat. The resulting cultures are reported to last as long as six months.

3. Culturing Paramecium A method somewhat similar to the hay infusion described above has been found successful in culturing *Paramecium*. To start, boil 1 liter of pond water (spring or tap water will suffice if pond water is not available). Add a handful of Timothy hay to the boiling water and continue boiling for 10 min. Cool the solution and let it stand for 2 days. Then introduce *Paramecium* into the medium.

4. Synthetic pond water preparation Another useful culture technique is the preparation of synthetic pond water requiring a specific composition of inorganic salts. The following technique is valuable in culturing *Blepharisma*.

Prepare *Brandwein's solution A* by dissolving the salts listed below in distilled water to make 1 liter of solution:

NaCl	1.20 g
KCl	0.03 g
$CaCl_2$	0.04 g
$NaHCO_3$	0.02 g
phosphate buffer (pH 6.0–7.0)	50 ml

Solution A should be diluted with distilled water, 1 part solution A: 10 parts distilled H_2O.

Rinse the finger bowls to be used for your cultures in hot water and then cold. Then prepare a 1% aqueous solution of powdered nonnutrient agar in solution A or distilled water. Heat this mixture until it is homogeneous and pour it into the finger bowls, creating a 1–2-mm layer of agar on the bottom of the bowls. Place eight rice grains in the still soft agar of each bowl.

To each bowl add about 50 *Blepharisma*, 15 ml of the *Blepharisma* growth medium, and 30 ml of dilute solution A. Each day for the following three days, add 15 ml of dilute solution A. You can expect the agar to separate from the glass in a few days.

In older cultures you may find very large *Blepharisma*, often cannibalistic.

5. **Culturing** *Euglena* A preparation of modified Kleb's solution is a good culture medium for *Euglena*. Dissolve the following salts in 1 liter of distilled water:

KNO_3	0.25 g
$MgSO_4$	0.25 g
KH_2PO_4	0.25 g
$Ca(NO_3)_2$	1.00 g
bacto-tryptophane broth powder (1-form)	0.01 g

To a glass battery jar add 100 ml of this solution, 20 boiled rice grains (boiled for 5 min), and 900 ml of distilled water. Allow this to stand 2 days. Introduce *Euglena* into the medium and store in indirect sunlight. Do not expose to direct sun for more than one hour per day. Inject the culture with *Euglena* again at 3, 6, and 9 days after the initial introduction. At the end of 2 to 3 weeks add 25 ml of the modified Kleb's solution in which an additional 10 mg of tryptophane powder has been dissolved.

EXERCISE

Introduce streptomycin into one of your cultures of *Euglena*.

■ **What are the effects?**
■ **Which structures within the organism are changed?**
■ **What might you add to the culture so that** *Euglena* **will still survive?**

6. **Vital stains** Vital stains are substances that will stain slowly but not kill the organism, allowing life activities (movement of undulipodia, for example) and details of living cell structure to be observed. The staining process is quite simple. Put a drop of the vital stain on a slide, letting it dry to an even film on the glass surface. (Extremely dilute stain is usually employed.) When you are ready, simply add a drop of the culture you wish to observe on the slide.

In order to learn the relative properties of various vital stains you may want to run a test of their effects on a variety of organisms. Using the listed series of stains on one group of protists, record your observations for each.

■ **What color does the nucleus stain? the vacuole? water expulsion vesicles? fat droplets?**
■ **Which stains reveal the undulipodia most clearly?**
■ **What general recommendations can you make for the relative advantages and usefulness of these vital dyes?**

	Amoeba	Para-mecium	Bleph-arisma	Vorti-cella	Mono-cystis
Methylene Blue					
Nile Blue Sulfate					
Methyl Green Acetic Acid					
Janus Green					

(Continued from page 53)

	Amoeba	Para-mecium	Bleph-arisma	Vorti-cella	Mono-cystis
Acetocarmine					
Neutral Red					
Congo Red					
Sudan III					
Lugol's Solution					

Additional information on vital stains will be found in the Appendix.

7. **Mating type determination** You may wish to expand your study of sexual reproduction (conjugation) by determining *mating types* of various ciliates. For further information refer to Chapter 18 (Genetics) and consult some of the references at the end of this chapter, such as Kirby (1950) and Wichterman (1953).

8. **Biological succession** A simple but dramatic demonstration of biological succession requires only the microscopic examination of similar samples taken daily from a jar of pond water or from a hay infusion (prepared as described in culture techniques). You can make standardized sampling counts of various protozoa found and chart the daily fluctuation of populations. Note the differences in species composition with time and with changing conditions as well as the population curves. (See Chapter 20, Ecology.)

The references suggested here will provide you with additional information on physiology, distribution, morphology, systematics, and identification of protists. These readings will greatly expand your interests in this field and should prove valuable in your further studies.

Suggested References

Allen, R. D., 1962. "Amoeboid movement." *Sci. American, 206*:112.

Corliss, J. O., 1959. "Comments on phylogeny and systematics of the Protozoa." *Syst. Zool.,* 8:169.

——, 1961. *The Ciliated Protozoa: Characterization, Classification, and Guide to the Literature.* New York: Pergamon press.

Florkin, M., and B. T. Scheer (eds.), 1967. *Chemical Zoology,* vol. 1, *Protozoa,* G. W. Kidder (ed.). New York: Academic Press.

Geise, A. C., 1971. "Photosensitization by natural pigments," in *Photophysiology,* vol. 6, A. C. Geise (ed.). New York: Academic Press, pp. 77–129.

——, 1973. *Blepharisma: The Biology of a Light-Sensitive Protozoan.* Stanford, Calif.: Stanford University Press.

Grell, K. G., 1973. *Protozoology.* New York: Springer Verlag.

Grieder, M. H., W. J. Kostir, and W. J. Frajola, 1958. "Electron microscope of *Amoeba proteus.*" *J. Protozool.,* 5:139.

Hadzi, J., 1953. "An attempt to reconstruct the system of animal classification." *Syst. Zool.,* 2:145.

Hall, R. P., 1953. *Protozoology.* Englewood Cliffs, N.J.: Prentice-Hall.

Hammond, D. M., and P. L. Long (eds.), 1973. *The Coccidia.* Baltimore, Md.: University Park Press.

Honigberg, B. M., *et al.*, 1964. "A revised classification of the phylum Protozoa." *J. Protozool.,* 11:7.

Hyman, L. H., 1940. *The Invertebrates: Protozoa through Ctenophora,* vol. 1. New York: McGraw-Hill.

Jahn, T., and F. F. Jahn, 1949. *How to Know the Protozoa.* Dubuque, Iowa: William C. Brown.

Jahn, T., and Votta, J. J. 1972. "Capillary suction test of the pressure gradient theory of amoeboid motion," *Science, 177*:636.

——, 1972. "Locomotion in Protozoa," *Ann. Rev. Fluid Mech., 4*:93.

Kirby, H., 1950. *Materials and Methods in the Study of Protozoa.* Berkeley, Calif.: University of California Press.

Kudo, R. R., 1966. *Protozoology,* 5th ed. Springfield, Ill.: Charles C. Thomas.

Lackey, J. B., 1959. "Zooflagellates," in *Fresh-Water Biology,* 2nd ed. New York: J. Wiley & Sons.

Levine, N. D., 1973. *Protozoan Parasites of Domestic Animals and of Man,* 2nd ed. Minneapolis, Minn.: Burgess Publishing.

Morholt, E., P. F. Brandwein, and A. Joseph, 1966. *A Sourcebook for the Biological Sciences.* New York: Harcourt, Brace Jovanovich.

Pitelka, D. R., 1963. *Electron-Microscopic Structure of Protozoa*. New York: Pergamon Press.

Sherman, J. W., and V. G. Sherman, 1970. *The Invertebrates: Function and Form; A Laboratory Guide*. New York: Macmillan.

Vickerman, K., and F. E. G. Cox, 1967. *The Protozoa*. Boston, Mass.: Houghton Mifflin.

Wichterman, R., 1953. *The Biology of Paramecium*. New York: Blakiston.

Wigg, D., E. Bovee, and T. H. Jahn, 1967. "The evacuation mechanism of the water expulsion vesicle ("contractile vacuole") of *Amoeba proteus*," *Protozool.*, *14*:104.

CHAPTER 4

PHYLUM PORIFERA

1. Introduction

There is nothing like a sponge—to parody Rodgers and Hammerstein. Nothing looks like a sponge, nothing acts like a sponge, nothing eats like a sponge. In fact, nothing is like a sponge. Find another living thing that feeds by filtering food through a thousand outside pores into a vast canal network whose cells do double and triple duty in a number of chores; and that can completely dissociate the fragments then wandering together again to make many small new organisms. And find in nature another example of such fragile beauty as the Venus' flowerbasket (Fig. 4.1), literally made of glass. These and other features make sponges unique. From the standpoint of *structure* (body organization and a spicule skeleton), *manner of feeding* (collar cells for food handling and water pumping), and *reproduction* (amphiblastula, gemmules, regenerative reorganization), truly there is nothing like a sponge.

Hence, some authors divide the multicellular animals into branches: sponges, the PARAZOA or "near animals" (literally, alongside of animals); and nonsponges or EUMETAZOA, the higher "true animals." The difference between the two branches is so clearly defined that it involves more basic characteristics than those dividing invertebrates from vertebrates. Sometimes a third branch is distinguished, the MESOZOA or "midanimals." However, many authors consider the MESOZOA a phylum within the EUMETAZOA rather than a separate branch.

The branch PARAZOA has just one phylum, the **PORIFERA**; therefore branch and phylum description are alike. A very ancient group with many fossils, **PORIFERA** contains about 5000 species, mostly marine. One family is found in fresh water (Fig. 4.2). Sponges are aquatic organisms attached to a substrate in shallow waters (Fig. 4.3), permeated by pores through which water passes into inner flagellated chambers where feeding occurs.

The *endoskeleton* (internal skeleton) consists of discrete *spicules* made of calcium carbonate (calcareous sponges), glass (siliceous sponges), or spongin (demosponges).

Reproduction is either sexual or asexual, sexual reproduction often including an unusual larval form, the *amphiblastula*. Asexual reproduction occurs by *budding, fragmentation* and *regeneration*, and *gemmule formation*. The last, occurring

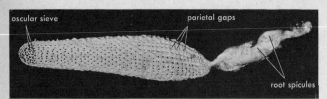

Figure 4.1. *Euplectella*, the Venus' flowerbasket, specimen 29 cm long.

chiefly in freshwater sponges, is the formation of specialized sponge cells into a resistant cystlike body—the gemmule—which can withstand freezing and desiccation, "hatching" into a growing sponge when environmental conditions improve (Fig. 4.4).

2. Classification

Kingdom ANIMALIA
   Branch PARAZOA
      Phylum **PORIFERA**
      Three classes are generally recognized:

1. CALCAREA, sponges with calcareous, 1-, 3-, or 4-rayed spicules
2. HEXACTINELLIDA, so-called glass sponges with siliceous, 6-rayed spicules
3. DEMOSPONGIAE, sponges with spicules of spongin (as in common bath sponges —Fig. 4.5) or silica, or both, or absent.

Much of the taxonomy of sponges is based on details of spicule structure because these hard parts are so easily studied (Fig. 4.6). This system agrees remarkably well with another more modern classification based largely on the steroid biochemistry of sponges. In this case, as in many such instances in various animal or plant groups, results of modern and classical methods of classification coincide, adding to our confidence in the scheme of classification based upon descriptive or morphological criteria.

3. Laboratory Instructions

For good results, the study of living sponges requires fresh material and excellent optics. Most students necessarily will be restricted to fixed material and prepared slides in their study of general body form and spicules.

I. *Scypha coronata* (*Grantia coronata*)

1. **Observations on living specimens** If you are fortunate enough to be able to watch a living calcareous (marine) sponge in a laboratory in its natural environment, try adding a small amount of powdered carmine or India ink to the water to demonstrate water currents passing through the animal.

Trace the direction of water flow.

■ **Is the incurrent or excurrent flow more rapid?**

■ **Is it advantageous to the sponge to have the *excurrent* flow be as rapid as possible?**

Observe under the dissecting microscope shape, color, mode of attachment, spicules

Figure 4.2.   Freshwater sponge on a branch.

Figure 4.3. Closeup of *Halaclona*, a marine Demospongian.

projecting from the surface, excurrent pore (*osculum*), incurrent pores (*dermal pores*).

Draw as many of these general features as you can observe in your own specimen.

2. **Preserved specimen of calcareous sponge** (Fig. 4.7) Repeat the above microscopic observations on preserved material. If living specimens are not available, make a sketch of general morphology of a preserved specimen.

3. **Dried specimen** (Fig. 4.8) Examine a dried sponge. With razor or scalpel make a careful cross section and longitudinal cut to show internal organization. Examine under a dissecting microscope and attempt to work out the main features of the canal system. In life the many *undulipodiated* (*flagellated*) *chambers* are lined with collar cells (*choanocytes*) whose undulipodia are directed toward the large central cavity or *spongocoel. Internal ostia* are canals that carry used water from undulipodiated chambers into the spongocoel via small openings (*apopyles*). Water first enters the sponge through the external dermal pores, then passes via in-current canals into the undulipodiated chambers where the essential feeding, excretory, and respiratory exchange takes place. Water then leaves via internal ostia and apopyles, passes into the spongocoel, and finally out the osculum.

The organization of *Scypha* is known as the *syconoid* form and is intermediate in complexity among the types of calcareous sponges (Fig. 4.9).

Locate the central cavity, which leads to the osculum, receiving water from each undulipodiated chamber. *Draw* a section to show the water path. Label key points.

Place a bit of calcareous sponge tissue in a glass dish, add bleach, examine the residue when bubbling ceases.

■ **What remains? Examine a drop of the residue under high power.**

Draw several spicules. In another dish add some 10 percent KOH to the sponge tissue and compare the results.

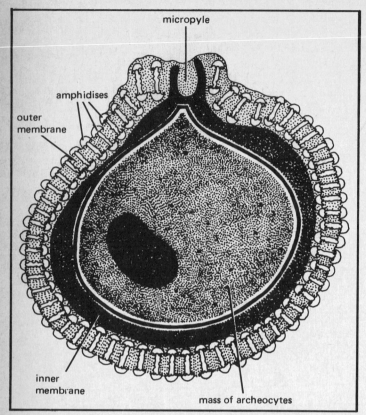

Figure 4.4. A durable asexual gemmule of a freshwater sponge (cross section).

Place some dilute acid on bits of sponge tissue in a glass dish.

■ **What do the results tell you about spicule composition in these sponges?**

4. **Stained sections on slides** The best way to observe structural detail is with well-stained, special preparations. Even then, normal cell appearance is seldom achieved because of distortion and shrinking. Living sponge cells viewed under the phase-contrast microscope are better for observing sponge cell types—a sight even few biologists are fortunate enough to see.

Study your slide section, then compare it with Fig. 4.10. Locate *choanocytes* and *amebocytes*, the two basic cell types in your sections. The amebocytes give rise to specialized cells, such as reproductive cells. (Where does sexual reproduction occur?) Look for masses of *spermatozoa* or developed *ova*. The various cell types often are not easily distinguished because they are not as highly

differentiated as in more complex animals. Other specialized cells are *scleroblasts*, which secrete spicules, and dermal cells or *pinacocytes*, which line the outer surface, incurrent canals, spongocoel, and entry pores.

■ **How are spicules normally arranged? (Compare your slide with a demonstration slide showing normal spicules.)**
■ **Review the function of the collar cell.**
■ **Identify the pathway that a drop of water with food particles takes through Scypha.**

In your textbook review the structure and stages of development of the amphiblastula larva.

II. Other Sponges

Examine dried or preserved specimens of as many sponge types as possible. Consult museum collections and textbooks for illustrations of living sponges to appreciate their variety and beauty.

1. *Leucosolenia* (Fig. 4.11) This is a highly simplified *asconoid* type of calcareous sponge, in which the spongocoel is at the same time a single, large flagellated chamber.

■ **How does Leucosolenia differ structurally from Scypha?**

2. **Venus' flowerbasket (Euplectella)**

■ **What class does this beautiful glass sponge represent? The fused skeletal elements form a continuous glassy network of extraordinary delicacy and complexity.**

3. **Horny or bath sponges**

■ **What is the skeletal material of these sponges?**

Members of this class (*leuconoid* sponges) have the most extensive network of chambers and ducts. Their organization should be contrasted with the *asconoid* and *syconoid* types. Leuconoid sponges are the most efficient in terms of water conduction, with a far greater total feeding surface. More convolutions and more flagellated chambers allow both a greater water-holding capacity

Figure 4.5.   Class Demospongiae, a bath sponge.

and a relatively more rapid exit of waste water out the osculum. Food-laden water enters a vast number of ostia, moves slowly as it enters, and spreads over a great number of flagellated chambers. Since all chambers end in a very small number of spongocoels and oscula, water is expressed rapidly. The simple and complex sponge types can be compared to an elementary hydrodynamic problem. In one case, water flows through ten feeder pipes, each 1 inch in diameter, which merge and exit through a single 10-inch pipe. In the more complex system, water flows through a hundred 1-inch feeder pipes and then out a 1-inch exit.

■ **Compare rate of water movement in the feeder systems (undulipodiated chambers) and in the exit pipes (osculum).**

■ **In which system does the water remain longer in the feeding area (allowing choanocytes more time to work), and in which is the water ejected farther (preventing contamination of the feeding zone)?**

This relative efficiency of water conduction apparently has enabled far more species of leuconoid than of asconoid or synconoid types to survive and to occupy a greater variety of habitats.

Review the three basic sponge types, their relative degrees of complexity, and the ecological importance of these differences.

■ **What special characteristics enable** *freshwater* **sponges to survive in their habitat?**

■ **In what respect is fresh water more demanding and challenging to an invertebrate's survival than sea water?**

Study *Spongilla* if available. It belongs to a subclass of the Demospongiae and contains spicules of silica plus binding threads of spongin. Usually it encrusts submerged sticks or stones.

Study a section drawing of a *gemmule* (Fig. 4.4).

■ **Is it a sexual or asexual form of reproduction?**

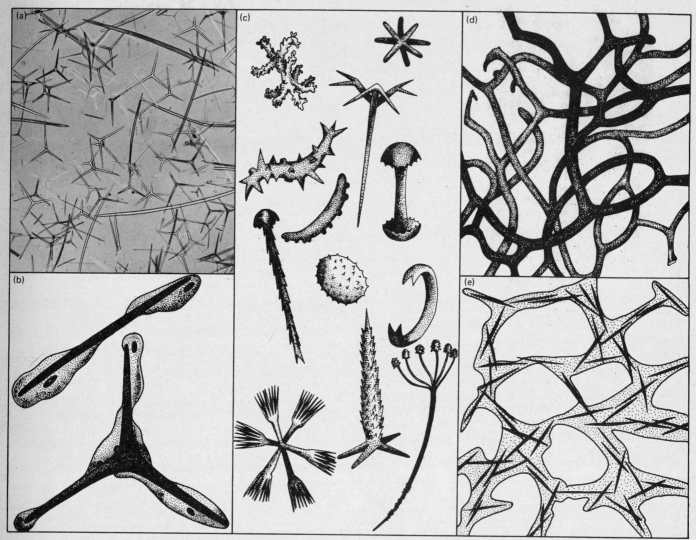

Figure 4.6. Spicules and spongin. (a) Assorted mon-axon and triaxon spicules of *Scypha*, ×125. (Courtesy CCM: General Biological, Inc., Chicago); (b) Develop-ing spicules with associated scleroblasts; (c) Siliceous spicules; (d) Spongin without spicules; (e) Spongin with spicules.

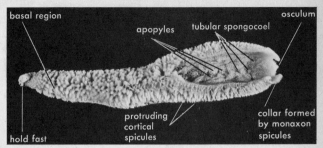

Figure 4.7. *Scypha coronata* with upper end cut to show spongocoel and macroscopic internal detail.

Figure 4.8. *Scypha*, gross and microscopic structural organization.

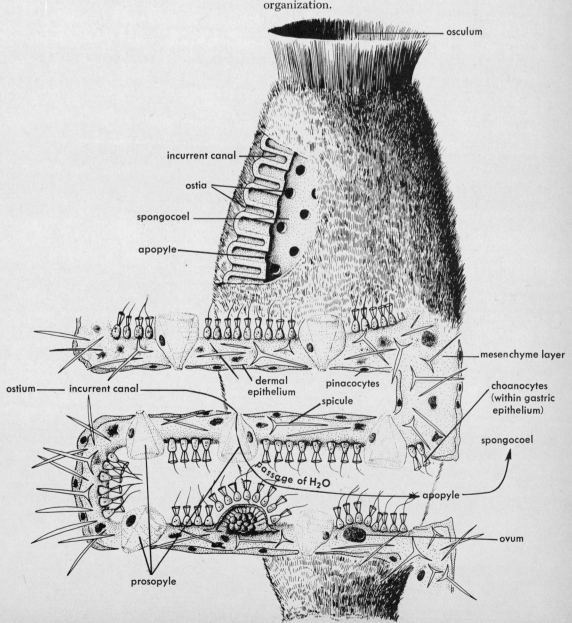

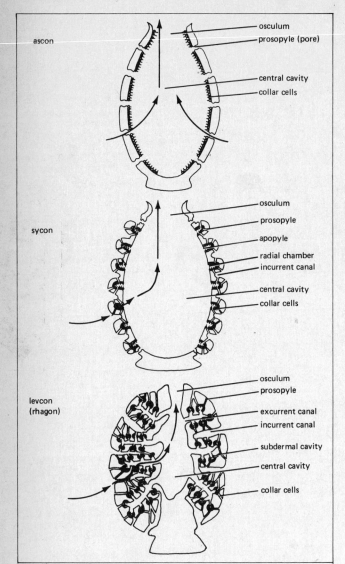

Figure 4.9. Types of canal systems of sponges. The sycon type is derived, theoretically, by the evagination of the wall of the ascon type of sponge into saclike chambers. Note how much each chamber of the sycon type of sponge resembles the single chamber of a simple sponge. The rhagon type, like the bath sponge, is more complex with an elaborate system of canals and undulipodiated chambers. The arrows indicate the course of water through the various types of sponges.

■ Of what advantage is a gemmule to the sponge groups possessing it?

■ Why do you suppose that gemmules are found only in freshwater sponges?

Return to the definition of the phylum and the basic sponge organization. Justify separation of sponges from all other animals.

■ Why is this phylum considered just above the protozoan level of organization, barely reaching the tissue stage of structural complexity?

■ Explain why sponges are considered to be an evolutionary "dead end" (not ancestral to any other groups) despite their primitive and simple organization.

III. Additional Studies

In addition to their remarkable powers of regeneration, sponges also have unusual powers of reassociation. When a sponge is broken into many separate fragments, these will reassociate forming one or more new aggregates. Workers found further that a mixture of cells from two different species will yield distinct aggregates—each of cells from only one sponge type.

1. **Reassociation**   Squeeze pieces of a fresh marine sponge, *Microciona*, through silk bolting cloth into a beaker of seawater be-

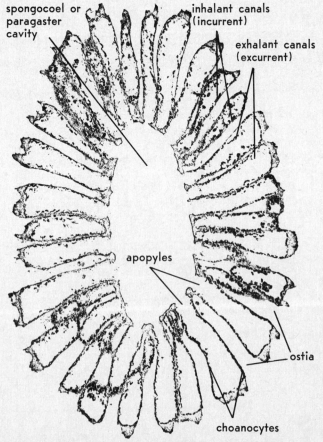

Figure 4.10.   Cross section of *Scypha*.

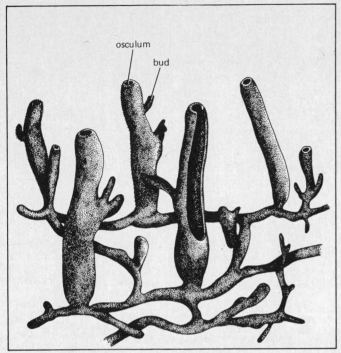

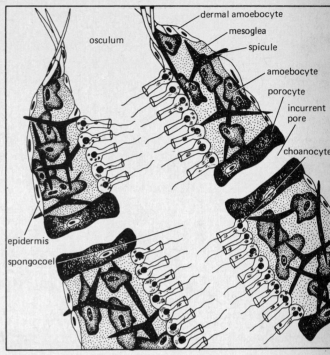

Figure 4.11. *Leucosolenia*: (*left*) colony form; (*right*) structural organization.

tween 3° and 5°C. (This will break the sponge into very small clumps of cells.) Then make a series of dilutions in seawater from your original suspension of sponge cells so that you have samples with decreasing cell densities (number of cells per unit volume of seawater). Next, place a syracuse watch glass in the bottoms of two small finger bowls and fill each bowl two thirds full of cool seawater. Put two slides on each watch glass. Very carefully pipette (with a Pasteur pipette) equal samples of the original suspension and the subsequent dilutions onto the slides in the finger bowls. Record the arrangement of the slides with the samples of decreasing cell density. The bowls should be covered and allowed to stand 24 hours—one bowl at room temperature, the other at a range of 3° to 5°C. At the end of this period, cautiously lift the slides from the bowls, cover with coverslips, and observe under the microscope.

■ **Describe the reaggregation that has taken place. Make a sketch of the typical clusters.**

■ **How does reassociation vary with decreasing cell density and temperature?**

2. *Microciona* and *Halaclona* (Fig. 4.3) Now press fragments of two different marine sponges, *Microciona* and *Halaclona*, through the silk bolting cloth onto one beaker (as above). Make dilutions as in the preceding exercise and follow the same directions for preparing finger bowls and for pipetting and storing the samples. After 24 hours record your observations.

■ **What has happened?**
■ **How might the segregation of the two cell types have occurred?**
■ **What other experiments would you suggest to test this phenomenon further?**

You may wish to collect several local freshwater sponges with which to try the cell-mixing experiment. In all cases you will find that the fresher the live material, the better your results will be.

Suggested References

Binder, G. P., 1923. "The relation of the form of a sponge to its currents." *Quart. J. Microscop. Sci.*, 67:293.

Brown, F. A., Jr. (ed.), 1950. *Selected Invertebrate Types.* New York: J. Wiley & Sons.

Buchsbaum, R., 1956. *Animals Without Backbones.* Chicago, Ill.: University of Chicago Press.

Dendy, A., 1926. "On the origin, growth, and arrangement of sponge spicules." *Quart. J. Microscop. Sci.*, 70:1.

Florkin, M., and B. Scheer (eds.), 1968. *Chemical Zoology*, vol. 2. New York: Academic Press.

Fry, W. G. (ed.), 1970. *The Biology of Porifera.* New York: Academic Press.

Hyman, L. H., 1940. *The Invertebrates*, vol. 1, *Protozoa Through Ctenophora.* New York: McGraw-Hill.

Jewell, M., 1959. "Porifera," in *Freshwater Biology*, 2nd ed., W. T. Edmondson et al. (eds.). New York: J. Wiley & Sons.

Jones, W. C., 1962. "Is there a nervous system in sponges?" *Biol. Rev.*, 37:1.

Levi, C., 1957. "Ontogeny and systematics in sponges." *Syst. Zool.*, 6:174.

Moscona, A., 1961. "How cells associate," *Sci. American*, 205:143–166.

Parker, G. H., 1914. "On the strength of water currents produced by sponges." *J. Exp. Zool.*, 16:443.

Rasmont, R., J. Bouillon, P. Castiaux, and G. Vandermeersche, 1958. "Ultrastructure of the choanocyte collar-cells in fresh-water sponges." *Nature*, 181:58.

Tuzet, O., 1963. "The phylogeny of sponges," in *The Lower Metazoa*, E. C. Dougherty, Z. N. Brown, E. D. Hanson, and W. D. Hartman (eds.). Berkeley, Calif.: University of California Press, pp. 129–148.

Van Well, P. B., 1948. "On the physiology of the tropical fresh-water sponge *Spongilla proliferans*: ingestion, digestion, and excretion." *Physiol. Comp. Oecol.*, 1:110.

Wilson, H. V., and J. T. Penny, 1930. "The regeneration of sponges (*Microciona*) from dissociated cells." *J. Exp. Zool.*, 56:73.

CHAPTER 5

PHYLUM CNIDARIA (COELENTERATA)

1. Introduction

What is often most impressive about the **CNIDARIA** is their stinging cells—the *nematocysts*. Indeed, anyone who has been stung by jellyfish or hydra will claim he knows the true "essence" of the phylum. But apart from the cnidarians' chemical warfare, actually a specialized form of predation, we find that this group of organisms has enormous structural variety and a wide range of adaptations, all built upon a common *basic body type*. Perhaps an entire course on basic zoological principles could be taught using the **CNIDARIA** as the sole source of material. Some of the biologically interesting aspects of these animals are the following:

1. **Tissue level of body organization CNIDARIA** do not possess organs, except for gonads and sensory structures, but they have two well-defined layers, *epiderm* and *gastroderm*, with a gelatinous layer of *mesoglea* sandwiched between. In the class ANTHOZOA (for example, *Metridium*), the mesoglea may be a true cellular tissue (mesenchyme or ectomesoderm) with considerable structural complexity. In *Hydra* (a member of the class HYDROZOA) it is reduced to a thin lining containing ameboid cells, located between the two tissue layers.

2. **Polymorphism (having different body forms)** This phenomenon is illustrated both by distinct developmental forms (life-cycle stages) and by striking differences between individual members (*zooids*) of a colony. In the jellyfish *Aurelia*, for example, the attached or *polyp* stage (in this group called the *scyphistoma*) develops by forming segments or buds that separate and become free-floating young *medusae* (specialized term: *ephyrae*). These develop into sexually mature adults, the familiar jellyfish form. Another type of polymorphism is seen in colonial animals such as *Physalia*, the Portuguese man-of-war, whose member zooids are so highly specialized for feeding, reproducing, and capturing prey that they appear to be separate parts or individual organs of a single animal.

3. **Life cycles** Cnidarian growth and development often involve a sequence of body forms, such as the sessile polyp and pelagic medusa. In some cases, environmental factors may influence this sequence. Varying degrees of dominance by one or the other stage pro-

duce many different life-cycle patterns. In *Obelia*, a representative type, both polyp and medusa are equally well developed. In *Hydra* and *Metridium* the medusa stage is entirely eliminated, the organisms leading an entirely sessile existence as an attached polyp. By contrast, in *Aurelia* the polyp stage is reduced and in *Pelagia* it is absent. Some species have a sequence of polyps or of medusae produced by budding.

4. **Evolution**   CNIDARIA are generally considered to be an evolutionary link connecting the **PROTOZOA** with the more complex **METAZOA** and are therefore an important group phylogenetically. They have contributed much to our understanding of the overall course of biological change. Review in your text and in the references (such as the iconoclastic but influential work of Hadzi, 1953 and 1963), the various (and conflicting) views on the origin of the higher metazoans. Other cnidarian features of general biological interest include the following:

5. **Symmetry**   *Radial* symmetry exists in this phylum, and *biradial* symmetry in the phylum **CTENOPHORA,** the comb jellies.

6. **Nematocysts** (Fig. 5.1)   These stinging capsules, arising from characteristic cells called *cnidoblasts*, characterize the phylum **CNIDARIA** probably more than any other single feature.

7. **Nervous system**   The nervous system consists of a network of interconnecting *nonpolar neurons* and epidermal sensory cells. This is a critical morphological advance over the sponges and protozoans, although it is a rudimentary system compared with that of the more complex metazoans.

8. **Reef formation** (Fig. 5.2)   Remains of corals and other anthozoans contribute in large measure to the formation of many miles of tropical coastline, fringing reefs, and atolls. The Great Barrier Reef of Australia is a 1500-mile-long example.

9. **Characteristics of embryology, regeneration, and growth**   Fundamental contributions to these topics have resulted from studies on *Hydra* and other coelenterates.

2. Classification

Phylum **CNIDARIA (COELENTERATA)**: aquatic, mostly marine organisms at the *tissue* level of organization; lacking definitive organ systems, head, or anus. Symmetry is radial or biradial. Tentacles surround the mouth, which leads to a saclike digestive cavity (*enteron*), with or without branches. Body forms a sessile, single or colonial polyp, or a floating medusa. Stinging cells of distinctive types characterize both polyp and medusa stages.

Figure 5.1.   The nematocysts of *Hydra*.

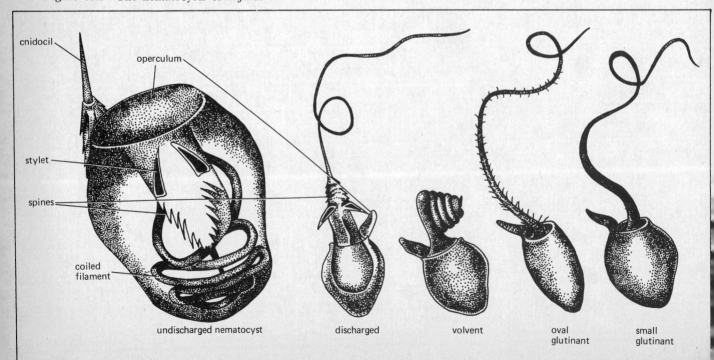

cnidocil
operculum
stylet
spines
coiled filament

undischarged nematocyst   discharged   volvent   oval glutinant   small glutinant

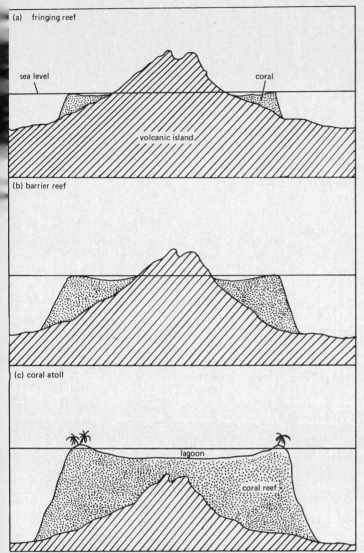

(a) fringing reef

sea level

coral

volcanic island

(b) barrier reef

(c) coral atoll

lagoon

coral reef

Figure 5.2. Development of a coral atoll. A fringing reef (*top*) is composed of coral that grows along the shore. The seaward side of the reef, rich in nutrients, grows rapidly, leading eventually to a barrier reef (*middle*). When the water rises, or the island sinks, the still-growing coral is all that is seen above the water (*bottom*). The resulting coral atoll may be a continuous reef, but it is usually divided by water channels extending through it from the ocean to the lagoon. Vegetation grows on accumulated debris.

The 10,000 or more cnidarian species are placed in three large classes, each based on overall body form.

Class 1. HYDROZOA, chiefly hydroids: the gastrovascular cavity is undivided and lacks nematocysts; a mesoglea which contains ameboid cells is present. Most hydroids possess a small medusa with a velum, but lack of stomadeum. Orders are the following: (1) HYROIDA (examples: *Hydra*,[1] *Obelia*), (2) TRACHYLINA (*Gonionemus*), (3) SIPHONOPHORA (*Physalia*), (4) MILLEPORINA (massive hydrocorals), and (5) STYLASTERINA (branched hydrocorals).

Class 2. SCYPHOZOA, true jellyfishes: these free-swimming medusae are bell-shaped, with four-part radial symmetry, a thick and gelatinous mesoglea, and no velum or stomadeum. The mouth is surrounded by gastric tentacles, there is a pouched gastrovascular cavity, and sense organs (*rhopalia*) on the edge of the bell. These sense organs and the gonads are the nearest approach to true organs in the phylum. The gonads are located in the gastric cavity. The polyp phase is reduced (as in *Aurelia*, for example) or absent entirely.

Class 3. ANTHOZOA, corals, sea anemones, and related forms: this class is predominantly sessile, with a flat oral disc from which hollow tentacles extend. The mouth leads to a stomadeum, then to an enteron divided into chambers by vertical septa bearing nematocysts, gonads, and, in some cases, special defensive (possibly digestive) tentacles (*acontia*). The mesoglea is well developed and highly cellular, forming a connective tissue. The heavy calcareous (epidermal) exoskeleton secreted by corals produces oceanic reefs.

Subclass 1. ALCYONACIA (8 septa with lateral projections; 8 tentacles): colonial only; includes the organ-pipe coral, soft corals, horny corals, sea fans (*Gorgonia*), sea pen (*Pennatula*), and sea pansy (*Renilla*).

Subclass 2. ZOANTHARIA (6 to many tentacles, never 8): includes the sea anemones (*Metridium, Anthopleura*), stony corals, and black corals.

We will study the following representative forms of each class:

HYDROZOA—*Hydra, Obelia, Gonionemus*
SCYPHOZOA—*Aurelia*
ANTHOZOA—*Metridium*

Remember that these samples are selected from among many and hence provide no more than a suggestion of the wide variety of adaptive variations in each class. Observe each critically and compare it with the others studied in order to obtain a balanced view of the whole phylum. "Take home" knowledge should include:

---

[1] The common name, hydra, is often used when referring to experimental work with various species of the genus *Hydra* or with similar genera, such as *Chlorohydra* and *Pelmatohydra*.

1. Basic cnidarian (coelenterate) organization.

2. Particular characteristics of each class, including basic form, specific structures, cell types, and mesogleal development.

3. Habitats in which the animal is normally found, and its life cycle. Wherever possible, add observations of behavior, such as motility, feeding reactions, response to selected stimuli, and study of regeneration and growth.

You should recognize how structural variation from the general body type permits adaptation of these structures and habits to specific environments.

3. Laboratory Instructions

I. Class 1. HYDROZOA; *Hydra* (Fig. 5.3)

Order 1. HYDROIDA

Hydra is an easily available example for study of cnidarian organization, including basic cell types and response to external stress. This organism is not a "typical" coelenterate, however. It is a fresh-water solitary polyp, without a medusoid phase and with a greatly reduced mesogleal layer.

1. **Gross structure of the living specimen** (Fig. 5.4)   Study a living hydra in a shallow dish. Add enough pond water to allow it free movement. Observe general body form under hand lens or dissecting microscope. Locate

Figure 5.3.   *Hydra*, whole mount.

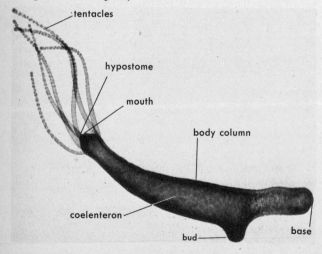

tentacles, nematocyst batteries, hypostome, mouth, base, and (if present) a bud and an ovary or testis. Observe the outer epiderm and inner gastroderm in your specimen. *Draw* an external view showing these structures.

2. **Behavior of the living specimen**   Observe reactions to the following:

a. *Motion*. Lightly tap the dish and rotate it. Note type of response.

b. *Touch*. Allow specimen to relax, then probe your specimen gently with a needle or fine point of a drawn-out glass rod, touching it at various points along the stalk and on the tentacles.

■ **What is the direction (*polarity*) of contraction (unidirectional or in both directions from the point of stimulus)?**

■ **Does this suggest something about the nature of the nervous system?**

■ **Which part is most responsive? Why?**

changes in pH, *salinity*. Add *dilute* acid, base, or salt solution.

■ **Any difference in the pattern of response?**

d. *Light*. Vary the local intensity of light.

e. *Presence of other hydras* (response to crowding or competition).

Now observe the following:

f. *Motility*, including "tumbling" and "inchworm" movements (Fig. 5.5). (Use a fresh specimen.)

■ **What might stimulate the response? Note location of specimen and check it after 2, 6, and 24 hours.**

g. *Nematocyst firing*. Place the specimen in a drop or two of water on a slide; cover and observe under the compound microscope (low power). Locate the intact nematocysts (each in a cnidoblast cell), which are found in rings or masses on the tentacles. Then tap the coverslip lightly or place a drop of safranin solution or fountain pen ink under the coverslip. Observe the discharge of nematocysts, which will appear like long threads (Fig. 5.1).

Locate and draw 2 different types of discharged cells.

h. *Feeding reaction*. Use fresh specimens starved about 48 hours. Observe *grappling, stinging,*

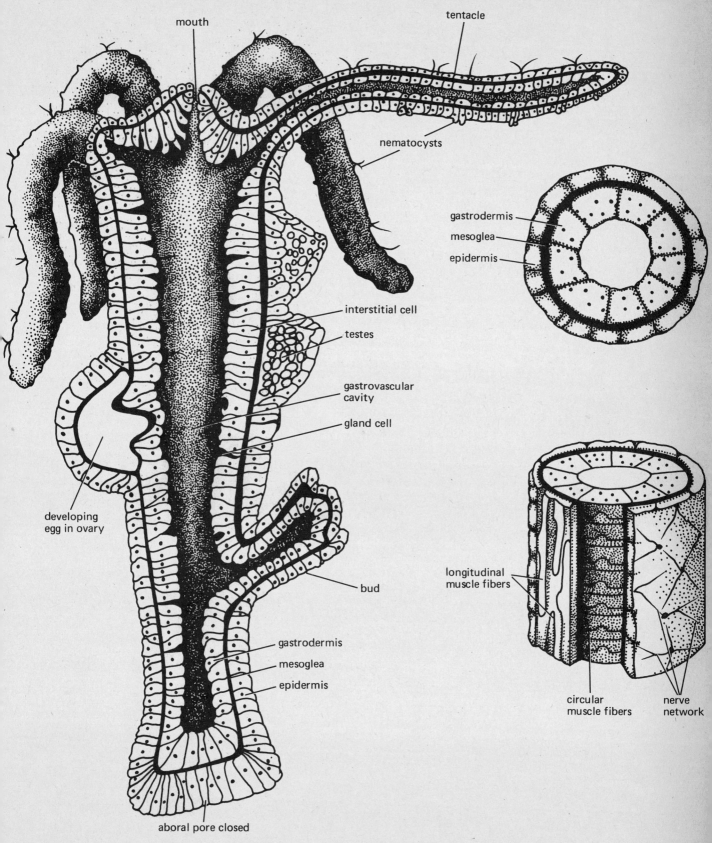

mouth

tentacle

nematocysts

gastrodermis

mesoglea

epidermis

interstitial cell

testes

gastrovascular cavity

gland cell

developing egg in ovary

bud

gastrodermis

mesoglea

epidermis

longitudinal muscle fibers

circular muscle fibers

nerve network

aboral pore closed

Figure 5.4. Structure of *Hydra*: (*left*) cut away to the entire *Hydra*; (*right*) (*top*) cross section, (*bottom*) section of trunk showing muscles and nerve net.

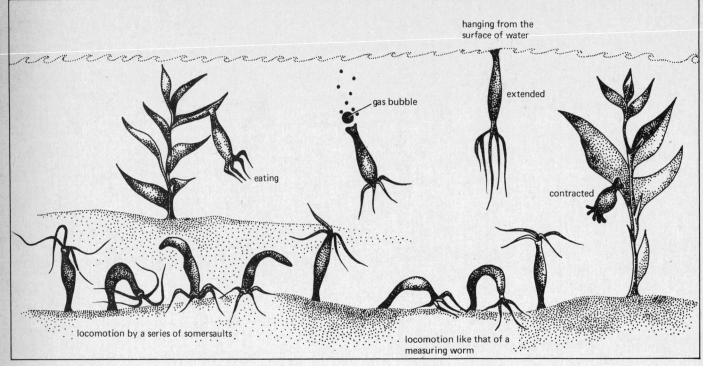

Figure 5.5.   Sketches show methods of *Hydra* locomotion.

and *engulfing* reactions. Use very small *Daphnia*, copepods, brine shrimp, bits of earthworm or of *Tubifex* worms, or minute scraps of almost any meat. Put the food particles close to the hydra, but be careful not to touch the tentacles. Record your observations.

Since no single preparation will show all of these activities to maximum advantage, share your observations with those of other students.

3. **Histology** (cross section) (Fig. 5.6) From a prepared slide, observe and draw the cellular detail seen in a selected (typical) section. Draw details in a wedge-shaped portion, *outline* the rest. Find: *epiderm*, with *epithelio-muscular* and *interstitial cells*; and *gastroderm*, with *epithelio-muscular*, *gland*, and *flagellated cells*. (Figs. 5.7 and 5.8). Try to locate *food vacuoles* in some gastrodermal cells.

■ **What additional structures are visible in a section through the hypostome? through the gonad (Fig. 5.9)?**

Diagram a longitudinal section cut from the hypostome and a tentacle down to the base of the animal. (Work out the structure from other sections you have studied.)

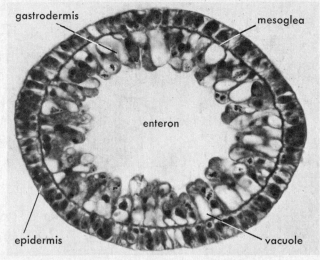

Figure 5.6.   *Hydra*, cross section through body column, ×500. (Courtesy CCM: General Biological, Inc., Chicago.)

4. **Questions**

■ **How does *Hydra* shorten? lengthen?**
■ **What type of *nervous system* controls this limited response?**
■ **What type of symmetry does *Hydra* show?**

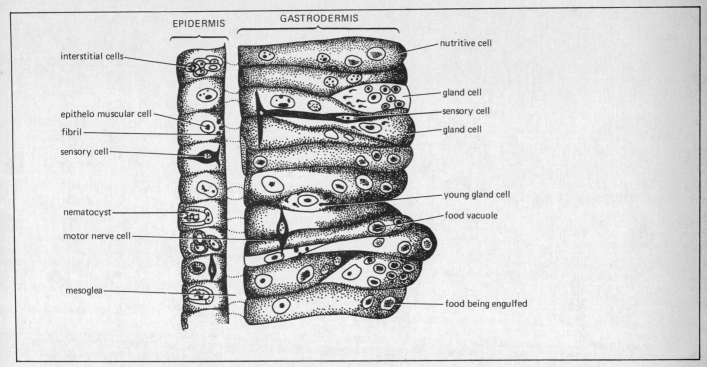

Figure 5.7. Longitudinal section of the body wall of *Hydra*, highly magnified to show the histology of the epidermis, mesoglea, and gastrodermis.

■ **What type of digestion is found?** (**Compare with a sponge and with a predaceous protozoan.**)

■ **How is waste material egested?**

■ **What is the life cycle?**

■ **How is the hydra adapted to a freshwater environment?**

■ **What other characteristics enable it to survive in its normal habitat?**

5. **Special projects**

a. Observe the distribution and habitats of *Hydra* in nature. Select a particular pond or series of sites, and keep careful records of collections and observations at regular intervals. *Look close.* You won't find the hydra at first, but bring back a few old leaves in pond water and they will "appear." Then check the pond site again *without disturbing the water.* The best way to find and to observe them is to:

b. *Culture* hydra in the laboratory from field collections. (For an easy method, see W.F. Loomis, 1959, *Ann. N.Y. Acad. Sci.,* 77:73.) Study the organism's life cycle, feeding behavior, growth pattern over a two-week period. *Artemia* (brine shrimp) are good for hydra. (To hatch *Artemia*

eggs in numbers, see the method by W.F. Loomis and H.M. Lenhoff, 1956, *J. Exp. Zool., 132*:555.)

c. Study regeneration of the tentacles, hypostome, and/or foot of hydra. Cut off the ring of tentacles with a fine pair of scissors, or cut off the hypostome and tentacles. Cut another hydra specimen in half. Observe twice daily for one week and record the rate at which these parts regenerate. Change the water daily.

d. Tie a piece of thread around the mid-region of hydra and see if you can observe differences in response by the two ends to various stimuli.

■ **What does this suggest about the concentration of nerve cells to the primitive nervous system displayed?**

■ **Is the behavior or response *adaptive* to the animal's survival?**

e. Observe the development of the sexual organs and of growth stages during *bud* formation (a form of asexual reproduction). Make slide preparations of the various stages of bud growth.

f. Investigate the feeding behavior of hydra as a study in chemoreception and food discrimination. (See Chapter 10.)

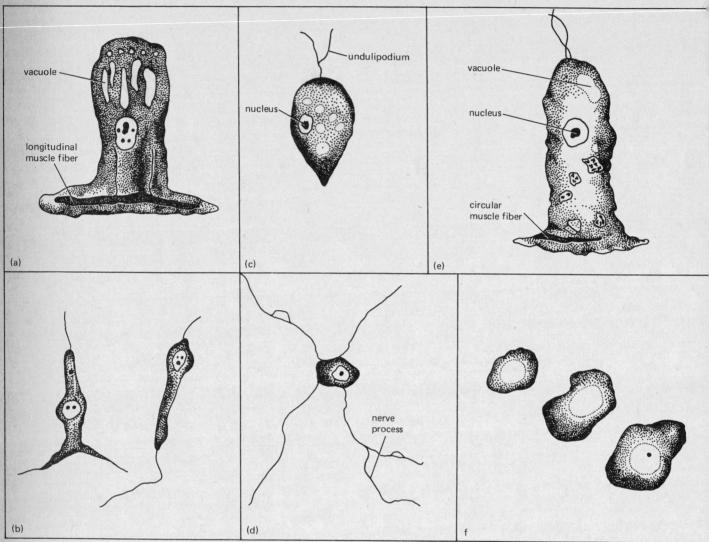

Figure 5.8. Principal cell types of the *Hydra*. (a) Epi-dermal muscular cell; (b) Sensory cells; (c) Gastro-dermal gland cell; (d) Conducting or motor nerve cell; (e) Gastrodermal nutritive cell; (f) Interstitial cells.

II. Class 1. HYDROZOA

Order 1. HYDROIDA; *Obelia* (Figs. 5.10 and 5.11)—a colonial marine hydroid

■ **In what way is *Obelia* a more representative cnidarian than *Hydra*?**

1. **Colony** (Fig. 5.10) Examine an intact colony under a hand lens, then look for greater detail by placing a small portion of the colony in water in a watch glass, or by studying a mounted, stained preparation. Using a dissecting microscope, study a *hydranth* (Fig. 5.12). Identify *tentacles, cnido-*

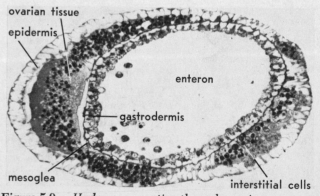

Figure 5.9. *Hydra*, cross section through ovarium.

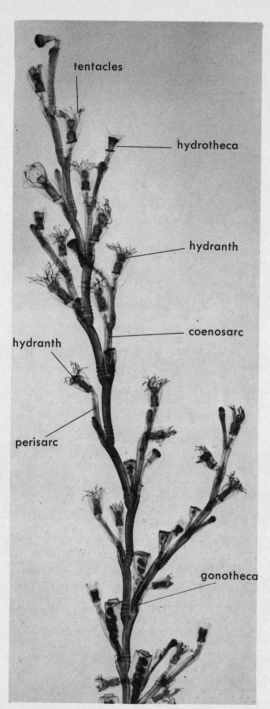

Figure 5.10.   *Obelia* colony.

blasts with their *nematocysts, hydrotheca,* the ringed outer *perisarc,* and the central *coenosarc.* Find a well-developed *gonangium* and locate the *gonotheca* (Fig. 5.13), *blastostyle,* and *medusa buds.*

   Sketch the general arrangement of a colony. *Draw* each of the two types of polyp

within this polymorphic colony and compare the functions of these forms.

■ **How does the individual *Hydra* perform these same functions of the colonial *Obelia*?**

■ **Is anything equivalent to the coenosarc, perisarc, hydratheca, or gonotheca found in *Hydra*?**

If live colonies of *Obelia* are available, maintain a population of several colonies in 2 gal of seawater at 24°–25°C. Use crushed *Artemia* for food.

■ **What becomes of the gonangia after their maturation?**

2.   MEDUSA   Study a medusa of *Obelia* (Fig. 5.14) or of some other hydroid, and locate the *bell* and the *manubrium* hanging from the *subumbrella* (ventral surface).

■ **Where is the mouth?**

From the mouth, food passes up to the *enteron* (gut) and out into the *radial canals* and *ring canal.* These structures are lined by gastrodermal cells in which *intracellular digestion* occurs. Digested food products are later passed to *amebocytes* (wandering cells), which transport them to other body cells.

   Between the tentacles are balancing organs, the *statocysts* (Fig. 5.8), containing a tiny particle (*statolith*).

■ **How do these organs work?**
■ **Why are they of such importance to the survival of medusae?**

The mesoglea is thick in all medusae. In fact, it comprises much of the volume of the organism.

■ **What is its function?**

Gonads containing ripe sperm or eggs are located in the four dark central structures near the manubrium base.

■ **How do sex cells escape from the parent animal?**

3.   **Life cycle**   Describe the steps in the *Obelia* life cycle.

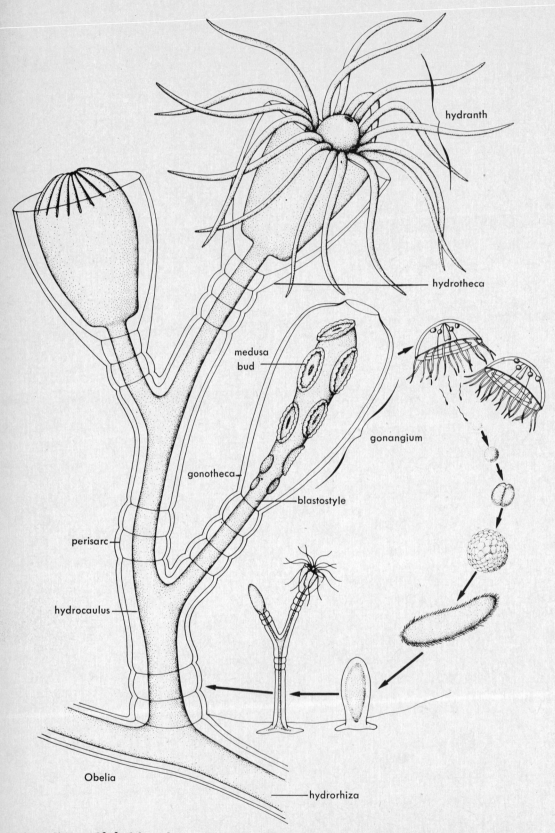

Figure 5.11. *Obelia* life cycle.

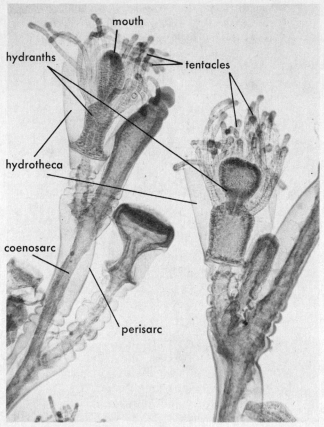

Figure 5.12.   *Obelia* hydranth.

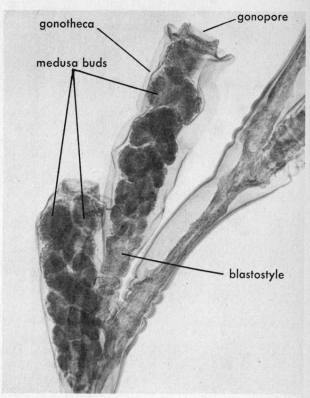

Figure 5.13.   *Obelia* gonangium.

III. Class 1. HYDROZOA

Order 2. TRACHYLINA; *Gonionemus* (Figs. 5.15 and 5.16)

1. **Anatomy**   Study the anatomy of a specimen under water and try to observe structures already described for the medusa of *Obelia*. Do not dissect this specimen if it is to be returned to the instructor. The *velum* will be particularly clear, as well as the large gonads.

Draw an oral view of *Gonionemus*.

■ **What is its normal habitat?**
■ **On what does it ordinarily feed?**

2. **Comparative material**   If available, observe museum specimens of other scyphomedusae, such as the jellyfish *Pelagia*, *Cyanea* (known to be up to 10 feet across!), or *Dactylometra*. Perhaps the most intriguing cnidarian is the colonial hydroid *Physalia*, the Portuguese man-of-war (order SIPHON-

Figure 5.14.   *Obelia* medusa, aboral view.

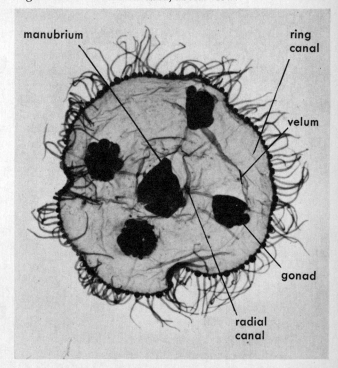

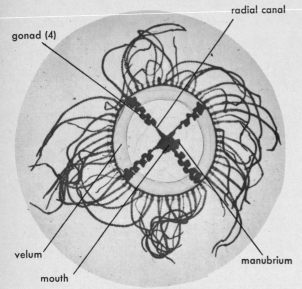

gonad (4)

radial canal

velum

mouth

manubrium

Figure 5.15. *Gonionemus*, oral view. (Courtesy CCM: General Biological, Inc., Chicago.)

OPHORA). The zooids that comprise the colony are so specialized and integrated within the colony that they appear to be parts or organs of a single animal. It probably represents the highest evolutionary development in which specialization of *whole individuals* rather than *cells* or subunits of a single organism is involved. Consider bee, ant, or termite colonies. Since the hive or colony members are polymorphic and interdependent, might not they be considered a loosely aggregated colony?

■ Do the zooids of *Physalia* develop from identical individuals?
■ How do we know that the resulting animal is derived from separate individuals and is not a single complex organism?

IV. Class 2. SCYPHOZOA; *Aurelia* (Figs. 5.17 and 5.18)

1. Anatomy Study the anatomy of an *Aurelia* medusa without dissecting it. (Return it intact later.)

■ Is a velum present?

Differentiate the SCYPHOZOA from the medusae of HYDROZOA.

Locate: *oral lobes, labial tentacles, gastric* or *enteric pouches* containing *gastric filaments, radial* and other *canals* (*interradial* and *aradial*), eight pairs of *lappets* at the edge of the *bell*, and *balancing organs.*

■ Can you find an *eyespot* in each of these balancing organs?

Draw a ventral view of *Aurelia*, giving details in one quadrant only.

■ How do jellyfish swim? feed? digest their food?

2. Life cycle Review the stages of the life cycle (Figs. 5.19–5.21).

■ Basically, how does it differ from that of *Hydra*? from *Obelia*?

3. Temperature effects If several live specimens of *Aurelia* are available, and some salt-water tanks for them, you can test the effect temperature changes has on the bell beat frequency (beats of the bell per minute). Start by introducing live *Aurelia* into three tanks of sea water, each set at a different (but physiologically tolerable) constant temperature. (Keep one of the tanks at the temperature you would expect of *Aurelia's* natural habitat.)

■ Do the organisms require a period of acclimation to the new temperature?

Count the bell beats for the individuals in each tank and prepare a graph by plotting bell beats per minute (if there is more than one individual in each tank) with water temperature shown on the x axis.

■ Compare the results in each tank. Are the differences consistent? significant? How do you account for them?
■ Would you expect the bell beat frequency of *Aurelia* living in the tropics to be the same as that of a species living in Arctic waters?
■ If an arctic *Aurelia* was transported to tropical waters, would you expect its bell beat frequency to become the same as that of the local (warm-water) species? Explain.

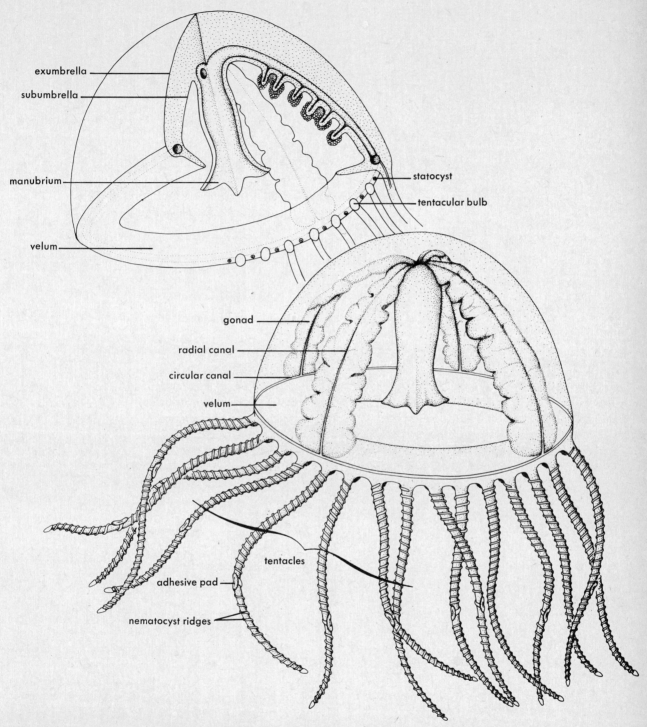

exumbrella

subumbrella

manubrium

velum

statocyst

tentacular bulb

gonad

radial canal

circular canal

velum

tentacles

adhesive pad

nematocyst ridges

Figure 5.16. *Gonionemus,* anatomical details.

V. Class 3. ANTHOZOA; *Metridium* (Figs. 5.22–5.26)

1. Anatomy On an intact specimen (Fig. 5.22) find: *oral disc, mouth, tentacles, body column,* and *pedal disc.*

2. Sections Split a fixed specimen lengthwise, from mouth to pedal disc. Locate the above structures, and in addition: the *gullet* with its *siphonoglyph* (one or two?), and the *enteron* lined by *septa* (or *mesenteries*) (Fig. 5.23).

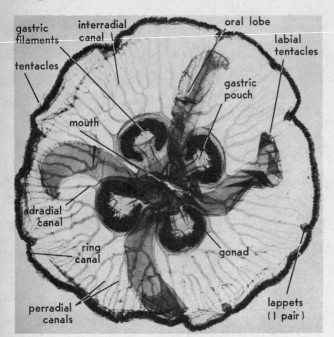

gastric filaments · interradial canal · oral lobe · labial tentacles · tentacles · gastric pouch · mouth · adradial canal · ring canal · gonad · perradial canals · lappets (1 pair)

Figure 5.17. *Aurelia*, adult, oral view.

At the edge of the septa are thick, convoluted septal filaments ending in *acontia*. The *gonads* resemble clusters of beads along the septa.

Point out each structure to your instructor so that he can check your understanding of the anatomy before you proceed with the next section.

Prepare and study several cross sections of *Metridium* (Figs. 5.25 and 5.26) to determine the different types of septal divisions. Find the *retractor muscles* on the septa.

■ Where are the body muscles found?
■ How does lengthening of the body column occur? shortening? withdrawal of tentacles?
■ How does the sea anemone avoid drying out when it is exposed to air and sunlight at very low tides?
■ Describe feeding and digestion in this animal.
■ What is the probable function of the various radial septa?
■ What is the life cycle of *Metridium*? of corals?

VI. Review Questions on CNIDARIA

■ How can we state that *Hydra* and *Metridium* have the same basic body type, in spite of the extreme differences in their structure?
■ Compare the life cycle of *Hydra*, *Obelia*, *Aurelia*, and *Metridium*, with particular reference to the relative importance in each of the polyp and medusa phases.
■ Which is the most highly developed (most complex) cnidarian you have studied?
■ Why do we speak of the cnidarian life cycle as a single developmental sequence, despite the fact that two phases (and also two types) of reproduction are involved? Why is the medusa considered to be the adult form?
■ *Define*: individual, aggregation, colony.
■ Compare the type of colony represented respectively by *Volvox*, *Obelia*, and *Physalia*. How do they differ?

VII. Special Projects

a. *Feeding reaction*. Use clean filter paper first, then bits of *Mytilus* (clam) juice or meat as test foods. Determine reactions to stimulation of tentacles, oral disc, mouth (lips), and column of living *Metridium* (aquarium or field specimens). Soak fresh filter paper in *Mytilus* juice, then test its effect on the same *Metridium* parts. Record the type and degree (speed, intensity) of response.

Compare these observations with similar feeding experiments on other polyps (hydroids or sea anemones).

b. What chemical substances or physical objects will cause the discharge of nematocysts? Cut-off several tentacles and place them in a depression slide; observe the arms under a dissecting microscope while you add the various test solutions.

c. Make a comparative study of nematocysts in as many species as you can study, such as *Hydra*, *Obelia*, *Aurelia*, *Metridium*, and *Physalia*.

d. Study the ciliary currents near the mouth of various test cnidarians by introducing particles of India ink, sand, or other foreign material. Compare this elimination response with the reaction to food particles. Record your results in chart form.

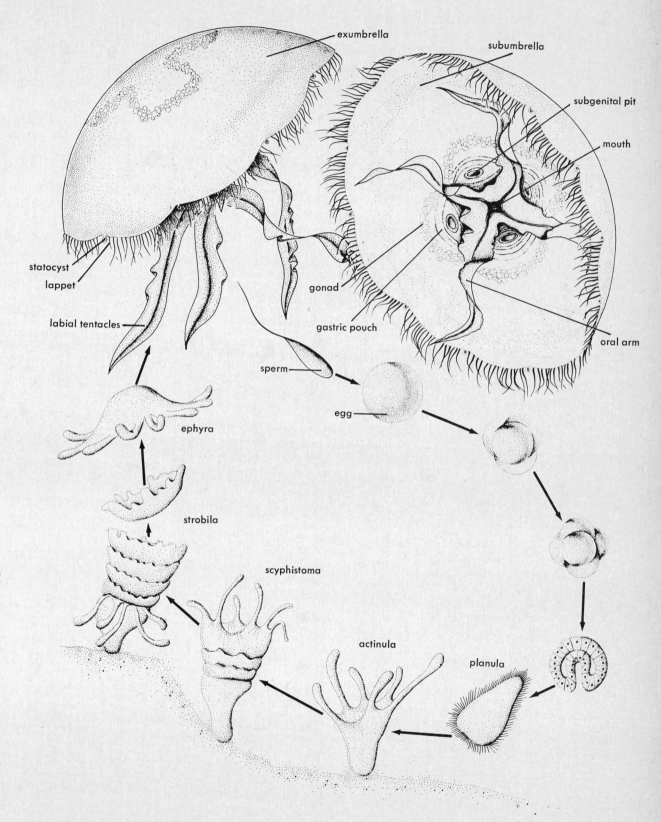

exumbrella

subumbrella

subgenital pit

mouth

statocyst

lappet

labial tentacles

gonad

oral arm

gastric pouch

sperm

egg

ephyra

strobila

scyphistoma

actinula

planula

Figure 5.18. *Aurelia* life cycle.

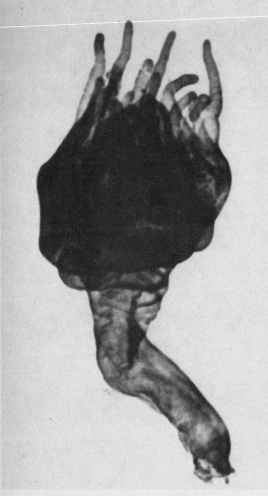

Figure 5.19.   *Aurelia*, scyphistoma stage.

Figure 5.20.   *Aurelia*, strobila stage.

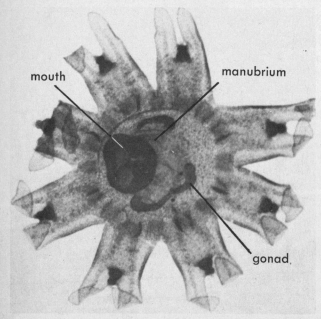

Figure 5.21.   *Aurelia*, ephyra stage.

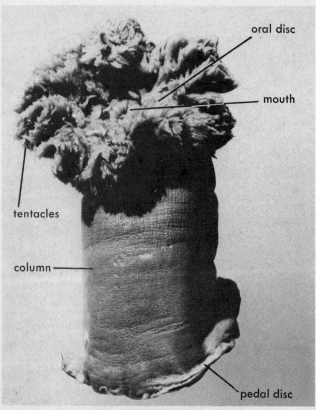

Figure 5.22.   *Metridium*, expanded.

Figure 5.23. *Metridium*, dorsoventral section.

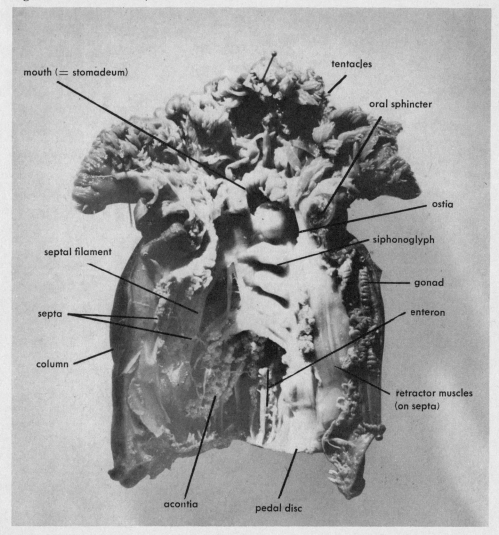

mouth (= stomadeum)

tentacles

oral sphincter

ostia

siphonoglyph

septal filament

gonad

septa

enteron

column

retractor muscles
(on septa)

acontia

pedal disc

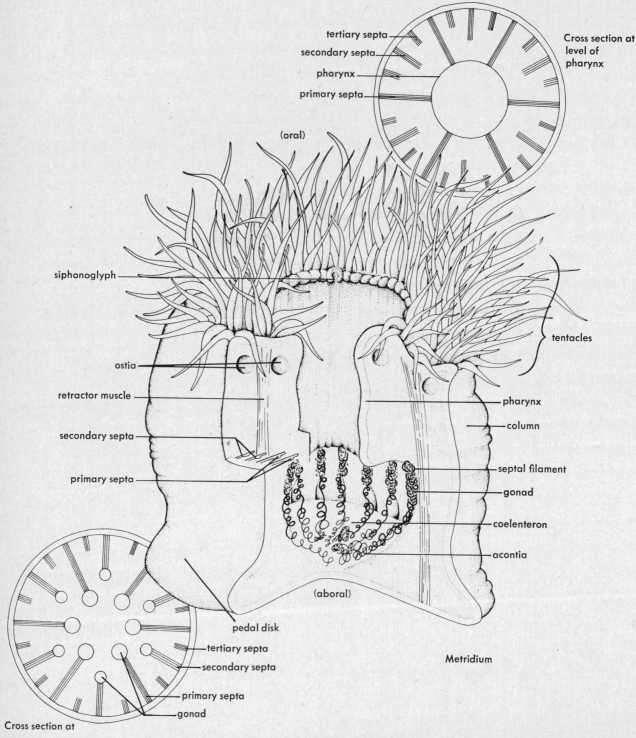

tertiary septa

secondary septa

pharynx

primary septa

Cross section at
level of
pharynx

(oral)

siphonoglyph

tentacles

ostia

retractor muscle

secondary septa

primary septa

pharynx

column

septal filament

gonad

coelenteron

acontia

(aboral)

Metridium

pedal disk

tertiary septa

secondary septa

primary septa

gonad

Cross section at

Figure 5.24. *Metridium*, semidiagrammatic illustration showing anatomical details.

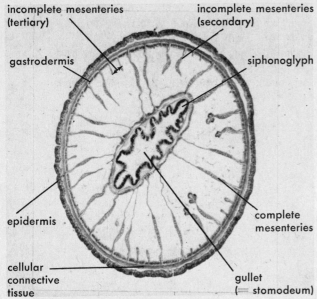

Figure 5.25.   *Metridium*, cross section at level of sto-
modeum.

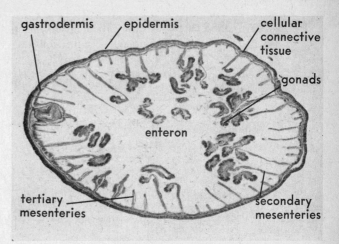

Figure 5.26.   *Metridium*, cross section at level of gonad.

Suggested References

Barnes, R. D., 1974. *Invertebrate Zoology*, 3rd ed. Philadelphia, Pa.: W. D. Saunders.

Batham, E. J., 1960. "The nerve net of the sea anemone, *Metridium senile*, the mesenteries and column." *Quart J. Microscop. Sci.*, 101:487.

Berrill, N. J., 1950. "Development and medusa-bud formation in the Hydromedusae." *Quart. Rev. Biol.*, 25:292.

Burnet, A. L., 1973. *Biology of Hydra*. New York: Academic Press.

Chapman, G., 1953. "Studies on the mesoglea of coelenterates." *Quart. J. Microscop. Sci.*, 94:155.

——, and L. G. Tilney, 1959. "Cytological studies of the nematocysts of *Hydra*," parts I and II. *J. Biophys. Biochem. Cytol.*, 5:69.

Crowell, S., *et al.* (eds.), 1965. "Behavioral physiology of coelenterates," *Am. Zool.*, 5:335.

Cuttress, C. E., 1955. "Systematic study of anthozoan nematocysts." *Syst. Zool.*, 4:120.

Darwin, C. R., 1896. *The Structure and Distribution of Coral Reefs*, 3rd ed. New York: Appleton-Century-Crofts.

Florkin, M., B. T. Scheer (eds.), 1968. *Chemical Zoology*, vol. 2, *Porifera, Coelenterata, and Platyhelminthes*. New York: Academic Press.

Garstang, W., 1946. "The morphology and relations of the Siphonophora." *Quart. J. Microscop. Sci.*, 87:103.

Hadzi, J., 1953. "An attempt to construct the system of animal classification." *Syst. Zool.*, 2:145.

——, 1963. *The Evolution of the Metazoa*. New York: Macmillan.

Hand, C., 1959. "On the origin and phylogeny of the coelenterates." *Syst. Zool.*, 8:191.

Horridge, G. A., 1956. "The nervous system of the ephyra larva of *Aurelia aurita*." *Quart. J. Microscop. Sci.*, 97:59.

Hyman, L. H., 1940. *The Invertebrates*, vol. 1, *Protozoa Through Ctenophora*. New York: McGraw-Hill. (The chapter on "Retrospect" in vol. 5 (1959) of this series summarized investigations on coelenterates from 1938 to 1958.)

Jones, C. S., 1949. "The control and discharge of nematocysts in *Hydra*." *J. Exp. Zool.*, 105:25.

Josephson, R. K., 1965. "The coordination of potential pacemakers in the hydroid Tubularia." *Am. Zool.*, 5:483.

Kramp, P. L., 1961. "Synopsis of the medusae of the world." *J. Marine Biol. Soc.*, 40:1.

Lenhoff, H. M., 1961. "Activation of the feeding reflex in *Hydra littoralis*," in *The Biology of Hydra*, H. M. Lenhoff and W. G. Loomis (eds.). Coral Gables, Fla.: University of Miami Press.

——, and W. F. Loomis (eds.), 1961. *Symposium on the Physiology and Ultrastructure of Hydra and Some Other Coelenterates*. Coral Gables, Fla.: University of Miami Press.

Lentz, T. L., and R. J. Barnett, 1965. "Fine structure of the nervous system of *Hydra.*" *Am. Zool.*, 5:341.

Loomis, W. F., 1959. "The sex gas of hydra." *Sci. American, 200*:145.

Pantin, C. F. A., 1965. "Capabilities of the coelenterate behavior machine." *Am. Zool.*, 5:581.

Payne, F., 1924. "A study of the fresh-water medusa, *Craspedacusta ryderi.*" *J. Morph.*, 38:387.

Rees, W. J. (ed.), 1966. *The Cnidaria and Their Evolution. Symp. Zool. Soc.* London no. 16. New York: Academic Press.

Yonge, C. M., 1958. "Ecology and physiology of reef building corals," in *Perspectives in Marine Biology*, A. A. Buzzati-Traverso (ed.). Berkeley, Calif.: University of California Press, pp. 117–135.

CHAPTER 6

PHYLUM PLATYHELMINTHES

The phylum **PLATYHELMINTHES** (G. *platys*, broad + *helmins*, worm) is composed of three classes: the free-living (that is, nonparasitic) TURBELLARIA, and two classes of highly specialized parasites, the TREMATODA (flukes) and the CESTODA (tapeworms). Generally the term "flatworm" is restricted to the class TURBELLARIA, in spite of its being the common name of the phylum. Flukes and tapeworms, along with certain other groups of parasitic worms, are called *helminths*, a convenient nontaxonomic term for these worm parasites.

The **PLATYHELMINTHES** demonstrate remarkable adaptations to both a free-living intertidal or fresh-water existence, and a parasitic mode of life. This phylum also displays (1) the organ level of complexity, (2) bilateral symmetry, and (3) the simplest form of body organization built largely of mesoderm. These three characteristics are such critical steps in the evolutionary sequence toward more complex organisms, that we shall consider separately their more widely extending implications before we review **PLATYHELMINTHES**.

1. Introduction to Higher Animals (Grade BILATERIA)[1]

Certain characteristics of free-living flatworms suggest that the phylum **PLATYHELMINTHES** is ancestral to the more complex animal phyla, and hence to all animals of greater complexity than a jellyfish or sea anemone. The increased specialization of mesoderm with the development of efficient organs is the most marked and significant distinction between flatworms and jellyfish or sponges. TURBELLARIA are the simplest animals yet discovered with a *brain* and *central nervous system* (CNS), a well-organized *excretory-osmoregulatory system*, and a complex *reproductive system*. Most body functions of flatworms are performed by specialized tissues or organs—a significant difference from the most com-

---

[1] The stages of complexity of body structures described in this section should suggest to the student the reasons why members of this and the following phyla are referred to as "higher animals."

plex cnidarians in which organs are more or less absent and highly specialized individuals (zooids) accomplish feeding, reproduction, defense, and other functions.

Accompanying bilateral symmetry in flatworms are improved sensory perception, increased muscular coordination, more rapid movement, and greater efficiency in processing food, excreting wastes, and controlling metabolic processes.

Associated with greater structural complexity is a pattern of life employing a *CNS-directed motility*, an active seeking of prey. This is a distinctly different pattern of behavior from that of coelenterates, which sting prey accidentally encountered in their nondirected swimming, floating, or sessile activities.

Much of the time in this course will be devoted to comparing structural and functional evolutionary changes in this and in succeeding animal phyla. We shall try to relate these changes to the animal's behavior in its particular habitat. Our purpose in doing so is threefold: *first*, to observe some of the various expressions of animal structure and function; *second*, to see how these demonstrate an increased efficiency and adjustment or *adaptation* to the environment; and *third*, to account for these adaptive changes.[2]

Structural or physiological complexity permits functional specialization. Specialization is associated with increased adaptation. Higher animals are derived. Recall that sponges have no such tissue, only wandering *amebocytes*. Cnidarians have *mesogleal* layers of varying complexity, from a thin jellylike layer to a true connective tissue. In **PLATYHELMINTHES**, mesoderm fills the space between body wall and gut with a mass of cells lacking clearly defined cell membranes. In more complex animals, mesodermal cells are clearly defined, forming tissues and organs of highly specific structure and function. Associated with this increase in complexity is the development of a *coelom*, a body cavity lined with mesoderm.

---

[2] *Caution*. The word adaptation is misleading if it is used to imply that an organism adjusts itself to the environment "on purpose." Indeed, this is not the case. The highly adaptive camouflaging spots of a leopard are *not* explained by its "needing" them—and thus developing them. Often we assume a simple cause and effect relationship between the environment and adaptation, between "need" and "response." However, it is well to remember that adaptation (the "response") is an *evolutionary* process that over a series of generations adjusts each species to its environment. The *mechanism* of adaptation by differential survival of natural variations in nature is a basic concern of this course and of much zoological research today.

Two other types of mesoderm organization occur in other phyla. **PLATYHELMINTHES** possess a *parenchymatous* (mesoderm-filled but not mesoderm-*lined*) body cavity. The platyhelminth organization; considered the most primitive type among the higher animals, is called *acoelomate* (without a true coelom). Further integration of organ systems is found in the phylum **ASCHELMINTHES** (see Chapter 7), with a body cavity that is *partially* lined with mesoderm. This type of mesoderm development is termed *pseudocoelomate* (false coelom).

In animals possessing a true coelom, the *eucoelomate* type, structures lying between the body covering (*epiderm*) and intestinal layer (*endoderm*) are enclosed by mesodermal sheets called *mesenteries* or *peritoneum*. One sheet underlies the body wall (hence: *somatic peritoneum*), another encloses the gut (hence: *splanchnic peritoneum*). The two sheets merge along the dorsal wall to form the *dorsal mesentery*, and in some instances form a *ventral mesentery* as well. Additional linings of mesentery enclose each organ, so that in all animals having a coelom, the organs are literally suspended within the body cavity by their mesodermal lining.

Each of these types of mesoderm organization characterizes large groups of animals of increasing structural complexity and integration. Mesoderm development is therefore used for a *supraphyletic* (above the phylum) taxonomic grouping—three *levels* within the *grade* BILATERIA (see Chapter 2):

Level 1. ACOELOMATA   lacking coelom
Level 2. PSEUDOCOELOMATA   partial coelom
Level 3. EUCOELOMATA   true coelom

2. Introduction to PLATYHELMINTHES

The acoelomate structure of flatworms allows a greater degree of tissue specialization than is possible in lower animals, permitting the *organ level* of complexity. Instead of a nerve *net*, flatworms possess a *central nervous system* with *ganglia* and *nerve trunks*. Sheets of opposing longitudinal and circular muscles underlie the epidermal layer of the body and control the body shape. The excretory (or, more likely, osmoregulatory) system, which is particularly characteristic of flatworms, consists of *flame bulbs* and *protonephridial tubules*, undulipodiated (ciliated) cells leading into tubes that collect body fluids and squeeze or drive them into a complex series of ducts and channels. Fluids pass afterwards to an excretory bladder and out one or more excretory pores. Flexible and coordinated motility is well developed in free-living TURBELLARIA; they possess cilia (usually on the ventral body surface) and secrete a coat of mucus on which they easily glide, assisted by muscular body movements. They can pass smoothly over the sharpest rocks or softest mud in their aquatic habitat. Some seek animal prey, others feed on vegetation (usually algae).

In their protected habitat inside the body of a vertebrate host, flukes and tapeworms have developed new specializations, including the loss of cilia and epiderm (although some immature stages retain these structures). They are covered by a tough, nonciliated tegument. In these internal and external parasites, directed motility seems to have been sacrificed for the capacity to adhere to and resist digestion (or other forms of destruction) by a host. Adaptations that provide a mechanism for "staying put" in the intestines, gills, or other protected niches in the vertebrate host include hooks, spines, suckers, and flaps of tissue (or combinations of these). Accompanying these structural changes are numerous physiological adaptations, as well as adjustments of the life cycle and reproduction to the exigencies of parasitism. These modifications attest to a remarkable degree of adaptation to a demanding environment. This is far from a state of "degeneration" that many persons consider parasites to occupy.

The digestive system among **PLATYHELMINTHES** is extremely variable. In fact, variations of this system partly determine the taxonomic orders of free-living flatworms, ranging from no gut (acoels) to many-branched gut (polyclad). In all cases, however, an *anus*, permitting one-way traffic of ingested food, is absent. The intestine of most flukes (TREMATODA) branches into two blind sacs, or *ceca*; tapeworms (CESTODA) have lost the system entirely and obtain food by absorption through the tegument. Dissolved materials pass directly into the interior parenchyma.

The reproductive system is perhaps the most impressive of all specialized structures in flatworms. Well developed in TURBELLARIA, it is extraordinarily complex in the flukes and tapeworms, which have essentially become reproductive machines. Since the chance of a parasite to transfer its progeny from one host to another is remote at best, high reproductive capacity is an essential adaptive mechanism for survival. In addition, the

TREMATODA often show highly bizarre, complex *life cycles* involving intermediate hosts, specialized larvae, and both sexual and asexual reproduction.

In tapeworms, sexual reproduction follows the mutual exchange of sperm between different worms or different segments of the same worm. Platyhelminths in general are *hermaphroditic* (*monecious*), that is, both sexes occur in the same individual.

■ **Speculate on the possible advantages of hermaphroditism to a parasite.**

Asexual reproduction (not involving gametes) occurs at the *larval* stage in the development of many flukes (trematodes) and among certain taenialike cestodes. This type of larval asexual reproduction is called *polyembryony* (the formation of several embryos from a single ovum).

A different kind of asexual reproduction occurs at the adult stage of most tapeworms. The body of these worms consists of a chain of segments (*proglottids*) budded out posteriorly from the anterior end (*scolex*). Each segment contains at least one complete sexual system, resulting in a duplication of sexual organs in a segmental train of identical units to form either a multiunit individual, or a colony if one chooses to call each segment a separate individual. In some heavy-bodied cestodes, the segments actually do break off and crawl about in the host intestine as if they were separate worms. Notice how the distinction between *reproduction* and *growth* may become modified or lost in considering the above question. (What is the distinction between the two among PROTOZOA?)

Polyembryony is characteristic of one of the two major groups of flukes (Order DIGENEA—a name denoting two hosts in the life cycle). A larva (*miracidium*) invades the tissues of a host, usually a snail, then buds off within itself individual germinal cells or clusters of cells that leave the initial stage and give rise to successive generations of structually different larval forms in the snail tissues (all derived from a single initial egg). The resulting embryos (*sporocysts* or *rediae*) develop, escape from the "parent" embryo into the tissues of the host, and proceed to form still another generation of larvae (*cercariae*), which finally leave the snail and seek the next host in this complex (digenetic) life cycle. In all digenetic flukes the general reproductive pattern is quite similar, but the specific hosts (at least two) and manner of development is characteristic of each species. Sexual reproduction occurs in the final host of these flukes, in which the worms mature, copulate, and produce eggs. Elucidation of trematode life cycles offers biological detective work in which there is considerable research interest.

Another group of TREMATODA, the order MONOGENEA (single-host life cycles), lacks polyembryony and develops directly on its specific host, usually a fish or other aquatic vertebrate.

Most of your time will be spent on fixed stained specimens. Remember, however, that the stained specimens are *animals*, not merely colored material on glass. Once very much alive, they are representative of living organism, some perhaps in your own body, and certainly in the bodies of multitudes of other animals and human beings. Those who have the opportunity to study and search for living examples of these parasites find that the common feeling of revulsion toward such a way of life is replaced by keen interest and continual surprise at the variety and extraordinary specializations among these organisms. Special projects, such as observation of regeneration in planaria or autopsy of hosts for parasitic worms, may particularly intrigue you.

3. Classification

Phylum **PLATYHELMINTHES**, flatworms: soft-bodied, flattened, usually elongated worms covered by epiderm or tegument, and unsegmented (except CESTODA). Digestive tract without an anus (TURBELLARIA and TREMATODA), or gut completely lacking (CESTODA). Excretory or osmoregulatory system consisting of flame cells and protonephridia leading to larger ducts and one or more excretory pores. Body acoelomate and filled with parenchyma. Nervous system consists of an anterior ganglion or nerve ring and one to four pairs of longitudinal nerve cords. No skeletal, circulatory, or respiratory organs. Usually hermaphroditic, with complex sexual apparatus; fertilization internal with yolky, shelled, microscopic eggs. Development direct in TURBELLARIA and MONOGENEA; indirect, with distinctive larval stages, in CESTODA and DIGENEA. Polyembryony in DIGENEA and some cestodes. Free-living or commensal (TURBELLARIA); parasitic (TREMATODA and CESTODA).

Class 1. TURBELLARIA, free-living flatworms: Ciliated epidermis with rhabdites and mucous gland; mouth ventral. Usually pigmented and with eyespots. Orders are the following:

(1) ACOELA[3] (no digestive tract); (2) RHABDOCOELA (single digestive tube, usually saclike); (3) ALLOECOELA (straight tube or with short branches); (4) TRICLADIDA (three-branched, usually with secondary short branches as, for example, in *Dugesia*; and (5) POLYCLADIDA (many-branched, mostly marine forms).

Class 2. TREMATODA, flukes: all parasitic; with tegument; suckers and/or hooks for attachment; mouth usually anterior, leading into a two-branched digestive tract; ovary single and testes usually paired.

Order 1. MONOGENEA: usually a single host, development direct (without distinct larval stages); large posterior sucker usually with hooks, anchors, or spines; parasitic on gills, mouth, pharynx, or anus of aquatic vertebrates.

Order 2. ASPIDOBOTHRIA: internal parasites whose entire ventral surface forms a specialized sucker divided into small compartments.

Order 3. DIGENEA: more than one host, the first a mollusk and the final host a vertebrate; two suckers; no hooks; polyembryony only in molluscan host (*Clonorchis, Fasciola*).

Class 3. CESTODA, tapeworms: parasites of vertebrates, usually intestinal; most segmented, with each segment (proglottid) lined with tegument, bearing one or two complete hermaphroditic reproductive systems; anterior scolex (plural: *scoleces*) with suckers and/or hooks for attachment. No digestive system or sense organs. Life cycle complex; no larval multiplication except in a few exceptional forms, such as *Echinococcus* and other relative of *Taenia* in which asexual multiplication of scoleces occurs in a larval cyst (*Taenia, Diphyllobothrium, Hymenolepis*).

4. Laboratory Instructions

I. Class 1. TURBELLARIA

Free-living flatworms will be represented by the freshwater planaria, *Dugesia tigrina* (Fig.

[3] Do not confuse Acoela with acoelomate or acoelom, descriptive terms for the *type of body cavity* in all platyhelminths.

6.1), or by some related form. It is an active, predaceous, responsive animal, extremely sensitive to chemicals, pH, light change, electrical fields, gravity, food, and similar stimuli.

1. **Living responses** Carefully record observations on a living specimen in pond water in a clean syracuse dish or watch glass. Follow its movements with a hand lens or dissecting microscope. Note its motility and reactions to such factors as food (bits of liver), jostling, and strong light; flip the worm upside down with a needle. Check reactions to a drop of weak acid or base, to 1 percent salt solution, or to other chemicals. (Remember that it may respond by disintegrating.)

2. **External anatomy** Observe the entire organism, its bilateral symmetry, and major axes. Locate the *dorsal* and *ventral* surfaces, *anterior* and *posterior* ends. Note the lobelike *auricles* (lateral extensions of the head), the *pharynx* and *pharynx sheath, mouth,* and *intestine.* Feeding it a liver paste mixed with carmine particles will give a startling demonstration of the digestive tract about 12–24 hours later, especially if white planaria are used. The *genital pore,* a small opening posterior to the mouth on the midventral surface, can be seen if the specimen is a lightly colored one.

3. **Detailed structure** Place the specimen on a slide in a drop of two of water, cover it with another slide or a strong coverslip, press slightly, and observe it under the high-dry

lens of your compound microscope. Look for detailed structure of the eye spots, cilia, pigment granules, and particularly the flame cells, detected by a faint flickering of the cilia inside the cell. The latter will be the reward of a persistent and diligent search, plus a certain amount of drying and compression of the specimen. After the specimen distintegrates (and it will), locate separated groups of cells and study their structure and the active movement of the ciliated cells of the ventral epidermis.

4. **Histology** (cross section) (Fig. 6.2) Study a stained cross section of a turbellarian. Use it to construct a three-dimensional view of the animal you have previously seen in surface view. Locate *rhabdites, gland cells,* and *cilia* in the *epiderm.* Find the *pharynx* (or *proboscis*), *pharynx sheath, intestine* (how many branches?), and endodermal *gastroderm* with gland and storage cells. Next observe the *parenchyma.* Look for the two ventral *nerve cords* near the midventral line. Below the epiderm find the *circular, transverse,* and *longitudinal muscle layers.*

Make a large drawing of your section, giving cellular detail for a small section only. Show only structures you can find in your own slide.

Figure 6.1. *Dugesia,* whole mount, stained to show intestine.

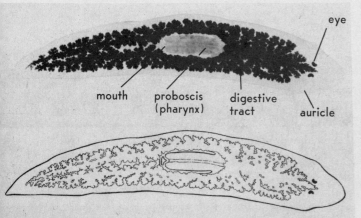

Figure 6.2. *Dugesia,* cross section through pharynx.

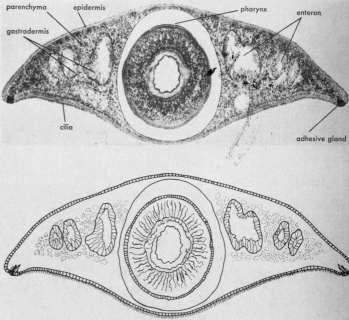

For a more complete understanding of this class, study other species from charts, diagrams, and text illustrations. Note variations and the appearance of other orders.

Check the structure of the reproductive system particularly, as it is extremely difficult to discern in *Dugesia*. Details of this system can be seen satisfactorily only in appropriate cross sections or in serial sections—one reason why there are so few experts on the taxonomy of TURBELLARIA.

5. **Special projects**

a. Search for planarians in fresh-water ponds, under leaves or stones, or on the lower surface of rocks turned over in the intertidal zone of a rocky coastal area. Try to locate representatives of each order of TURBELLARIA. Bring living planarians back to the laboratory in fairly large jars containing the original water in which they were found. *Keep them cool and do not crowd* with other specimens or organic material. Classify them to order and record observations on their feeding and responses to various stimuli. Special experiments can be developed to test reactions to such stimuli as an electric field, varying light intensity, chemicals or drugs, and various types of food. Keep careful notes on rhythms of activity, feeding preferences, combativeness, mating, or other forms of activity of undisturbed animals. Such observations could prove to be original and valuable scientific contributions.

b. Study regeneration or replacement of body parts. Plan your experiment in advance after review of references such as Bronsted (1955) suggested at the end of this chapter. Careful preparation and follow-through during the experiment will pay off in results—regeneration instead of disintegration. Use a sharp scalpel, or a chip of a razor blade mounted on a wooden applicator stick to cut selected large specimens of *Dugesia* or some other planarian. Place specimen (starved for a week) in a drop of water on a *clean* slide, place slide on ice cube to slow it down. Appropriate cuts should allow you to produce planaria with 2 heads, or 2 tails, or a double or multiple heads at one end. Maintain your operated specimens in a covered container kept filled to a uniform level with clean pond or tap water allowed to stand for two days in a cool, dark, protected place. *Do not feed them during the experiment.* Keep the

worm fragments in separate labeled containers (petri dishes or small jars). If partial cuts are made to test formation of multiple heads, these must be recut periodically to prevent healing over. Examine the worms at regular intervals for a two-week period; draw their outlines periodically to show zones of regeneration.

■ **Is regeneration a form of reproduction?**

c. Study conditioned response (a form of learning) in planarians. Review pertinent references at end of chapter. Expose a planarian to strong light.

■ **What is its reaction?**

Next give the planarian a mild electric shock (using an electrical stimulator hooked up to the aqueous medium).

■ **How does it respond?**

Now repeatedly expose the planarian to the strong light followed immediately (only a few seconds later) by the mild electric shock. You may wish to use several planarians, giving each one different numbers of exposures to the light–shock sequence. For example:

planarian	light–shock exposures
A	25
B	75
C	100

Now shine the strong light on each of the planarians without giving the electric shock afterwards. Repeat this many times.

■ **What are their reactions?**
■ **What minimum number of exposures to the light–shock sequence will "train" the planarian?**
■ **Does conditioning or training improve with increased numbers of light–shock exposures?**
■ **What is a simple definition of conditioned response?**
■ **Make a graph by plotting the number of light–shock exposures on the x axis and on the y axis the number of conditioned responses (out of a constant number of trials) for each planarian.**

Cut up one or several of the conditioned planarians and feed the segments to untreated individuals.

■ How do these new individuals respond to the strong light alone? Are these individuals more easily trained (i.e., show fewer light–shock treatments)?

Further work can include regeneration studies on conditioned individuals. Do the flatworms regenerated from the head and tail segments of a conditioned flatworm have the same ability to be trained?

II. Class 2. TREMATODA, flukes

This class includes many hundreds of parasitic species, from a half millimeter to a meter in length. Generally, they are leaf-shaped (hence: *fluke*) and characterized by a complex hermaphroditic reproductive system, expressed in the order DIGENEA by a weirdly complex life cycle (as mentioned earlier).

Two species of digenetic flukes are used in most introductory laboratories—the human liver fluke, *Clonorchis sinensis* (Fig. 6.3) and the sheep liver fluke, *Fasciola hepatica* (Fig. 6.4). The latter is atypical in that all major organ systems have multiple blanches, making it difficult to study. *Clonorchis*, a more representative fluke, is easier to study structurally. However, many other species might also be used. Examination of lungs, intestine, and bladder of a few freshly killed frogs will dramatically demonstrate the abundance of flukes in nature.

1. **Adult structure** (*Clonorchis*, stained whole mount) Find the *suckers* (anterior *oral sucker* surrounding the mouth, and *ventral sucker* or *acetabulum*), then the *pharynx* and intestinal branches (*ceca*). Study the reproductive system carefully with the aid of Fig. 6.3. The male system consists of two *testes, sperm ducts, seminal vesicle, cirrus* or *penis* (often enclosed in a cirrus pouch), and a *genital pore* (common exit of both reproductive systems). The female organs, somewhat more complex, consist of an *ovary* connected by an egg tube or *oviduct* to a fertilization and yolk accrefion chamber, the *oötype*. Connected to the oötype is a *seminal receptacle* (for storing sperm received during copulation) and a *vitelline duct* bearing yolk material from the yolk or *vitelline glands*. Fertilization occurs in the oötype, after which yolk is rapidly

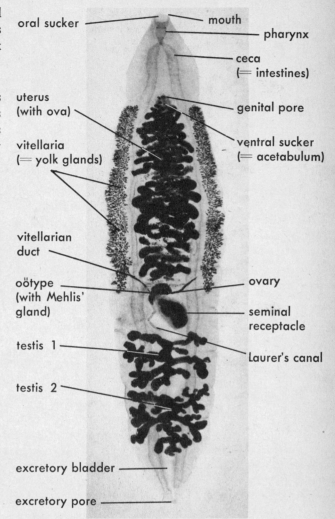

Figure 6.3. *Clonorchis*, whole mount, stained to show reproductive system.

pressed onto the egg in a stamping-mill process, in some forms producing a finished egg every few seconds. The oötype is usually surrounded by a rather diffuse *Mehlis' gland*, often difficult to see. From the oötype, the newly fertilized, yolk-laden egg moves into the *uterus*, a large coiled structure characteristic of digenetic flukes. Here, the egg shells are hardened and darkened by a chemical tanning process, and eggs are shoved up the coils of the uterus by newer eggs pushing out of the oötype, until they finally emerge out of the genital pore, often in a near-continual stream of eggs.

Locate the terminal *excretory pore* and, if it is not obscured by other organs, the *excretory bladder*.

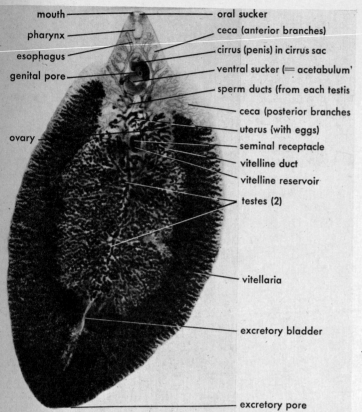

Figure 6.4. *Fasciola hepatica*, whole mount, stained to show internal structure.

■ **Why are flame cells not visible in your specimen?**

Review the fluke structures and work out the reproductive system so that you can follow the process of sperm and egg formation, copulation, fertilization, yolk accretion, and egg passage from ovary to genital pore.

■ **What happens to the eggs after they pass out the genital pore?**

2. **Larval structure** If available, study living or stained preparations of larval stages: *miracidium, sporocyst, redia, cercaria,* and *metacercaria.*

■ **In which of these stages does asexual reproduction occur? by what process?**

3. **Special projects**

a. Collect freshwater snails such as *Helisoma, Gyraulus, Physa, Lymnaea,* or mudflat brack-ish-water snails like *Cerithidea,* and examine them for larval stages of flukes. Crush or break open the snail shells in pond water in a syracuse dish so that you can remove the visceral whorl and search it for parasites. Any infection will usually be readily apparent, as the larval flukes, when present, are generally abundant. Under the dissecting microscope look for lashing, tailed cercariae, elongate or saclike aprocysts, or the larger, slowly moving rediae, often colored yellowish and distinguished by an oral sucker and a darkly pigmented gut. Find the germ balls and later developmental forms inside the sporocysts and rediae. Developing cercariae should be easy to discern. Identify as many different types of larvae as you can, especially different types of cercariae (representing different species infecting the snail). Consult the parasitology textbooks in your reading list for additional examples (e.g., Cheng, 1973; Olsen, 1973).

b. Autopsy (necropsy) of nearly any vertebrate will yield parasites. Frogs, salamanders, fish, snakes, carnivorous or fish-eating birds, bats, and rodents are all excellent hosts. Nothing creates an appreciation of parasitism more quickly than the sight of a *living* fluke, tapeworm, or nematode—especially when you find it yourself in an animal you have collected.

III. Class 3. CESTODA, tapeworms

Tapeworms are always good for a recoil—especially those said to measure "79 feet long" by certain casual eyewitnesses. Most tapeworm parasites of animals are only several inches long, and they are just as interesting and a good deal easier to study. Human tapeworms of the genera *Taenia* or *Diphyllobothrium* are comparative giants. They *may* grow to 40-foot lengths, but 10 to 15 feet is more common. Refer to your text for the life cycle of cestodes and, for additional details, consult Cheng, Hyman, Olsen, and Smyth from your reading list.

1. **Structure of adult Taenia or Dipylidium** (double-pored dog tapeworm) Observe the gross anatomy of your preparation.

■ **Can you identify the worm from the shape of the segment or appearance of the scolex on your slide?**

The small anterior end is the *scolex* (Fig. 6.5) which should be examined first. Locate the *suckers* (how many?) and an anterior projection or *rostellum*. In many species (such as *Dipylidium caninum from dogs*) this structure will be lined with one or more rows of hooks.

■ **What is the function of the scolex?**
■ **Where is the tapeworm mouth?**

Study a *mature* segment (Fig. 6.6) in which are found the fully developed, intact male and female reproductive systems. (These systems will be destroyed or distorted by pressure of eggs in the *ripe* or *gravid* segment.) Find the male structures, including the scattered *testes, sperm ducts, seminal vesicle, cirrus pouch,* and *genital pore*. The female system in the same segment will show the *vagina* (slender tube that passes from the genital pore to the *seminal receptacle*, passing above the center of the divided *ovary*). It is seen as a fine, down-curving line. As in flukes, eggs from the ovary move down an *oviduct* and into the *oötype*. Sperm, having entered the system through the vagina after copulation, migrate from their storage point in the seminal receptacle, through the oviduct, into the oötype, where fertilization occurs. Eggs are then quickly coated with yolk, and shell material is pressed over them by contraction of the oötype. This material comes from the compact *vitelline gland* located below the ovary. In other groups of tapeworms, the vitelline gland may be distinct follicles either scattered throughout the segment or in discrete marginal bands. (These criteria are means for distinguishing *orders* of tapeworms.) Finished eggs are quickly pushed along into the uterus or into the parenchyma, where they gradually fill and form it into a shape characteristic for each species.

■ **How does the uterus of *Taenia* compare with that of *Clonorchis*?**
■ **What traits distinguish scolex, immature, mature, and gravid segments of *Taenia* from *Dipylidium*?**

Draw the mature segment, label the parts, and be prepared to tell the function of each organ.

■ **How do eggs escape from the uterus?**
■ **How do they get into the intermediate host?**
■ **How do the larvae reach their final host?**

Study a *ripe* or *gravid* segment of *Taenia* (Fig. 6.7). The reproductive organs are practically obliterated by egg-filled branches of the uterus. Under high power, examine some eggs mounted in water, using a coverslip. Try to distinguish the six-hooked larva (the *oncosphere* or *hexacanth*) inside the shell. This larva hatches from the shell after being stimulated to do so in the stomach of its *intermediate host*, which became infected by accidentally ingesting eggs or segments from feces of the *final host*. The released larva then penetrates the intermediate host's intestinal wall, invades the body tissues, and forms a *cysticercus* (or *bladder worm*) in muscle or other tissues.

■ **What are the intermediate hosts of *Taenia solium* and *Taenia saginata*?**
■ **How does the *final host* (man among them) become infected?**

Another larval form, the *cysticercoid*, occurs in the life cycle of those tapeworms that utilize an *invertebrate* intermediate host, usually an insect. The cysticercoid of *Hymenolepis diminuta* (a common rat tapeworm), for example, lives in grain beetles. They infect the rat when the latter feeds on beetle-contaminated grain. And how did the beetles get infected?
2. **Larval structure** Study a slide of a stained *Taenia* cysticercus (or bladderworm). It is found, as we know, in the flesh of an appropriate vertebrate intermediate host. Locate the *bladder, neck, scolex,* and, if possible, the *rostellum* with a double row of *hooks* that characterizes *Taenia solium*. *Taenia saginata* has a similar cysticercus but lacks the hooked rostellum.

3. **Special projects**

a. Slit open carefully and examine intestines of rabbits, cats, or dogs (from animals destroyed at the local pound), sheep (from a slaughter-

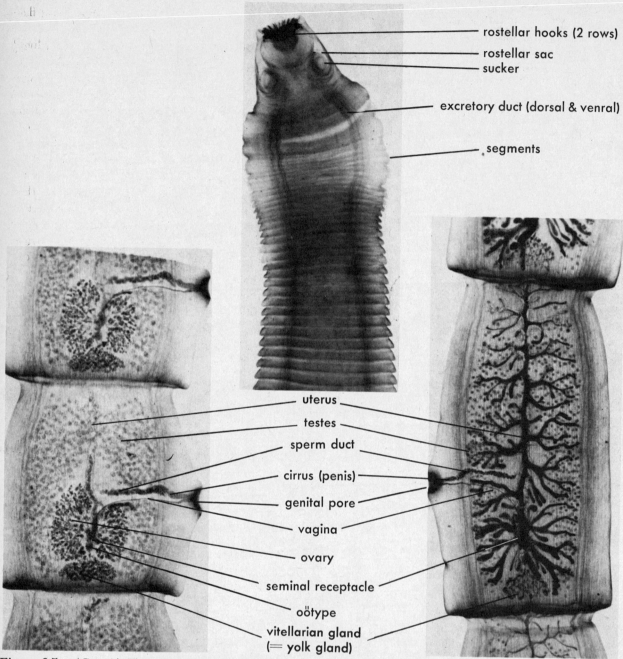

rostellar hooks (2 rows)
rostellar sac
sucker

excretory duct (dorsal & venral)

segments

uterus
testes
sperm duct
cirrus (penis)
genital pore
vagina
ovary
seminal receptacle
oötype
vitellarian gland
(= yolk gland)

Figure 6.5. *(Center) Taenia pisiformis*, scolex.
Figure 6.6. *(Left) Taenia pisiformis*, mature segments.
Figure 6.7. *(Right) Taenia pisiformis*, gravid or ripe segments.

house), bass, trout, pike, or other predaceous fish. Using techniques for processing the tapeworms you can find in the reading list, prepare your own stained, mounted, and labeled slides.

b. Set up an experimental tapeworm life cycle, using available tapeworm ova or larvae. If a

naturally infected rabbit can be obtained, feed cysts from the liver of this animal in a bit of tissue to an experimental dog. The final host (dog) need not be sacrificed unless adult worm specimens are desired. Eggs can be recovered from the feces and fed back to uninfected rabbits to complete the cycle. The cat and rat share a similar *Taenia* tapeworm. A simpler life cycle, employing grain beetles (genus *Tribolium*), laboratory rats, and the rat tapeworm, *Hymenolepis diminuta*, can be studied with relatively little difficulty. Because of the ease of transmission of this parasite under controlled conditions, *Hymenolepis* is used in many laboratories for a number of research projects designed to study tapeworm development or physiology.

IV. General Questions

1.   What are some likely morphological intermediate stages in the transition from free-living to parasitic flatworms?

2.   What special turbellarian structures are lacking in the parasitic flatworm groups? How is this absence correlated with function? What structures characteristic of flukes and tapeworms are absent in free-living flatworms?

3.   Is a tapeworm a colony or an individual? Compare it with *Volvox* and *Physalia*.

4.   What primary adaptations would you consider necessary for parasitism in general? Would you include physiological as well as structural modifications? If so, which ones?

5.   Are helminth parasites degenerate organisms? If not, what do you feel would be a more appropriate term?

6.   How would you explain the fact that *Clonorchis*, with a relatively small adult worm, has a reproductive capacity comparable with that of *Taenia*, which has larger adult size and far greater egg-producing capacity?

7.   Speculate on the possible advantages there are in a *multihost* parasite life cycle compared with a *single-host* cycle.

Suggested References

Best, J. B., 1963. "Protopsychology." *Sci. American, 208*(2).

Brand, R. von, 1973. *Biochemistry of Parasites*, 2nd ed. New York: Academic Press.

Bronsted, H. V., 1955. "Planaria regeneration." *Biol. Rev., 30*:65.

Brown, F. A., Jr. (ed.), 1950. *Selected Invertebrate Types*. New York: J. Wiley & Sons (Chapters by L. H. Hyman).

Buchsbaum, R., 1956. *Animals Without Backbones*. Chicago, Ill.: University of Chicago Press.

Cheng, T., 1973. *General Parasitology*. New York: Academic Press.

Dawes, B., 1946. *The Trematoda*. London: Cambridge University Press.

Florkin, M., and B. T. Scheer (eds.). *Chemical Zoology*, vol. 2, *Porifera, Coelenterata, and Platyhelminthes*. New York: Academic Press.

Hyman, L. H., 1951. *The Invertebrates: Platyhelminthes and Rhynchocoela*. vol. 2. New York: McGraw-Hill, pp. 52–219. (The chapter on "Retrospects" in vol. 5 (1959) in this series summarizes research on Turbellaria from 1950 to 1958.)

Jennings, J. B., 1957. "Studies on feeding, digestion, and food storage in free-living flatworms." *Biol. Bull., 112*: 63.

——, 1962. "Further studies on feeding and digestion in triclad Turbellaria." *Biol. Bull., 123*:571.

McConnell, J. V., 1967. *A Manual of Psychological Experimentation on Planarians*. Mental Health Research Inst., Univ. of Michigan, Ann Arbor, Mich.

Noble, E. R., and G. A. Noble, 1971. *Parasitiology. The Biology of Animal Parasites*, 3rd ed. Philadelphia, Pa.: Lea & Febiger.

Olsen, O. W., 1974. *Animal Parasites: Their Life Cycles and Ecology*, 3rd ed. Baltimore, Md.: University Park Press.

Ruebush, T. K., 1941. "A key to the American fresh-water turbellarian genera, exclusive of the Tricladida." *Trans. Am. Micr. Soc., 60*:29.

Schell, S. C., 1970. *How to Know the Trematodes*. Dubuque, Iowa: Wm. C. Brown Co.

Schmidt, G. D., 1970. *How to Know the Tapeworms*. Dubuque, Iowa: Wm. C. Brown Co.

Smyth, J. D., 1966. *The Physiology of Trematodes*. San Francisco, Calif.: W. H. Freeman.

Yamaguti, S., 1958. *Systema Helminthum.*, vol. 1, *The Digenetic Trematodes of Vertebrates*, pts. I and II. New York: Interscience Publishers.

——, 1959. *Systema Helminthum.*, vol. 2, *The Cestodes of Vertebrates*. New York: Interscience Publishers.

CHAPTER 7 · PHYLUM ASCHELMINTHES

1. Introduction

Phylum **ASCHELMINTHES** is an assemblage of six different classes,[1] the largest of which is the class NEMATODA—the threadworms (G., *nema*, thread) or roundworms. These ubiquitous small worms outnumber all other groups of multicellular animals except insects and mites.

The *pseudocoelom* is one important characteristic shared by all aschelminths. The *cuticle*, together with its underlying *hypoderm*, a cellular or syncytial (multinucleated protoplasmic mass without apparent cell walls) lining, and a mesodermal muscle layer combine to form the outer wall of the body cavity. The intestine, lacking a mesodermal covering, forms the inner margin of the body cavity. The pseudocoel (space between the intestine and the body wall) is filled with fluid and the tubes of the reproductive system. This incomplete or false coelom, as previously reviewed, represents an evolutionary stage between the ACOELOMATA (which phyla?), in which the body cavity is filled with undifferentiated mesodermal cells, and the EUCOELOMATA (all the so-called higher phyla), in which the body cavity is completely lined by mesoderm. The advantages of a coelom will be described in later chapters.

■ **Can you suggest an advantage of the pseudocoelom over the acoelomic condition?**

A second major advance in **ASCHELMINTHES** is a *complete digestive system*: mouth, intestine, and anus. Cellular specialization in this system allows ingested food to be processed along a one-way passage, with unassimilated matter finally ejected out the anus, rather than having it returned and passed back out the oral opening, as in TURBELLARIA and most TREMATODA.

■ **What are the advantages in terms of digesting and processing food in this type of intestinal tract?**

The cuticle is not only an extraordinarily effective protection against harmful external substances (or digestive enzymes in the case of internal parasites) but also serves as an exoskeleton, allowing an extensive area for muscle attachment.

Some aschelminthes may reproduce by *parthenogenesis* (development from an unfertilized

---

[1] These six classes are combined into one phylum because of certain common features. Some authorities consider each of these classes as a separate phylum.

egg). In some NEMATODA, ROTIFERA, and GASTROTRICHA, males may be entirely lacking, and successive generations of females produced parthenogenetically. Subsequently, both sexes then may be produced, followed by normal sexual reproduction, producing a resistant or "winter egg." In other species, no males have ever been discovered.

Still another remarkable characteristic of the phylum is a tendency toward constancy of cell number (*eutely*), in which the organs and often the entire animal are composed of a precise and relatively small number of cells. The number may be constant not only for the species but also for the larger taxonomic group and can be used as part of its morphologic definition. (For example, the central nervous system in *Ascaris* and many other nematodes consists of 162 cells.)

2. Classification

Classes in the phylum **ASCHELMINTHES**[2] include:

Class 1. NEMATODA: roundworms or threadworms (discussed below).

Class 2. NEMATOMORPHA (or GORDIACEA): "Horsehair worms," somewhat nematode-like; larvae parasitic in insects, adult worm emerges from its arthropod host in fresh or salt water, then reproduces and spends a brief free-living life.

Class 3. ROTIFERA: rotifers, microscopic "wheeled animals" (look for them as abundant denizens of your pond-water cultures); free-swimming, sessile, and tube-building forms known (Fig. 7.1).

Class 4. GASTROTRICHA: microscopic aquatic animals, with spiny cuticular skin and ventral undulipodia.

Class 5. KINORHYNCHA (or ECHINODERA): minute, marine, and mud-dwelling jointed worms, similar to gastrotrichs and rotifers.

Class 6. PRIAPULIDA: burrowing marine worms, thick-bodied, with a large spined anterior proboscis. These were once considered a class of **ASCHELMINTHES,** but are now thought to constitute a separate phylum.

I. Class NEMATODA

Characteristics of NEMATODA include an elongate, pointed body, round in cross section

(hence: roundworms), and a tough, impermeable cuticle.[3] A complete digestive system, separate sexes, tubular reproductive system, and a pair of "lateral lines" (containing nerve trunks and excretory tubes) are additional nematode features.

Biologically, nematodes are the most important and successful of the aschelminths. Parasitic nematodes infect more hosts in a greater variety and number than do any other groups. Humans, for example, are parasitized by pinworms, by *Trichinella, Trichuris, Strongyloides, Ascaris, Dracunculus* (guinea worm), numerous filarial worms, and several species of hookworms (Fig. 7.2). The incidence of infection with some worms may be almost 100 percent in certain areas. Sanitation-conscious Americans have one of the highest rates of trichinosis in the world, acquired from hogs fed raw garbage or slaughter-house leavings (Fig. 7.9). Probably every reader of these words has been infected by pinworms at least once in his lifetime. While most of us are familiar with nematodes as internal parasites of humans and other animals, most nematodes are free-living and microscopic. They occupy a great number and variety of habitats in soil, mud, and fresh or salt water. Soil nematodes, plant parasites of the first magnitude, cause millions of dollars in agricultural damage annually. Aquatic nematodes are important members of the biological food chain at the microscopic level, as they represent a significant portion of the total mass of living things. The majority, however, are still undescribed and unknown to science.

3. Laboratory Instructions

I. *Ascaris lumbricoides*[4]

1. **External anatomy** (Fig. 7.3) Observe a fixed or living specimen of *Ascaris*. Find the *head*, with three *lips* or bulbous bosses surrounding the *mouth*. (Examine from the an-

---

[2] Their varied nature and the broad, perhaps unnatural, assemblage in which these classes are placed causes some irreverant observers to term them "ash-can helminths."

[3] If free-living existence preceded parasitism, as is logically assumed, one can see how the resistant cuticle would serve the *parasitic* nematode just as well as it did its free-living ancestor. This condition is sometimes called *preadaptation.*

[4] *Ascaris* is not a typical nematode; it is merely large and easily available, and therefore commonly used in beginning zoology laboratories. Like *Taenia* and *Fasciola, Ascaris* is a giant, with most of its bulk filled with a tremendously elongated, highly coiled reproductive system capable of producing 200,000 eggs daily.

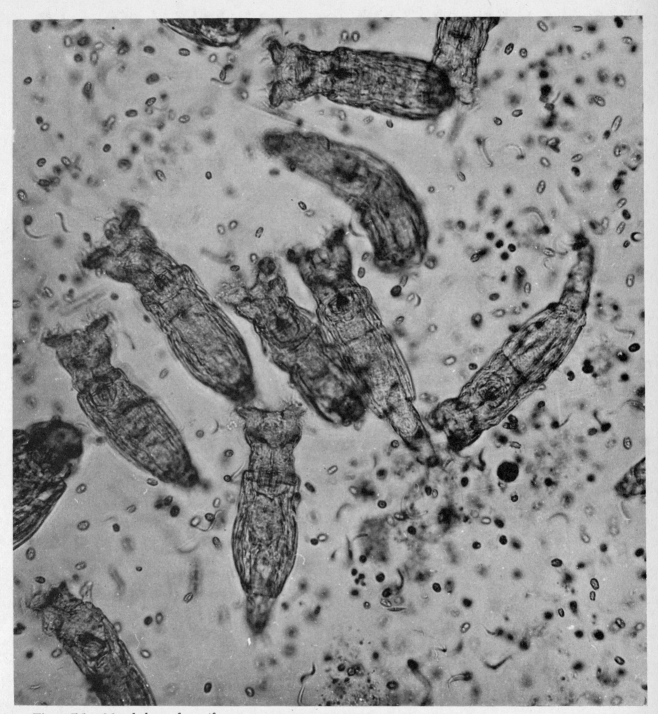

Figure 7.1.   Morphology of a rotifer.

terior end; look directly down on the head in good light using a strong hand lens.) The head can be examined under a dissecting microscope after cutting it off close to the lips and mounting it face up on a slide. This type of mounting is often used in nematode identification. The largest lip is the dorsal; the other two are the ventrolateral lips.

The male, considerably smaller than the female, is identified by its hooked or curved tail, which sometimes shows paired copulatory *spicules* (possibly serving as sperm

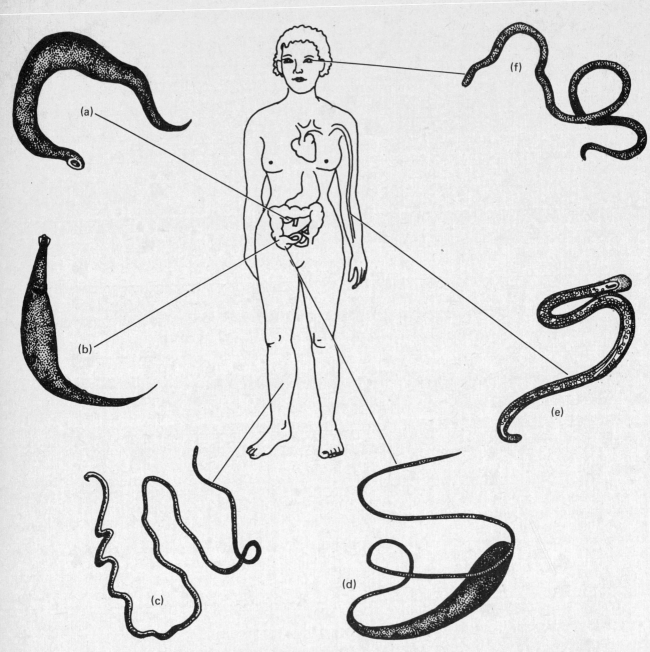

Figure 7.2. Roundworms that parasitize man. (a) Hookworm, *Necator americanus*, from intestine; (b) Pinworm, *Enterobius vermicularis*, from cecum; (c) Guinea worm, *Dracunculus medinensis*, from subcu- taneous tissues of appendages; (d) Whipworm, *Trichuris trichiura*, from cecum; (e) Microfilaria of *Wuchereria bancrofti*, from blood; (f) Adult eye worm, *Loa loa*.

guides) (Fig. 7.4), which protrude from the *cloaca* (the combined digestive and genital opening). Locate the *lateral lines* and the smaller *dorsal* and *ventral lines*, which mark the *dorsal* and *ventral nerve trunks*. Through the cuticle of a female worm, one can usually see the numerous fine coils of the reproductive system surrounding the intestine and filling most of the pseudocoel. The female genital pore can usually be seen one third of the way back from the head on the midventral surface.

II. *Trichinella spiralis* (Figs. 7.7, 7.8, and 7.9)

Study a section of trichina-infected pork or other meat. Locate a cyst and observe it under high power. Note the coiled *larva*, the surrounding *cyst wall* enveloped in host tissue (a much-modified host muscle cell, and possibly inflammatory cells around that), and perhaps evidence of calcification—a host tissue reaction that eventually kills the larva.

■ How did the parasite get to this site? Where are the *adult* worms found? How did they get there?

■ Why do we say that man, the *final* host, is also an *intermediate* host? Is this also true of the pig?

■ Why is man a "dead end" for completion of the life cycle?

■ Compare the life cycles of *Trichinella* (Fig. 7.9) with those of *Ascaris*, hookworm (*Ancylostoma*), and pinworm (*Enterobius*).

III. Free-Living Nematodes

1. **Observations** Study examples of living nonparasitic nematodes. Genera commonly available include the soil nematodes *Rhabditis* and *Cephalobus*, and the "vinegar eel," *Turbatrix aceti*, commonly found at the bottom of cider vinegar barrels. Place a few worms on a slide in water or in 5 *percent methylcellulose* to slow their rapid thrashing movement.

Rhabditis maupasi is commonly found in the excretory system or coelom of the earth-

Figure 7.8. *Trichinella spiralis* larva digested out of cyst, ×300.

worm. These are actually larval stages of a worm that lives on decaying matter in soil. Culturing these with one or two decaying earthworms will yield a rich nematode harvest. Although members of the genus *Rhabditis* are not *obligate parasites*, this species is a free-living form capable of spending the larval part of its life cycle inside the earthworm. Other species demonstrate similar degrees of parasitism (such as dwelling in ulcerating sores, feeding on the bacteria there).

■ How might this degree of parasitism evolve from this *facultative* type to a completely dependent or *obligate* type?

Observe the direction and orientation of movement of the soil nematode. The alignment and location of muscle cells along the body wall of nematodes indicate that their movement is possible only in one plane. The worms on your slide lie on their sides, so the observed undulatory *lateral* movement is actually a *dorsoventral* swimming action. Lightly press the coverslip down or let the preparation dry slightly to stop the worm's movement. Study the detailed structure of lips and head, pharynx (note the valvular bulb), intestine, and anus. Observe the clarity and relative simplicity of the reproductive system as compared with that of *Ascaris*.

■ Do you see evidence of *ovoviviparity* (hatching of eggs inside the uterus)?

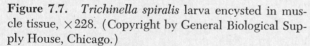

Figure 7.7. *Trichinella spiralis* larva encysted in muscle tissue, ×228. (Copyright by General Biological Supply House, Chicago.)

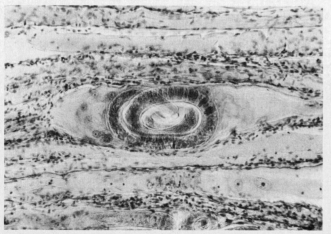

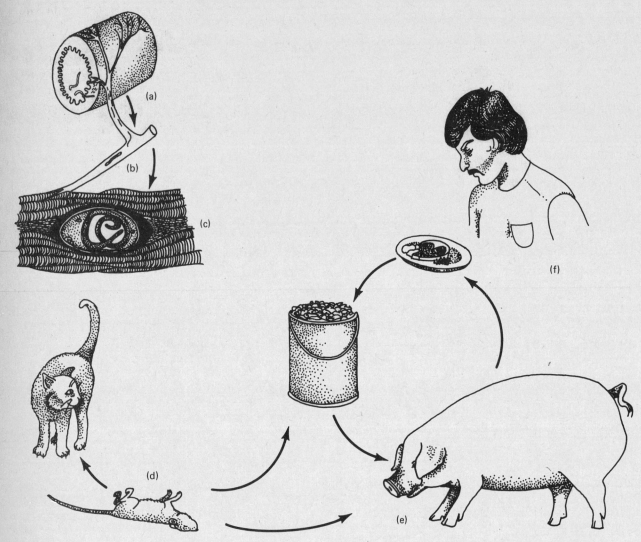

Figure 7.9.   Life cycle of *Trichinella*. (a) Adult female becomes embedded in small intestine of host and produces larvae; (b) Larvae enter lymph or blood vessels; (c) Larvae encyst in muscle; (d) Rodents act as additional or reservoir hosts, becoming infected by ingesting infected pork scraps; (f) Humans can become infected by eating poorly cooked or raw infected pork, liberating the larvae, which become adults in the intestinal mucosa, and which subsequently release larvae that enter the circulation and penetrate muscle fibers to produce *trichinosis*.

■ **Are special sensory structures visible about the head?**

■ **What would you consider to be the chief morphological differences between** *parasitic* **and** *free-living* **nematodes on the basis of your observations? How would these differences help each organism to adapt to its particular environment?**

2. Special projects

a. Make a microscopic study of free-living or plant-parasitic nematodes from soil, plant, or water samples. Study the samples microscop-

ically. To obtain larger numbers of worms, simple concentration and culture techniques can be employed.

One common procedure (described in most parasitology texts) is the use of the Baermann funnel, a simple apparatus that allows worms to swim out of a soil sample through a fine screen or cloth into a funnel containing warm water, connected to some rubber tubing. A pinch clamp on the tubing into a receptacle controls release of the worms collected in the bottom of the funnel. Another suitable method is the use of a tier of successively finer wire-mesh strainers that allow only nematodes and smaller organisms to pass into the sample chamber. Perhaps a more dramatic demonstration would be your own nematode culture:

Pour salt agar (0.5 percent NaCl, 1.5 percent agar, and distilled water) into a sterile petri dish to a depth of about 4 mm. Allow it to harden and place small bits of rich soil on the surface. Incubate at room temperature for periods varying from two or three days to a week or more, depending on the richness of the culture desired. If the soil contains nematodes, you will soon observe swarms of larvae at all stages of development on the surface and tunneling into the agar. Samples can be removed for detailed study at any time. *Rhabditis* will be one of the most common forms. If you try culturing worms from various types of soil you will be convinced of the abundance and variety of soil nematodes.

Try adding a piece of earthworm tissue to another agar plate. A flourishing culture derived from nematode larva living within the earthworm should appear in three or four days. Observe and record the life-cycle changes and relative abundance of the various stages of nematodes on your culture over a fixed period of time. Estimate from periodic samplings the population size and rates of growth and development of the colony (see Chapter 18).

■ **What further observations should you be able to make?**

A variety of cultures from different soil or water samples will enable you to observe many nematode species with a wide assortment of interesting structures and adaptations. You will find this a relatively little known but a potentially rich area of biological observation and research.

IV. General Questions

1. What structural characteristics of free-living nematodes permit or increase the likelihood of successful parasitism? For example, what "preadaptations" to parasitism may such nematodes have?)

2. What traits found in parasitic nematodes are of special importance in their adaptation to this specialized existence?

3. Name some of the traits or structures that help to account for the abundance, wide distribution, and ecological versatility of nematodes.

4. Compare adaptations to parasitism among the *protozoa, trematodes, tapeworms,* and *nematodes.*

5. What general characteristics unite the six classes of **ASCHELMINTHES** into a single phylum?

Suggested References

Brand, T., 1966. *Biochemistry of Parasites.* New York: Academic Press.

Brown, F. A., Jr., 1950. *Selected Invertebrate Types.* New York: J. Wiley & Sons.

Chandler, A. C., and C. P. Read, 1961. *Introduction to Parasitology.* New York: J. Wiley & Sons.

Cheng, T., 1973. *General Parasitology.* New York: Academic Press.

Chitwood, B. G., and M. B. Chitwood, 1950. *An Introduction to Nematologym,* 2nd ed. Baltimore, Md.: Monumental Printing Company, sec. I.

Florkin, M., and B. T. Scheer (eds.), 1969. *Chemical Zoology,* vol. 3, *Echinodermata, Nematoda, and Acanthocephala.* New York: Academic Press.

Goodey, T., 1951. *Soil and Freshwater Nematodes.* New York: J. Wiley & Sons.

Harris, J. E., and H. D. Crofton, 1957. "Structure and function in the nematodes: internal pressure and cuticular structure in *Ascaris.*" *J. Exp. Biol.,* 34:116.

Hyman, L. H., 1951. *The Invertebrates,* vol. 3. *Acanthocephala, Aschelminthes, and Entroprocta.* New York: McGraw-Hill.

Lee, D., 1965. *The Physiology of Nematodes.* Edinburgh, Scotland: Oliver and Boyd.

Olsen, O. W., 1974. *Animal Parasites; Their Life Cycles and Ecology,* 3rd ed. Baltimore, Md.: University Park Press.

Noble, E. R., and G. A. Noble, 1971. *Parasitology. The Biology of Animal Parasites,* 3rd ed. Philadelphia, Pa.: Lea & Febiger.

Rogers, W. P., 1962. *The Nature of Parasitism.* New York: Academic Press.

Yamaguti, S., 1961. *Systema Helminthum,* vol. 3. *The Nematodes of Vertebrates,* pts. I and II. New York: Interscience Publishers.

CHAPTER 8

PHYLUM ANNELIDA

1. Introduction

There are about 8000 species of segmented worms in the phylum **ANNELIDA.** This coelomate group contains the earthworms (class OLIGOCHAETA), marine bristleworms (class POLYCHAETA), leeches (class HIRUDINEA), and a small group of tiny marine annelids thought to be primitive (class ARCHIANNELIDA).

The annelid represents an evolutionary stage midway in complexity between the primitive metazoan and the vertebrate level of organization. It is one of the simplest of the *eucoelomate* animals (possessing a true coelom, completely lined with sheets of mesoderm), yet it possesses *organ systems* characteristic of all the so-called higher organisms.

Earthworms, the representatives of this phylum usually selected for classroom study, are abundant in moist soil and easily available to every student. Although demonstrating the *annelid basic body type,* earthworms possess distinctive adaptations enabling them to thrive in an environment of moist soil and decaying organic matter.

2. Classification

Phylum **ANNELIDA**: Bilaterally symmetrical worms characterized by a ringlike series of similar (metameric) *segments* containing paired setae, nephridia, ventral ganglia, and other duplicated structures; a complete digestive tract, large coelom, thin nonchitinous cuticle, closed circulatory system. Widely distributed, mostly marine, but also in fresh water and soil. About 8000 species.

Class 1. OLIGOCHAETA, earthworms and allies (Figs. 8.1 to 8.14): external and internal segmentation, head and parapodia reduced, or parapodia lacking, few setae per somite. Gonads in anterior section, monoecious (hermaphroditic), clitellum secretes an egg cocoon, development direct (no distinct larval form). Occur chiefly in moist soil and fresh water. Soil-dwelling forms with complex alimentary canal adapted for the ingestion of soil and digestion of organic matter. About 4000 species.

Class 2. POLYCHAETA, bristleworms, sandworms, clamworms, sedentary worms (Figs. 8.15 to 8.21): numerous somites with internal and external segmentation; well-developed parapodia with many setae; head with tentacles, no clitellum. Usually dioecious (sexes in separate individuals), gonads appear only at certain periods, trochophore larva. Chiefly marine, consist of active predators and tube-builders or attached forms feeding by tentacles. About 4000 species.

Class 3. HIRUDINEA, leeches (Figs. 8.22 and 8.23): large posterior sucker, often a smaller one at anterior end. Highly muscular, yet extremely distensible coelom filled with connective or muscle tissue. Thirty-four somites but with many external divisions (*annuli*). No tentacles or parapodia; setae usually also lacking. Hermaphroditic, cocoons formed, direct development. Active predators or bloodsucking ectoparasites. About 300 species.

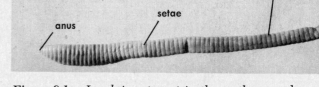

Figure 8.1. *Lumbricus terrestris*, the earthworm, dorsolateral view.

3. Laboratory Instructions

I. Class OLIGOCHAETA; *Lumbricus terrestris*[1]

1. **Observations** If a living worm is available, observe its general appearance, segmentation, epidermal texture, and surface color dorsally and ventrally. Find the anterior and posterior ends—mouth and anus. Study its method of locomotion. Test its reactions to stimuli such as moisture, mild acetic acid, dilute salt solution, and touch. Place the worm in a dish and study it under a dissecting microscope. Find the dorsal blood vessel (note the direction of blood flow); locate the small lateral bristles or setae and watch their movements. If possible, periodically observe a number of worms over several days. Accurately record your observations on methods of feeding, production of castings, copulation, and cocoon formation.

■ **Can you suggest a series of experiments to test more fully the habits and reaction patterns of earthworms?**

2. **External anatomy** (Figs. 8.1 through 8.4) For anatomical study use a specially

prepared specimen, one killed and fixed in formalin, which hardens tissues. More preferable is an anesthetized living worm (immersed in water or placed on wet toweling saturated with chloroform).

Note the general *body form, segmentation, clitellum, mouth, prostomium,* and *anus* (Fig. 8.1). Under a hand lens observe the shape of the mouth and anal openings (Figs. 8.2 and 8.3). Count the segments from I[2] (prostomium plus peristomium) to the nonannulated clitellum. On the anterior ventral surface, locate *male gonopore* on XV, and the openings of the female *oviducts* on XIV (Fig. 8.4). Between these openings and the clitellum, find the ventral or *genital ridges*, marked by *genital setae*. Compare the genital setae with *lateral setae*. By careful use of a

Figure 8.2. *Lumbricus*, anterior end, ventral view.

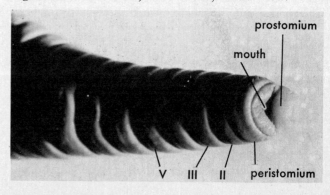

---

[1] Should you be given specimens of other earthworm genera, such as *Eisenia* or *Helodrilus*, look for minor differences in arrangement and size of genitalia and the other organs compared with those of the common *Lumbricus terrestris*.

[2] Roman numerals are customarily used to indicate the earthworm segments.

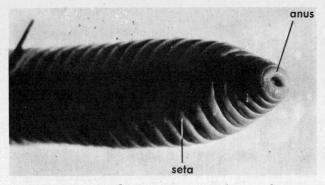

Figure 8.3. *Lumbricus*, posterior end, ventral view.

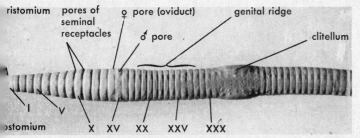

Figure 8.4. *Lumbricus*, ventral view showing position of external body openings. (The symbols ♂ and ♀ are used to indicate male and female, respectively.)

hand lens, try to find the ventral openings of the *seminal receptacles* between IX and X, and X and XI.

Draw a ventral outline showing the structures seen by you.

3. **Internal anatomy** (Figs. 8.5 to 8.14)
a. *General layout.* Carefully place your specimen

Figure 8.5. Anterior view of earthworm, showing general arrangement of internal organs.

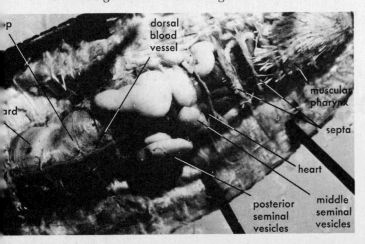

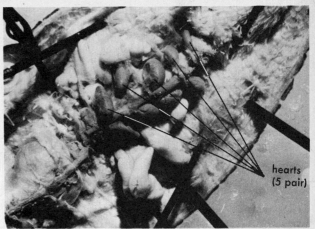

Figure 8.6. Same as Figure 8.5, with reproductive organs dissected away to show 5 pairs of hearts.

dorsal side up in your dissecting pan. Add enough tap water to submerge the worm. Make a shallow dorsal incision a few segments posterior to the clitellum. Then cautiously make a continuous shallow cut just to the right of the median dorsal wall, continuing to the anterior end of the worm. Do not cut deeper than the thickness of the outer body wall or the scissors will injure organs lying below the thin epidermis. Carefully separate the flaps by inserting a scalpel under the cut edges. Keep the blade as close as possible to the body wall. Cut away adhering tissues and septal walls, moving toward the anterior until the full length is cleared along each side. Pin back the exposed flaps. Place the pins firmly, pointed obliquely away from the specimen. For orientation, observe the general layout of *pharynx, esophagus, hearts,* and *dorsal blood vessel, genitalia* (mostly *seminal vesicles*), *crop,* and *gizzard* (Figs. 8.5, 8.6, and 8.7), and crosswalls that mark the internal segmentation. These form *septae* with paired *nephridia* running through them. By placing pins every five segments to serve as markers, it will be unnecessary to make repeated counts of segments. Keep your dissection clean by using an eyedropper to wash away debris.

b. *Reproductive system.* First observe the exposed reproductive organs, the most conspicuous of which in the male are three pairs of sperm reservoirs where spermatozoa mature—the *seminal vesicles* (Fig. 8.5). These are large trilobed structures overlying the esophagus in IX, XI,

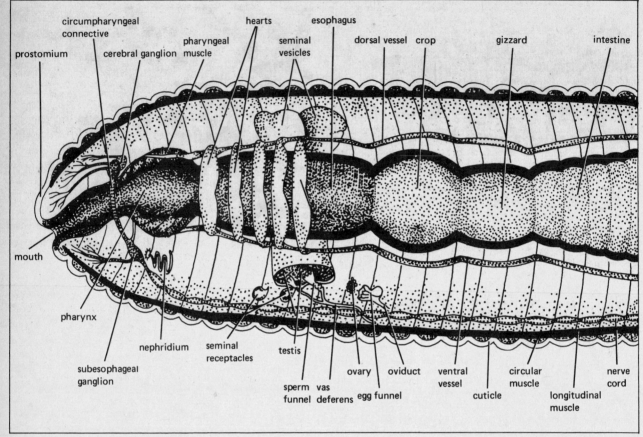

Figure 8.7. Diagram of anterior region of an earthworm with lateral body wall removed to show internal anatomy.

and XII; the *posterior*, usually the largest, forms a folded, bilobed sac. The *middle* and *anterior* vesicles are considerably smaller. Occasionally one vesicle may be larger than its partner. These vesicles all join basally (IX and X) where they form four *testis sacs* (X and XI), which enclose *sperm funnels* and the minute testes.

During copulation the funnels discharge spermatozoa from the seminal vesicles through *sperm ducts* into a single *vas deferens* at XII. Sperm then pass out the male pores (previously seen in XV) and cross over to the *seminal receptacles* through the pores between IX and X, and X and XI. *Both* worms receive sperm as this is a process of mutual cross fertilization between hermaphroditic organisms. Most of the named structures can be found by carefully pushing the seminal vesicles aside with a blunt probe, or by removing them from one side as in Figs. 8.8 and 8.9. To locate the indistinct testes and sperm funnels will require careful

searching under a dissecting microscope and teasing away the teste sacs. Finding these organs is a technical challenge and you will experience a sense of achievement if you succeed in working out the morphological and functional relationships of these structures within the reproductive system.

The female system consists of paired *ovaries* (XIII), the *ovarian funnels* (XIV), and *oviducts* (XIV), ending in the female openings seen in XIV. These tiny structures are best observed with a hand lens or dissecting microscope after careful removal of the seminal vesicles (Figs. 8.8 and 8.9). The seminal receptacles (or *spermathecae*)—two pairs of small lobular whitish sacs in IX and X—lie lateral to or under the anterior and the middle seminal vesicles.

Study your text and charts of copulation and cocoon formation; carefully work out the functioning of the entire reproductive system, and learn the highly specialized habits asso-

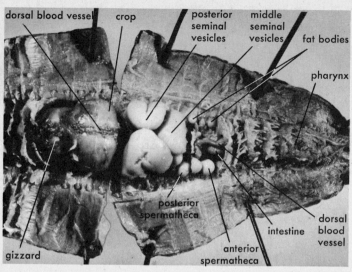

Figure 8.8. Dissection, anterior end, showing general arrangement of reproductive organs with spermatheca and with anterior vesicles removed from left side.

ciated with it. A model of the earthworm, if available, will help to relate the interacting parts.

■ **How do the specialized reproductive organs aid in the survival of *Lumbricus* in its terrestrial environment?**

Remember that a large proportion of annelids are marine polychaetes and lack such terrestrial or freshwater adaptations as cocoon formation.

c. *Circulatory system.* Use an anesthetized living worm. Much of this system cannot be seen in gross dissection, but you should be able to locate the following: *dorsal hearts*—five pairs (VII–XI) (Figs. 8.6 and 8.7); *dorsal blood vessel* (above digestive tract, Figs. 8.6 and 8.7); *ventral* or *subintestinal vessel* (look under intestine); and *body wall* vessels.

Study models or charts of the circulatory system of *Lumbricus*. It is a *closed* system (meaning?) in which the blood flows anteriorly through the dorsal vessel, through the hearts, and into the ventral vessel posteriorly; then out into the body wall *capillaries*; and from them via *lateral-neural vessels* to the *subneural vessel*; finally returning to the dorsal blood vessel through numerous *parietal vessels*.

Diagram this circulatory pathway, showing direction of flow and names of primary vessels.

d. *Digestive system.* The digestive system—*mouth* and *buccal cavity* (I–II), *pharynx* (II–V), *esophagus* (VI–XIV), *crop* (XV–XVI), *gizzard* (XVII–XVIII), *intestine*, and *anus*—is a continuous, highly specialized, food-handling tube. The animal can be visualized as a tube within a tube, that is, an intestine within the outer body wall. The intestine is enclosed by the *inner wall of the coelom* (the *splanchnic mesoderm* or inner *peritoneum*) and the *outer* wall of the coelom (the *somatic mesoderm* or outer *peritoneum*). The latter structure is at the same time the innermost lining of the *body* wall, lying just below the epidermis and outer muscle layers. The space between these two mesoderm layers is the coelom, in which are found the elements of the reproductive, excretory, and other organ systems (each wrapped in a coating of inner peritoneum). Specialization of the digestive tract in *Lumbricus* is an adaptation for processing large quantities of soil and detritus. The ingested soil is passed through the mouth and buccal cavity, the well-developed muscular pharynx, and the esophagus—a pas-

Figure 8.9. Intestine and seminal vesicle removed from left side and right posterior seminal vesicle displaced, disclosing the 4 testes sacs in which sperm funnels are found. Examine closely for ovaries and egg funnels as well.

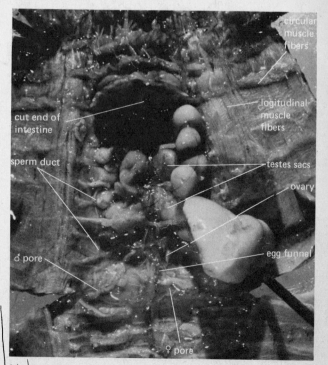

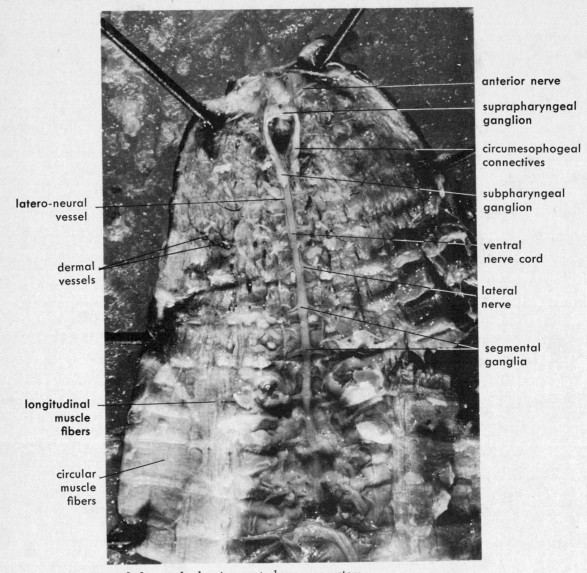

anterior nerve

suprapharyngeal ganglion

circumesophogeal connectives

subpharyngeal ganglion

ventral nerve cord

lateral nerve

segmental ganglia

latero-neural vessel

dermal vessels

longitudinal muscle fibers

circular muscle fibers

Figure 8.10.   Anterior end dissected, showing central nervous system.

sageway leading to the storage and digestive area. In X and XI, the esophagus may have three paired glandular swellings (*calciferous glands*) for calcium and iron regulation. Immediately posterior is the thin-walled *crop* for storage, and a thick, muscular, grinding *gizzard*, followed by the long *intestine* for food digestion and absorption. Specialized cells (*chloragogue cells*) along the outer intestinal wall appear to function for glycogen synthesis and breakdown of proteins to ammonia and urea. The *typhlosole*, a fold extending into the lumen of the intestine, greatly increases the intestinal surface area (Fig. 8.12). The alimentary canal ends at the *anus*, through which soil and un-

used food material is ejected. Piled up fecal *castings* are often seen on lawns. These castings represent a substantial amount of tunneling and turnover of soil, an important contribution to aeration and the loosening of millions of tons of surface soil. Charles Darwin gave a fascinating account of this in a book on the earthworm's role in enriching the soil (*The Formation of Vegetable Mould, Through The Action of Worms*, 1881).

Draw an outline of the dissected worm; show the complete alimentary canal with specialized portions in their correct locations.

e. *Central nervous system* (see Figs. 8.9, 8.10, and 8.11). Carefully remove the alimentary canal

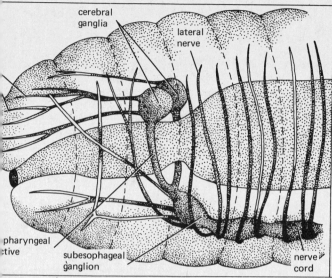

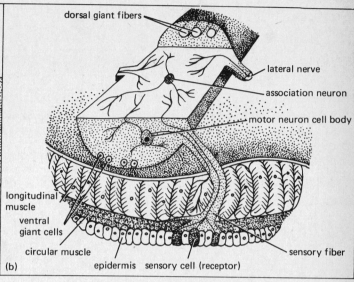

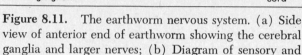

Figure 8.11.   The earthworm nervous system. (a) Side view of anterior end of earthworm showing the cerebral ganglia and larger nerves; (b) Diagram of sensory and motor·neurons of the ventral nerve cord of an earthworm, showing their connections with the epidermis and the muscles to form a reflex arc.

from the anterior esophageal region and look for the white *nerve cord* below it. This cord, with its string of *segmental ganglia*, is metameric (segmented) and the chief component of the *central nervous system*. Find the *ventral* (or *subpharyngeal*) nerve *ganglion* in each segment, with *lateral nerves* coming out from each. Dissect out a portion of the cord in the midbody, place it in a drop of water on a slide, cover, and examine under low power. How many lateral nerves come from each ganglion?

Figure 8.12.   *Lumbricus,* cross section through posterior region.

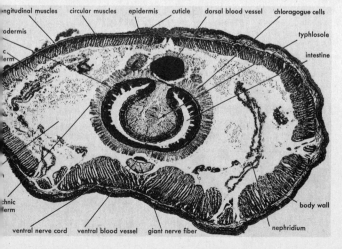

Trace the rest of the nerve cord anteriorly. Near the mouth, find where this nerve separates into two *circumpharyngeal connective nerves* that encircle the anterior end of the pharynx and unite dorsally to form a "brain" (*cerebral ganglia*).

■ **Can you find any nerves branching from these ganglia or from the connective nerves?**

Diagram the nervous system of your specimen. Show location of *ganglia, lateral nerves,* and connectives; note their location and corresponding somite numbers (include only the first 15 segments in your diagram).

f. *Excretory system.* Locate the paired *nephridia* that pierce the septal walls along each side of the animal, one pair per segment. The *nephrostome,* or ciliated opening of each nephridium, drains waste fluids from the coelom. It can be seen just anterior to each septum, with the rest of the nephridium located just posterior to the septal wall as a winding tube (with numerous associated fine blood vessels). The tube connects posteriorly to a bladder-like portion, which empties to the exterior through a *nephridiopore.* Carefully cut out a septal wall in the posterior half of your specimen. Keep the nephridium attached, mount it on a slide in a drop of water, cover, and examine under low

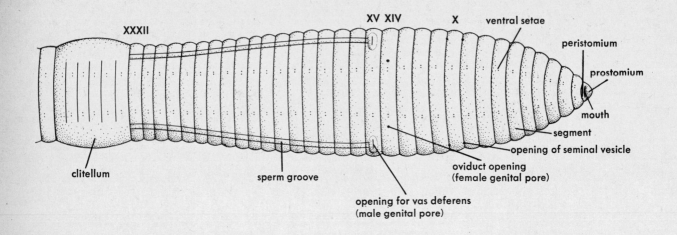

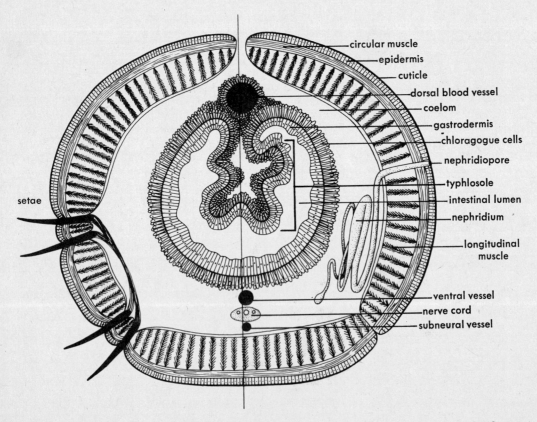

Figure 8.13. *Lumbricus,* semidiagrammatic illustration. (*top*) ventral view of anterior end; (*bottom*) cross section through posterior region.

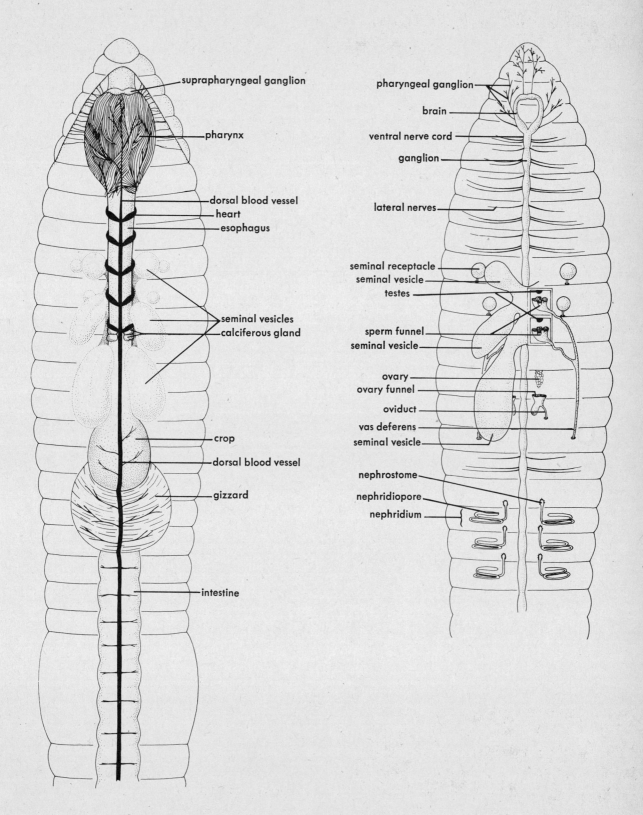

Labels (left diagram):
- suprapharyngeal ganglion
- pharynx
- dorsal blood vessel
- heart
- esophagus
- seminal vesicles
- calciferous gland
- crop
- dorsal blood vessel
- gizzard
- intestine

Labels (right diagram):
- pharyngeal ganglion
- brain
- ventral nerve cord
- ganglion
- lateral nerves
- seminal receptacle
- seminal vesicle
- testes
- sperm funnel
- seminal vesicle
- ovary
- ovary funnel
- oviduct
- vas deferens
- seminal vesicle
- nephrostome
- nephridiopore
- nephridium

Figure 8.14. *Lumbricus*, semidiagrammatic illustration: (*left*) internal anatomy; (*right*) digestive system removed to depict reproductive, excretory, and central nervous systems.

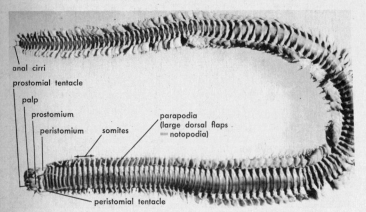

Figure 8.15.   *Neanthes*, dorsal view.

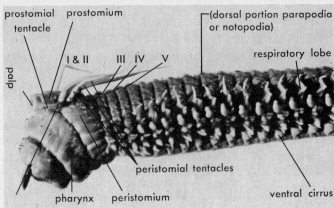

Figure 8.17.   *Neanthes*, anterolateral view.

power to observe its general organization. Then study the nephridium under higher magnification.

■ **What further details can you discern?**

Study of a living specimen will permit you to see the great activity in the nephridium. The ciliary action of the nephrostome takes up coelomic fluid and moves it into the nephridial tube. Parts of the tube wall also possess cilia for maintaining rapid flow of coelomic fluid.

■ **What occurs in the tubule?**
■ **What is the function of the blood vessels encircling the nephridial tube?**

Specially stained permanent preparations of these organs may be available for study.

Draw a nephridium to show its principal features.

Figure 8.16.   *Neanthes*, anteroventral view.

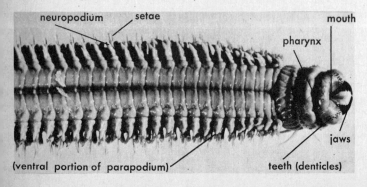

4. **Histology**   In a stained cross section of the earthworm, locate the structures shown in Figs. 8.12 and 8.13, try to correlate each structure with organs you have seen in longitudinal view in your dissection. Use high power to observe details of cellular structure. In the body wall find: *cuticle, epiderm* (with glandular and epithelial cells), *circular muscle layer, longitudinal muscle layer* (appearing feathery in cross section), and *somatic peritoneum* forming the outer lining of the coelom. Find on the intestine the *chloragogue* layer, which serves as part of the *inner* coelomic lining. Locate the *submucosa* of the intestine, with its *connective tissue, muscle tissue* (how many layers?), and *blood vessels*. Finally, observe the inner layer or *mucosa*, with *connective tissue* and *ciliated columnar cells* which form the innermost lining of the intestine.

Locate: *dorsal, ventral, subneural,* and *lateral-neural* blood vessels now viewed in cross section. Review the part each plays in overall circulation.

Study a section through the ventral nerve cord; find the three clear *giant fibers*, other *nerve fibers*, and occasional nerve *cell bodies* (nuclei). The nerve cord is enclosed in a layer of muscle cells and blood vessels. Review the function of these structures. Correlate structures you have just seen in microscopic cross section with those you have previously seen macroscopically by dissection (see Figs. 8.11 and 8.14).

5. **Special experiments**   The earthworm, one of the most intensively studied inverte-

brates, offers excellent material for experiments on development, growth, regeneration, and response to various environmental stimuli or insults. Observations on habits and reactions among living worms have already been mentioned with respect to the reproductive cycle.

a. *Embryology* of earthworms can be studied by dissecting various embryonic stages from cocoons. A cocoon is a membranous girdle, secreted by the clitellum, which hardens after it has slipped off the worm. As the cocoon is shifted anteriorly on the worm, it picks up several ova and a number of spermatozoa from the pores that empty into the space between the cocoon and body wall. Fertilization occurs within the cocoon. The fertilized eggs or *zygotes* are enclosed by the cocoon when it shrinks up around the eggs and the ends pinch up after the cocoon has been shed into the soil.

Collect cocoons from a laboratory earthworm culture in damp soil, or try to find some yourself in protected rich soil with earthworm castings.

Set up a time schedule and dissect out developing embryos at different stages. Examine them under a dissecting microscope (keeping the worms in water) or under your compound microscope if the worms are small enough. Compare these stages with newly emerged juvenile worms.

b. *Regeneration.* Growth or replacement of parts is also a fascinating study. Use your own ingenuity to set up experiments designed to answer such questions as the following:

    i. Does somite replacement occur?

    ii. Which end can regenerate?

    iii. How many segments are capable of replacement?

    iv. What can you conclude concerning the relationship between *structural specialization* and the *capacity to regenerate*?

Annelids other than *Lumbricus* may be used, sometimes to considerable advantage. The fresh-water oligochaete, *Tubifex tubifex*, can be reared in large numbers in the laboratory. They are small and easily handled. Allow about two weeks for regenerative changes to occur.

Draw the stages of development or change as you see them, and maintain accurate records of your *procedures*, *observations*, and *results*.

c. *Histology.* To provide a three-dimensional sense of the tissues seen in very thin section on your stained slides, prepare your own sections by hand from a preserved earthworm, cutting transverse sections one somite thick from various points along the body length. A new razor blade is the best tool. Select a worm with a relatively undistended crop and intestine so that other organs will not be forced against the body wall. Sections can be cleared for better observation by soaking them in *chloral hydrate* or *oil of wintergreen*.

Compare your observations with those from stained tissue sections. Be sure to try several transverse sections so that you can work out the changes between anterior, genital, and posterior sections.

Be prepared to compare tissue layers and basic body type of the earthworm (for **ANNELIDA**) with those of hydra (for **CNIDARIA**) and of the frog (for **CHORDATA**).

d. *Other studies*

Can you devise other studies employing earthworms or other annelids to test such activities as:

    i. Response to an electric field.

    ii. Response to various drugs.

    iii. Method of locomotion.

    iv. Manner of feeding, including peristalsis, ciliary activity, movement of setae, and circulation of coelomic fluid.

Smaller oligochaetes, such as *Aeolosoma* or *Dero*, may prove useful for some of these studies.

II. Class POLYCHAETA; *Neanthes virens* (*Nereis virens*)

This class, containing most members of the phylum, includes both relatively simple and extremely specialized and bizarre worms (among the more specialized are the sedentary polychaetes adapted for mud or tube dwelling). They all retain recognizable annelid characteristics, especially with regard to internal structure. Though their anatomy in general is much like that of the earthworm, they lack the special adaptations for terrestrial existence seen in *Lumbricus*.

Polychaetes, on the other hand, have developed many special structures absent in oligochaetes that enable them to live in the marine environment. Some, like the clamworm, *Neanthes virens* (Figs. 8.15 to 8.19), have paddlelike setae for swimming and respiration, and a formidable

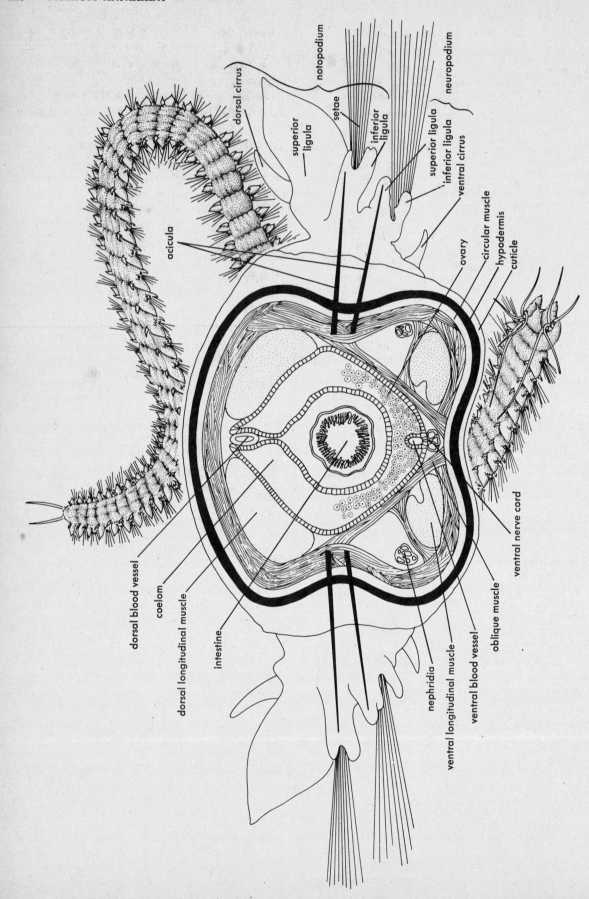

Figure 8.18. *Neanthes*, cross section through midposterior region.

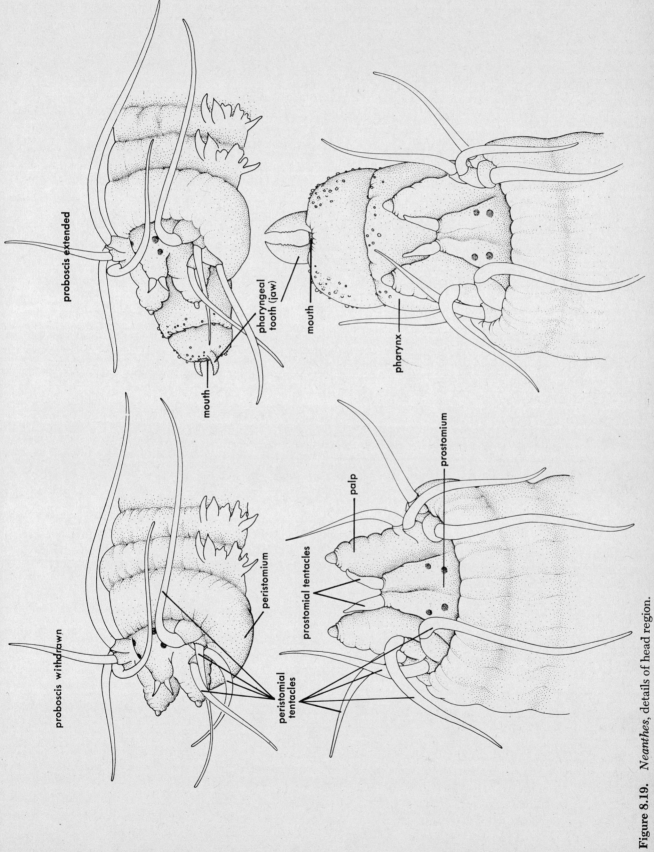

proboscis extended

pharyngeal
tooth (jaw)

mouth

mouth

pharynx

proboscis withdrawn

peristomium

peristomial
tentacles

palp

prostomium

prostomial tentacles

peristomial
tentacles

Figure 8.19. *Neanthes*, details of head region.

pair of hooklike jaws (Figs. 8.16 to 8.19), that can evaginate and withdraw for capture and ingestion of soft-bodied prey. Other polychaetes are adapted for making tubes in mud or sand. Each of these species constructs a highly specific type of tube, some of which may be cemented to rocks or shells. Feeding and respiration are accomplished by a writhing mass of anterior undulipodiated tentacles.

1. **External anatomy** The *parapodia* and lateral swimming and respiratory organs are prominent. Examine the structure of *head* and *jaws*. If your specimen was killed in an extended position, identify the *pharynx* with its *denticles* and *jaws*.

Draw the head structures you can see in dorsal view. Identify and label structures found on the *prostomium* and *peristomium*. Include one of the following views in your drawing:

a. Head (if pharynx is everted): *jaws, denticles, pharynx, prostomium* with *palps* and stubby *prostomial tentacles, eyes, peristomium* (consisting of segments I and II), with four pairs of *peristomial tentacles* (Fig. 8.17).
b. Head (if pharynx is retracted): jaw and pharynx withdrawn but *mouth* visible below peristomial tentacles; other structures as noted above (Fig. 8.19).

■ What are the sensory functions of tentacles and palps?

Remove a parapodium, mount it on a slide in water, cover, and examine under the dissecting microscope. Note the two *lobes*, a dorsal *notopodium* and a ventral *neuropodium*. Each has a spine or *cirrus* (plural: *cirri*) and a cluster of bristles or setae around a single supporting needle-like *aciculum*. The notopodia have many fine blood capillaries coursing through them. This thin, exposed surface is clearly respiratory.

■ Why can one assume this to be true?
■ How might it be proved?

Draw and label the above structures of a parapodium.

This annelid demonstrates *homonomous segmentation*—division of the body into a series of nearly equal (homonomous) segments. Later we will compare this with a more advanced type found in higher arthropods, *heteronomous segmentation*, in which successive segments are markedly different.

■ Why are polychaetes (especially worms like *Neanthes*) thought to be relatively primitive annelids?
■ What is meant by the "tube within a tube" body plan?

2. **Histology** In a stained cross section of *Neanthes* locate the structures shown in Fig. 8.18.

■ How do these structures differ from those of *Lumbricus*?
■ How are these differences related to the food habits and way of life of each worm?

3. **Review of polychaetes** Examine various polychaetes from available laboratory specimens and charts, from your text, and references listed at the end of the chapter. You may be surprised. They form an unexpectedly varied and interesting assemblage.

■ Relate the variation in polychaete structure to the widely diverse habitats in which you would expect to find these worms.

Study the drawings of *Arenicola* (Fig. 8.20) and *Amphitrite* (Fig. 8.21). Note the large blunt anterior end, the annulations of the skin (not indicative of true segmentation—why?), and the vestigal parapodia of *Arenicola*. *Amphitrite* has a thicker body at the anterior end where a crown of tentacular filaments are found. Setae are restricted to the anterior end of the body.

■ In what kind of habitat would you expect to find *Arenicola*? *Amphitrite*?
■ How do these two species differ from *Neanthes*? from *Lumbricus*?

III. **Class HIRUDINEA;** *Placobdella* (Fig. 8.22) or *Hirudo* (Fig. 8.23)

Not a popular organism with most people, leeches are nonetheless interesting and highly adapted animals. The leech was used for bloodletting in medical practice for many centuries. In some places even today leeches are still used for this purpose.

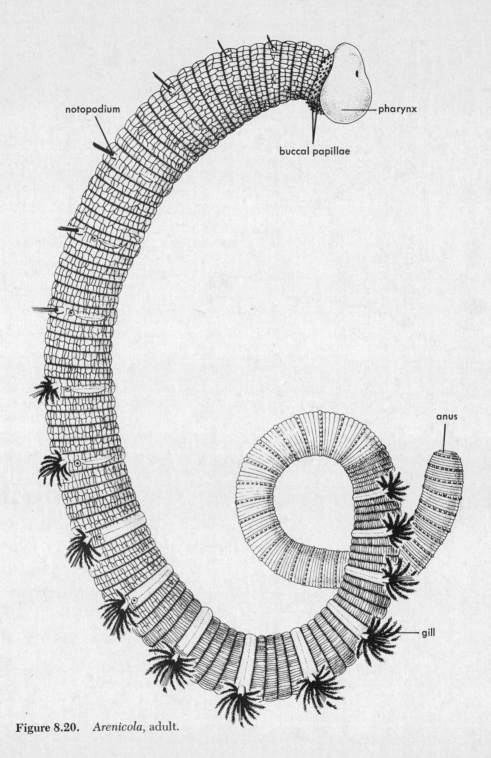

Figure 8.20.   *Arenicola*, adult.

Although chiefly known for their blood-sucking habits, many leeches are freshwater and marine predators, feeding voraciously on snails, worms, and other aquatic soft-bodied organisms. Some are scavengers, others ectoparasites (meaning?). Bloodsucking species have an enormously distensible, branched gut, enabling them to feed to repletion on those rare occasions when they find a host and then to store the bloodmeal for long periods of abstinence, often several months. Some leeches have become adapted to a terrestrial habitat, particularly in warm, moist, tropical areas where they are disagreeable and often dangerous, attacking men and animals in great numbers.

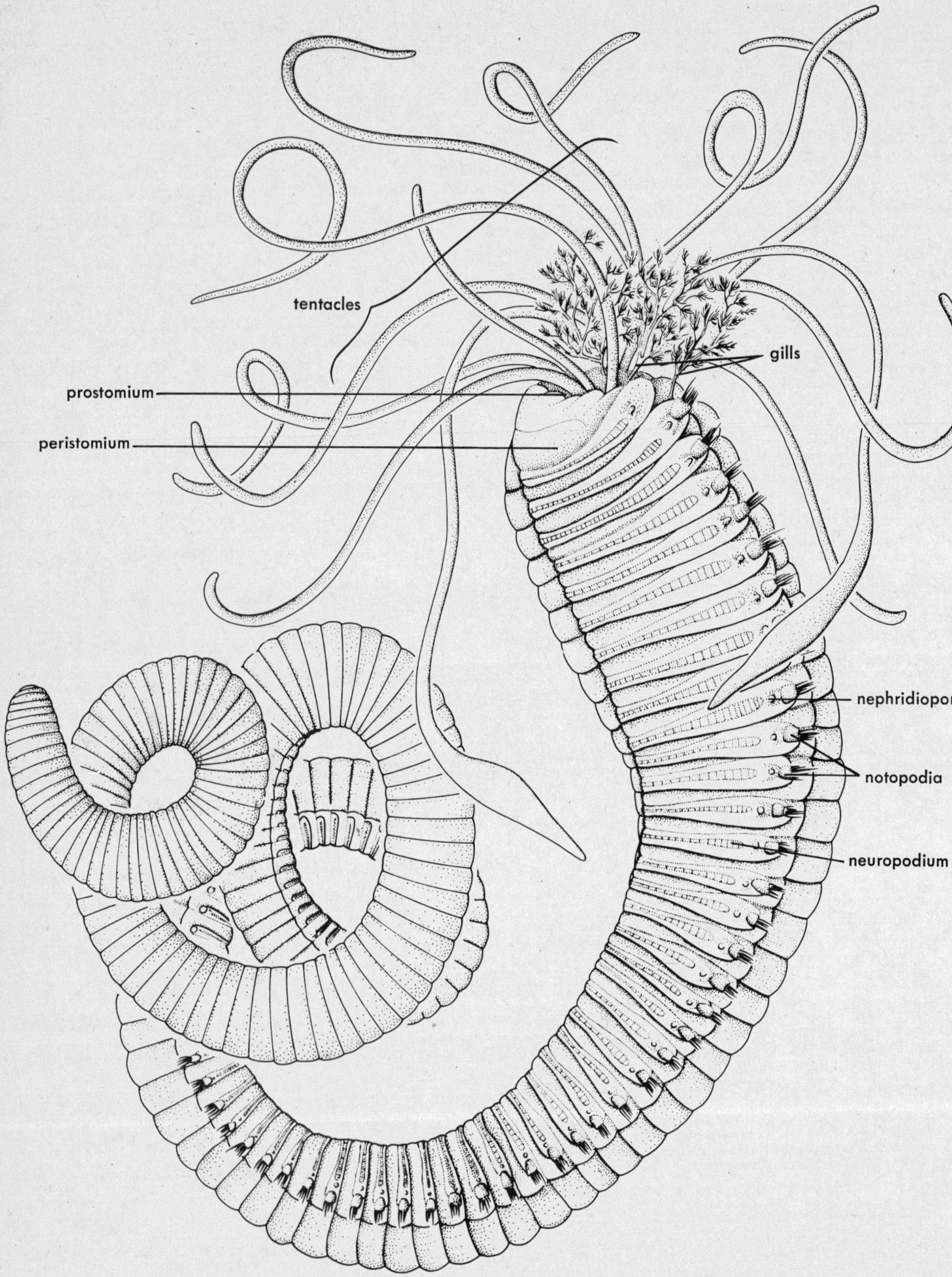

tentacles

gills

prostomium

peristomium

nephridiopore

notopodia

neuropodium

Figure 8.21. *Amphitrite*, adult.

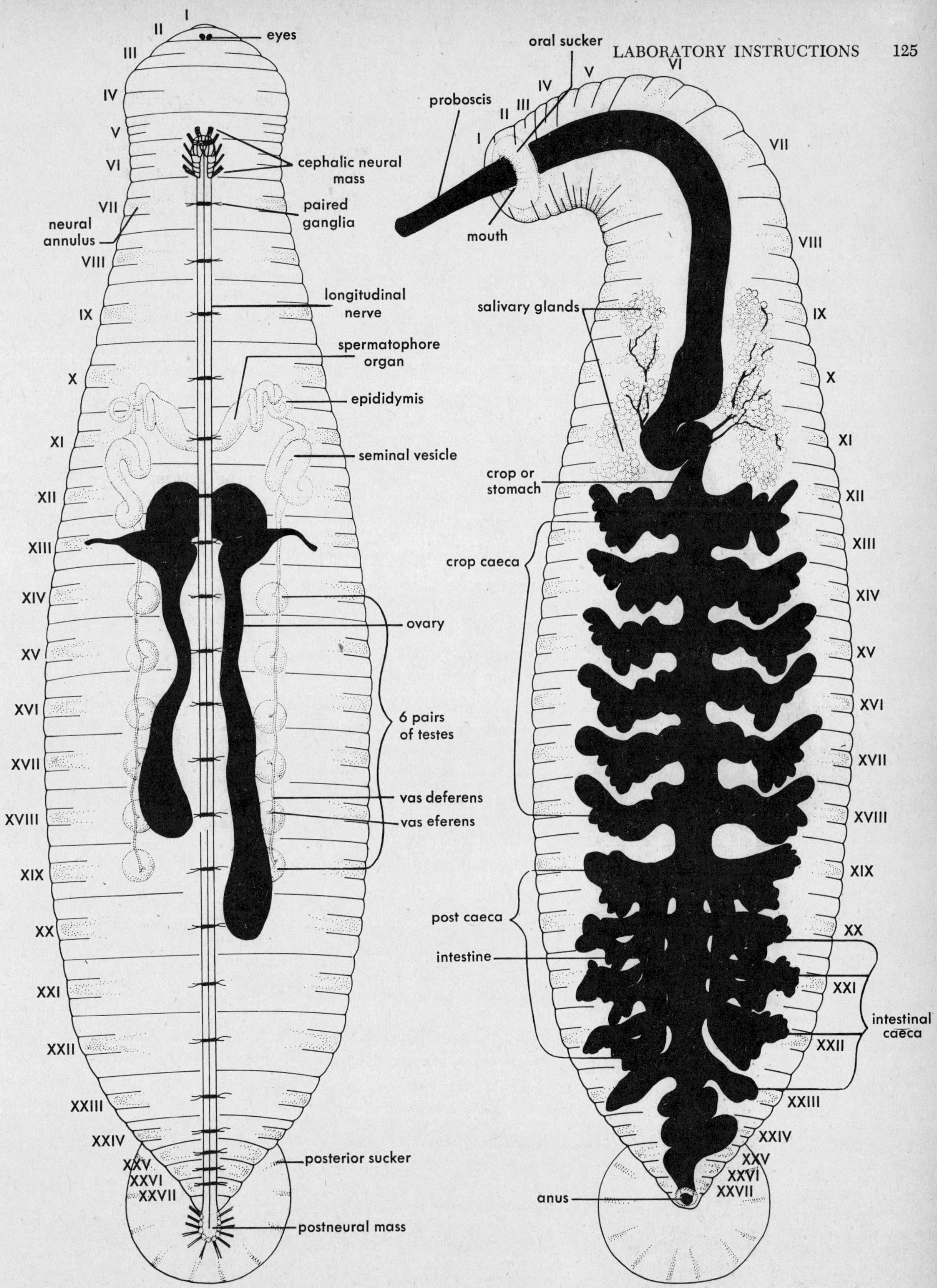

Figure 8.22. *Placobdella*: (*left*) internal anatomy of nervous and reproductive systems; (*right*) internal anatomy of digestive system.

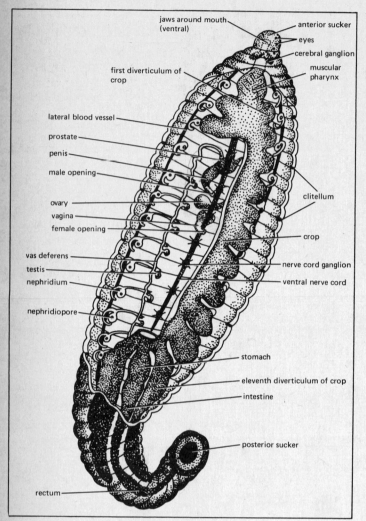

jaws around mouth (ventral)
anterior sucker
eyes
cerebral ganglion
muscular pharynx
first diverticulum of crop
lateral blood vessel
prostate
penis
male opening
clitellum
ovary
vagina
female opening
crop
vas deferens
nerve cord ganglion
testis
ventral nerve cord
nephridium
nephridiopore
stomach
eleventh diverticulum of crop
intestine
posterior sucker
rectum

Figure 8.23. Diagrammatic dorsal view showing the segmentation and internal anatomy of the leech (*Hirudo medicinalis*). Part of the crop is cut away on the left side to show the ventral nerve cord and reproductive organs.

1. **External anatomy**   Observe general appearance and the external annulations.

■ How does this *external* annulation differ from true segmentation? (See Fig. 8.22.)
■ How can you determine which are the true segments in leeches? Locate the anterior and posterior ends, the mouth and anus.
■ What roles do the suckers play?
■ Do the mouth parts show special adaptation for bloodsucking?

2. **Internal anatomy**   Make a shallow incision along the middorsal line. Carefully remove tissue attached to the body wall. Pin back the body wall to expose the internal organs.

■ Which system is most prominent?
■ How does the external appearance of this system differ from that of *Lumbricus*?

Identify the muscular pharynx and crop. This latter structure generally has 11 pairs of *diverticula* (ceca) that extend laterally. The crop is long and connects to the intestine, which also has several lateral diverticula and terminates in the anus. The latter opens dorsally and anterior to the posterior sucker.

The ventral nerve cord, reproductive system, several nephridia, and some blood vessels can be seen by removing the digestive tract.

■ Why do you suppose the digestive system is so prominent? How is this related to the leech's way of life?
■ What special adaptations do leeches possess for success as bloodsucking ectoparasites?
■ What characteristics of leeches demonstrate a relationship with other annelids?

Study laboratory demonstrations or charts for internal anatomy and structures not apparent in your own specimen.

Suggested References

Bahl, K. N., 1947. "Excretion in the Oligochaeta." *Biol. Rev.*, 22:109.

Barnes, R. D., 1974. *Invertebrate Zoology*, 3rd ed. Philadelphia, Pa.: W. B. Saunders.

Brinkhurst, R. O., and B. G. Jamieson, 1972. *Aquatic Oligochaeta of the World.* Toronto: Toronto University Press.

Brown, F. A., Jr. (ed.), 1950. *Selected Invertebrate Types.* New York: J. Wiley & Sons.

Buchsbaum, R., 1956. *Animals Without Backbones.* Chicago, Ill.: University of Chicago Press.

Chapman, G., and G. E. Newell, 1947. "The role of the body fluid in relation to movement in soft-bodied invertebrates; I—The burrowing of *Arenicola.*" *Proc. Roy. Soc.* (London), *1348*:431.

Dales, R. P., 1957. "The feeding mechanism and structure of the gut of *Owenia fusiformis.*" *J. Mar. Biol. Assoc.* (*U.K.*), 36:81.

———, 1963. *Annelids.* London: Hutchinson University Library.

Gosner, K. L., 1971. *Guide to Indentification of Marine and Estuarine Invertebrates: Cape Hatteras to the Bay of Fundy.* New York: Interscience.

Gray, J., 1939. "Studies in animal locomotion, VIII. The kinetics of locomotion of *Nereis diversicolor.*" *J. Exp. Biol.*, *16*:9.

———, and H. W. Lissman, 1938. "Studies in animal locomotion, VII. Locomotory reflexes in the earthworm." *J. Exp. Biol.*, 15:506.

Grove, A. J., 1925. "On the reproductive processes of the earthworm *Lumbricus terrestris.*" *Quart. J. Micr. Sci.*, 69:245.

Herland-Meewis, H., 1964. "Regeneration in annelids." *Adv. Morphol.*, 4:155.

Lavarack, M. S., 1963. *The Physiology of Earthworms.* New York: Macmillan.

McConnaughey, C., and D. L. Fox, 1949. "The anatomy and biology of the marine polychaete *Thorcophelia mucronata.*" *Univ. Calif. Publ. Zool.*, 47:319.

MacGinitie, G. E., 1939. "The method of feeding of *Chaetopterus.*" *Biol. Bull.*, 77:115.

Mann, K. H., 1962. *Leeches (Hirudinea); Their Structure, Physiology, Ecology and Embryology.* New York: Pergammon Press.

Meglitsch, P., 1967. *Invertebrate Zoology.* New York: Oxford University Press.

Prosser, C. L., 1934. "The nervous system of the earthworm." *Quart. Rev. Biol.*, 9:181.

Ramsay, J. A., 1949. "The osmotic relations of the earthworm." *J. Exp. Biol.*, 26:46.

Sherman, I. W., and V. G. Sherman, 1970. *The Invertebrates: Function and Form. A Laboratory Guide.* New York: Macmillan.

Sutton, M. F., 1957. "The feeding mechanism, functional morphology, and histology of the alimentary canal of *Terebella lapidaria.*" *Proc. Zool. Soc. London*, 129:487.

Wells, G. P., 1945. "The mode of life of *Arenicola marina L.*" *J. Mar. Biol. Assoc. (U.K.)*, 26:170.

Wolf, A. V., 1940. "Paths of water exchange in the earthworm." *Physiol. Zool.*, 13:294.

CHAPTER 9

PHYLUM ARTHROPODA

1. Introduction

Approximately three fourths of all animal species are in the phylum **ARTHROPODA.** Arthropod body organization must therefore be an extraordinary successful one. Great structural flexibility and high capacity to form new species are also implied by these numbers. The success of the phylum is obvious, whether measured by numbers of individuals, species, or total mass structural variety, adaptability, or evolutionary plasticity.

The arthropod basic body type is characterized by: (1) bilateral symmetry; (2) segmentation; (3) hardened exoskeleton, usually chitinous; (4) jointed appendages; (5) strong tendency toward *tagmosis* (fusion of blocks of segments to form major regions—head, thorax, and abdomen) and toward *heteronomous metamerism* (formation of specialized segments and appendages); (6) discontinuous growth, usually occurring immediately after shedding the exoskeleton (molting); (7) no distinct trochophorelike larval form in early development, such as in annelids and mollusks;[1] (8) ceph-

alization (increased size and specialization of brain and central nervous system); (9) tendency toward reduction of coelom and formation of hemocoel; and (10) retention of certain annelidlike characters (dorsal heart with ostia; nerve ring around esophagus; ventral ganglia paired in each segment but modified by fusion of ganglia and segments in more advanced forms).

The *major* groups of arthropods are classified according to their *segmentation, tagmosis*, and *appendages*. The first part to form embryologically, the head, contains the most specialized appendages and is probably the most useful key to relationships.

One of the most interesting aspects of arthropod biology is the extraordinary impact that the chitinous exoskeleton has had on the form, function, adaptability, and evolution of the group. This is discussed with respect to insect evolution as an illustration. Not only the term **ARTHROPODA** (meaning jointed feet), but many other characters—the manner of growth, circulatory and respiratory systems, size, musculature, and even habitat—can be related to this tough, jointed, hollow skeleton. Limitations too—small size, short life span, restricted brain size—can be traced to the skeletal structure. Keep the relationship between exoskeleton and evolution of the phylum in mind

---

[1] Although a nauplius larva is found in CRUSTACEA, and a wormlike larva in higher insects, both are later stages in the embryological sequence.

during your survey of this enormous group. Test its applicability as you become better acquainted with the examples reviewed here.

2. Classification

A comparative study of **ARTHROPODA** is difficult, since its major elements became diversified before the earliest clearly recognizable fossils were deposited in the Cambrian period, 500 million years ago. Arthropod evolution is so vast a field that innumerable examples are required to get a real feeling for patterns of change and for groups that have evolved during this long period. The embryology of higher arthropods lacks phylogenetically illuminating early stages, which makes basic relationships even more difficult to trace.

We will have time to study very few examples, but these will provide a brief introduction to some major patterns of arthropod adaptation. Check all names used against the classification outlined in Chapter 2; add to the outline generic names referred to here that are not listed there.

Two great groups (subphyla) of arthropods are generally recognized: the **MANDIBULATA,** jawed or mandibulate arthropods; and the **CHELICERATA,** arthropods whose first appendage bears clawlike pincers.

A huge assemblage is included in the subphylum **MANDIBULATA,** chiefly in the classes CRUSTACEA and INSECTA. In the subphylum **CHELICERATA** is the large class ARACHNIDA (scorpions, spiders, mites, ticks, and relations); the extinct tribolites (TRILOBITA); and the extinct giant eurypterids with the still-existent horseshoe crabs (MEROSTOMATA); and the peculiar marine sea spiders (PYNCOGONIDA).[2]

In the class CRUSTACEA, subclass MALA-COSTRACA, is an assemblage of water dwellers typified by crayfishes (*Cambarus*), lobsters (*Homarus*), and shrimps, all of which are included in the order DECAPODA (malacostracans with ten pairs of walking legs). Mantis shrimps, beach fleas, and sowbugs represent other major groups of the subclass MALACOSTRACA.

Other important subclasses in the class CRUSTACEA include barnacles (CIRRIPEDIA) and copepods (COPEPODA). In the subclass BRANCHIOPODA are found the most primitive crustacea, showing very little specialization of the segments or appendages; examples are the fairy shrimps and brine shrimps (order ANOSTRACA). Perhaps no animals

can compare with the widespread copepod, (such as the marine *Calanus*) in overall abundance or importance in the ecological food chain of the sea.

Our study of the crayfish as a single example of the entire class CRUSTACEA must therefore cover much biological territory, a great period of geological time, an enormous area of the earth's surface (about four sevenths—all the water-covered surface of the globe), and a rich evolutionary divergence in numbers and kinds.

The two remaining classes of **MANDIBULATA,** the INSECTA (or HEXAPODA, meaning six-legged) and MYRIAPODA (centipedes and millipedes Fig. 9.1), contain many familiar examples. Despite the fact that INSECTA is by far the largest class of all, we will have time to study only one or two species. Perhaps 800,000 to a million species of insects have already been described. Some entomologists believe that this includes less than half of the actual number of living insect species. We will examine the common lubber grasshopper, *Romalea*, and the American cockroach, *Periplaneta*. (These examples are selected chiefly because they are easily procured.)

Whether there are one or two million species, insects represent a biological success story. Insect history differs markedly from that of the class CRUSTACEA in one important ecological respect: insects have become terrestrial. They have bridged the great gap between water breathing and air breathing. (Terrestrial isopods still require a moist habitat for survival.) Occupation of land opened a wealth of new environments in which insects spread rapidly, as early as the Pennsylvanian

---

[2] Sometimes the PYNCOGONIDA, here considered a class of **CHELICERATA,** is recognized as an additional subphylum of arthropods. These "sea spiders" are strange, slow-moving, marine, somewhat spiderlike creatures. The extinct TRILOBITES here listed as a class in the chelicerate line, also are sometimes classified as a distinct subphylum. The subphylum **ONYCHOPHORA,** a phylogenetically important but rather rare and small group of caterpillarlike animals, shows important evolutionary relationships with the annelids. (Review your text discussion of *Peripatus*.) Some authors list these organisms as a separate *phylum*, indicating uncertainty over the exact position of this ancient group. Other small subphyla within the **ARTHROPODA** are accorded a higher taxonomic level by various authors because of their extremely unusual structure (or simply because we do not have enough evidence to place them phylogenetically precisely where they belong). These include the subphylum **PENTASTOMIDA**—wormlike parasites of vertebrates (possibly derived from mites and considered by some specialists to be a class of chelicerates)—and the TARDIGRADA, "water bears"—microscopic, soft-bodied, largely unsegmented, eight-legged creatures, sometimes placed in the chelicerate subphylum.

Figure 9.1. The segmentation of a millipede (*left*) is reminiscent of the annelid form, while the jointed appendages on each segment (*right*) are one of the characteristics of an arthropod.

period (late Carboniferous), 320 million years ago, and most rapidly during the Cretaceous period, about 100 million years ago. In doing so, however, their adaptations for air breathing and resistance to desiccation as well as osmoregulatory limitations seem to have prevented reoccupation of the marine environment from which their ancestors presumably arose. (Although a small number of larval insects live in fresh water, how many *adult* insects have you seen inhabiting ponds and streams? Some water beetles carry bubbles of air under water, an indication of the difficulty adult insects have in returning to an aquatic environment.)

Concurrent with the arthropod transfer to the terrestrial habitat was the development of flowering plants, which provided insects with a great variety of shelter, food, and protection. In turn, insects came to play a significant role in the distribution and structural modifications of these plants, many of which possess highly specialized structures for pollination by a specific insect group—a type of evolutionary partnership, illustrating *coevolution.*

Insects demonstrate to an unusual degree the evolutionary process called *adaptive radiation,* in which a basic structural or functional modification permits rapid occupation of a new environment. This exploitation of previously unavailable habitats is followed by diversification—formation of new species—and then a new spread into more specialized niches within the new environment. If another evolutionary modification appears in one of the ecologically or geographically isolated groups, a new wave of habitat occupation may carry these animals into still different environments. A succession of such major structural and functional changes or adaptive breakthroughs, each producing a wave of spread and specialization, is what is meant by adaptive radiation.

When arthropods transferred from an aquatic to a terrestrial environment, their jointed exoskeleton—durable, impervious, and light in weight—determined a series of profound morphological and physiological changes that introduced adaptive radiations into the numerous habitats afforded by land and air.

The first of the major changes related to the exoskeleton, and influencing the evolution of insects, was probably the *tracheal system,* an extensive network of air-conducting tubes branching from the exoskeleton to every cell of the organism. The presence of tracheae, along with the strength, lightness, extensive area for muscle attachment, and other advantages of the exoskeleton, permitted early scorpionlike arthropods to become air-breathing land dwellers. The reduction of the number of walking legs to six freed the head appendages for specialization in feeding, chiefly by their modification to form *jaws. Flight* was made possible by membranous wings, a highly organized muscular system, a rapid metabolic rate, small size, and relative indestructibility—all features made possible by the exoskeleton. Vast new ecological realms were made available by this remarkable development. [No other invertebrates and only reptiles, birds, and certain mammals (such as bats) have evolved the power of flight.]

Metamorphosis, related to mode of growth and ultimately to the exoskeleton, provided still further specialization: the protected *egg* stage for critical and delicate early development; the worm-like *larva* for feeding and growth; the *pupa,* another protected stage for the transition from larva to adult; and the flying *adult,* adapted for reproduction, occupation of a different ecological niche, and distribution. Metamorphosis meant specialization of each body form to functional needs at each stage of the life cycle, such as development, growth, reproduction, and distribution of the species.

Each of these major stages of insect evolution was marked by adaptive radiation, made possible by important morphological and physiological changes, and molded by the structural limitations and advantages afforded by the chitinous skeleton.

Although the ant may one day inherit the world, it will *not,* as pictured in science fiction tales, become huge in size. Total size is restricted in insects for two reasons. First, tracheae can work efficiently only in small volumes, as the air moves in and out of them chiefly by diffusion. Very large insects simply could not exchange gases rapidly enough by tracheal respiration. Second, just after each molt the insect skeleton is soft and flexible—the period of *discontinuous growth.* Being unsupported by exoskeleton during this vulnerable period, the body must be supported by the external medium. This is a manageable problem to a lobster—but what would happen to a lobster in *air* during the molting period? Its body would be permanently distorted by its own weight. Imagine what would happen to the monster grasshopper pictured in science fiction! However, instead of becoming huge, the ant and his innumerable counterparts might simply multiply to the fantastic degree possible for these organisms and occupy the total available environment. In this event, insects would prove an excessively burdensome competitor to man for food and space—far more dangerous to us than ten-foot man-eating insects would be.

Review the body type of the phylum **ARTHROPODA** as exemplified first by crustaceans and then by insects.

■ **How do these differences permit occupation of distinct environments and subsequent divergence of the two groups?**

The second major division of **ARTHROPODA,** the subphylum **CHELICERATA,** is comprised of the spiders, scorpions, ticks, mites, and allied forms. Here the general arthropod pattern is usually modified for predation as well as for existence on land. Many chelicerates are equipped with poison claws or glands, and their mouthparts are usually adapted for sucking out the juices or soft tissues of their prey. Although chelicerates are highly resistant to desiccation, few except mites have spread to the variety of habitats utilized by insects, perhaps because of their more specialized food habits. The two-sectioned body consists of *head* (*cephalothorax*) and *abdomen.* Six pairs of jointed, segmented appendages are hinged onto the cephalothorax. These include one pair of clawed *chelicerae* (containing the poison fangs and ducts, or claws), one pair of *pedipalpi* (six-jointed, leglike structures to hold or crush prey, specialized for sperm transfer in males), and *four pairs* of seven-jointed walking legs. True antennae and mandibles, characteristic of the **MANDIBULATA,** are entirely absent. The respiratory system also has specialized features: *book lungs* in some spiders, *book gills* in king crabs, *tracheae* in the abdomen of other spiders, *direct diffusion* through the cuticle in mites.

The class MEROSTOMATA includes king crabs and extinct eurypterids (huge scorpionlike predators of Paleozoic seas). ARACHNIDA include nine somewhat familiar orders, including many feared and often maligned examples: SCORPIONIDA—scorpions (Fig. 9.2); PEDIPALPI—whip scorpions; ARANEIDA—spiders (divided into two suborders; trapdoor spiders and tarantulas in one, and black widows, funnel web spiders, hunting spiders, orb weavers, crab spiders, and jumping spiders in the other); SOLPUGIDA—sun spiders; PSEUDOSCORPIONIDA—false scorpions; PHALANGIDA—harvestmen or daddy longlegs; and ACARINA—ticks and mites.

Chelicerate mouth structure, absence of wings, and predaceous food habits in general restrict distribution of members of this subphylum, at least in comparison with that of insects. Spiders, ticks, and mites, however, do occupy a vast area and many types of terrestrial habitats.

Through marked reduction of size and simplification of body structure, mites have in fact developed a different widespread adaptive radiation. It has enabled them to become abundant soil dwellers and to feed on a wide variety of materials from organic debris to blood. Although mites represent an extension of the arthropod type into numerous ecological niches, these tiny creatures are among the least known groups of animals.

Other classes of **CHELICERATA** are the extinct TRILOBITA; the PYCNOGONIDA or sea

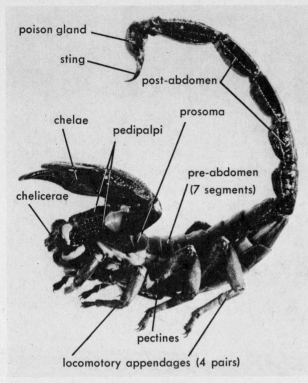

poison gland

sting

post-abdomen

chelae

pedipalpi

prosoma

chelicerae

pre-abdomen
(7 segments)

pectines

locomotory appendages (4 pairs)

Figure 9.2.   Scorpion, lateral view.

spiders; and the MEROSTOMATA. The latter group is represented by an animal that is a good candidate for a crossword puzzle, the subclass XIPHOSURA. Along the east coast beaches one can see these bizarre creatures, with small bodies in an oversize armored shield ending in a spiny tail. The single surviving member is *Limulus* (Fig. 9.3), the king crab, a primitive reminder of ancient Paleozoic seas once also inhabited by trilobites and predaceous 10-foot eurypterids, all related merostomes.

Our detailed study of the phylum **ARTHROPODA** must be confined to just two or three examples: the decapod malacostracan crustacean *Cambarus*, the crayfish; the grasshopper *Romalea*; and, if time permits, the cockroach *Periplaneta*.

Illustrated here are common examples of some of the more important arthropod groups within the class CRUSTACEA. They represent a wide range of types starting with one of the most primitive, *Artemia* (Fig. 9.4), the brine shrimp, with little specialization of its appendages. The abundant water fleas *Daphnia* (Fig. 9.5) and *Cyclops* (Fig. 9.6) are widespread members of nearly every aquatic environment, part of the so-called microcrustacean fauna. They are primary members of the food chain supporting nearly all larger aquatic animals. *Argulus* (Fig. 9.7) is specialized for at-

tachment (note suckers) and is a common fish ectoparasite. *Lepas* (Fig. 9.8) and *Balanus* (Fig. 9.9) are barnacles, specialized for permanent attachment to a substrate, and the latter is well adapted for protection against wave action and periodic exposure to intertidal desiccation. The familiar sowbug (pill bug) *Oniscus* (Fig. 9.10) is adapted for terrestrial existence. *Homarus* (Fig. 9.11), the succulent lobster, represents an advanced crustacean with a high degree of modification of its appendages, with specialized mouthparts and powerful pincers.

3.   Laboratory Instructions

I. Cambarus

1.   **Classification**
Phylum **ARTHROPODA**
Class CRUSTACEA
Subclass MALACOSTRACA
Order DECAPODA
Family Astacidae
Genus *Cambarus procambarus* (in western United States the Pacific form, *Pacifastacus*, may be used).

2.   **General observations** The crayfish, a common inhabitant of freshwater streams and ponds, often emerges at night and makes characteristic burrows and mounds in wet fields or gardens. Along with its marine relatives, lobsters and crabs, the crayfish feeds chiefly on decaying organic matter. (It is also locally known as the crawfish.)

Observe the living crayfish in an aquarium. Watch it locate food with its antennae, tear off chunks with its large pincers (*chelipeds*), tease and macerate it with the *mandibles*, and finally pass the smaller pieces to the five pairs of mouthparts (*maxmillae* and *maxillipeds*) that handle and ingest the morsel.

Observe walking, swimming, and respiratory movements.

■ Which legs are involved in each process?
■ How are the antennae used?
■ How does the animal right itself when turned over?

Pick your crayfish up by the dorsal shield (*carapace*). Hold it firmly!

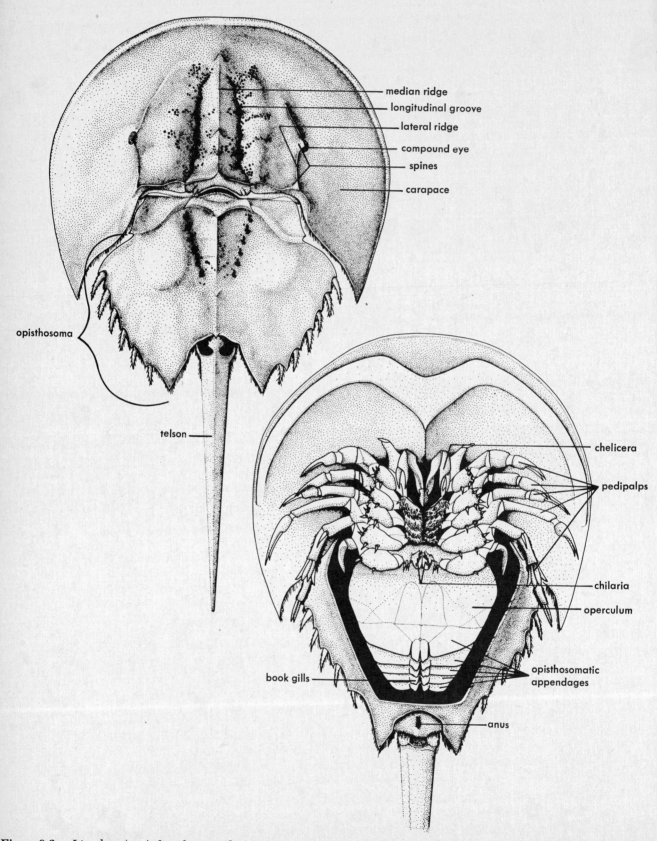

Figure 9.3. *Limulus*: (*top*) dorsal view; (*bottom*) ventral view.

■ Why does the animal buck and flip its tail down so strongly?

■ What is the function of this movement when the animal is in water?

■ How do you suppose that crayfish digs in soft earth?

Observe the respiratory currents of a crayfish resting in a shallow pan of water and note how they are produced. Add a few drops of India ink near the animal and watch the particle flow into the respiratory stream.

SUGGESTION

In many areas of our southern and prairie states, it is possible to watch foraging and digging crayfish ("crawdads") at night. Seek them out with a flashlight in marshy or moist open fields and watch their pattern of movement, digging, and rapid escape. Dig a few from their tunnels and observe their action in a laboratory or home aquarium. Baked or boiled crayfish represent another biological aspect of these organisms well worth investigating.

3. **External anatomy** (Figs. 9.12 and 9.13) If your specimen is preserved in formalin, avoid fumes by rinsing it overnight in tapwater before dissection.

a. *Segmentation.* Observe the *tagmata*, or grouping of segments, into three general regions: (1) *head* (somites I–V, with *antennules, antennae, mandibles,* and two pairs of *maxillae*); (2) *thorax* (=pereion) (somites VI–XIII, with three pairs of *maxillipeds,* one pair of *chelipeds,* and four pairs of walking legs or *pereiopods*); and (3) *abdomen* (=pleon) [somites XIV–XIX with five pairs of swimmerets or *pleopods,* and a *uropod* and *telson* (=tail)].

■ Are all appendages jointed?

■ How do the pincers operate? (Better use a pickled specimen.)

■ What are the advantages of a hollow exoskeleton?

■ How does surface area for muscle attachment compare between a hollow exoskeleton and a solid endoskeleton?

■ What are the *disadvantages* of an exoskeleton?

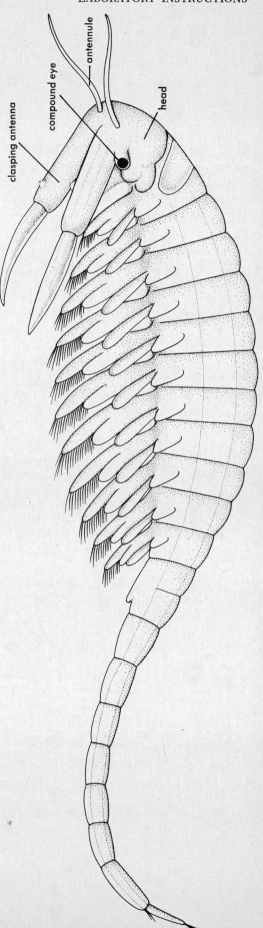

clasping antenna

compound eye

antennule

head

Figure 9.4. *Artemia,* lateral view.

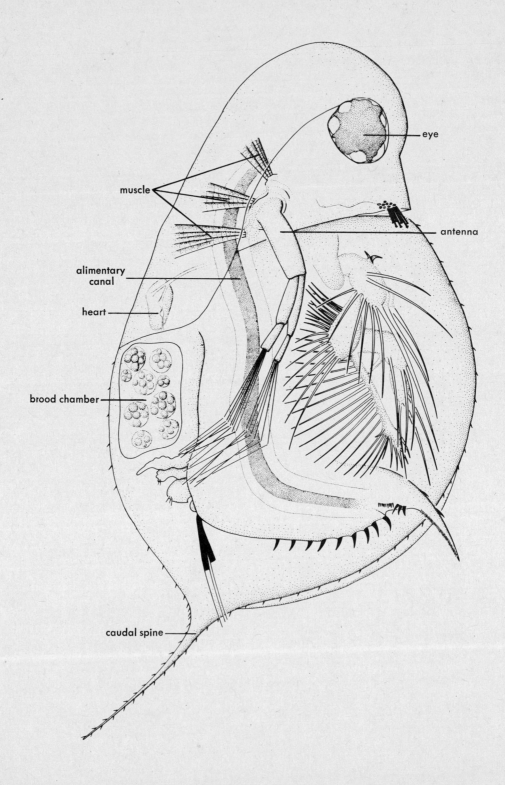

eye

muscle

antenna

alimentary
canal

heart

brood chamber

caudal spine

Figure 9.5.   *Daphnia*, lateral view.

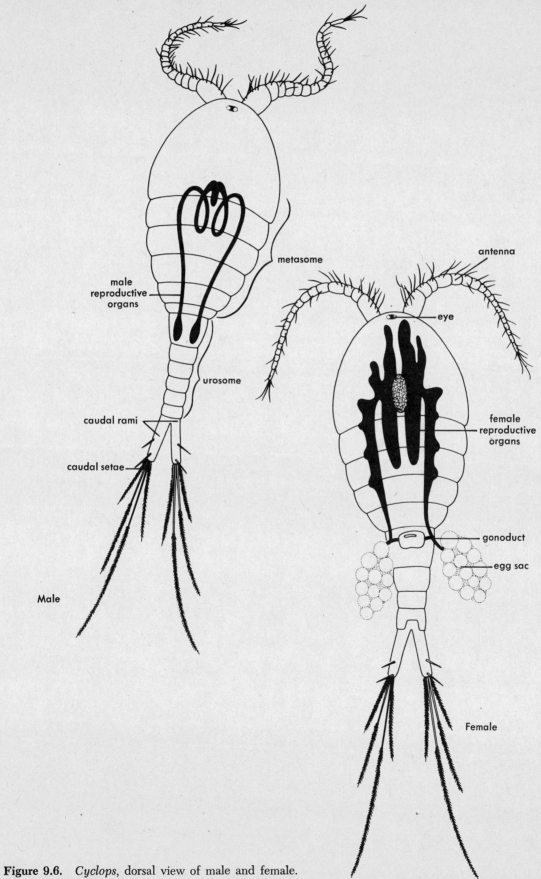

Figure 9.6. *Cyclops*, dorsal view of male and female.

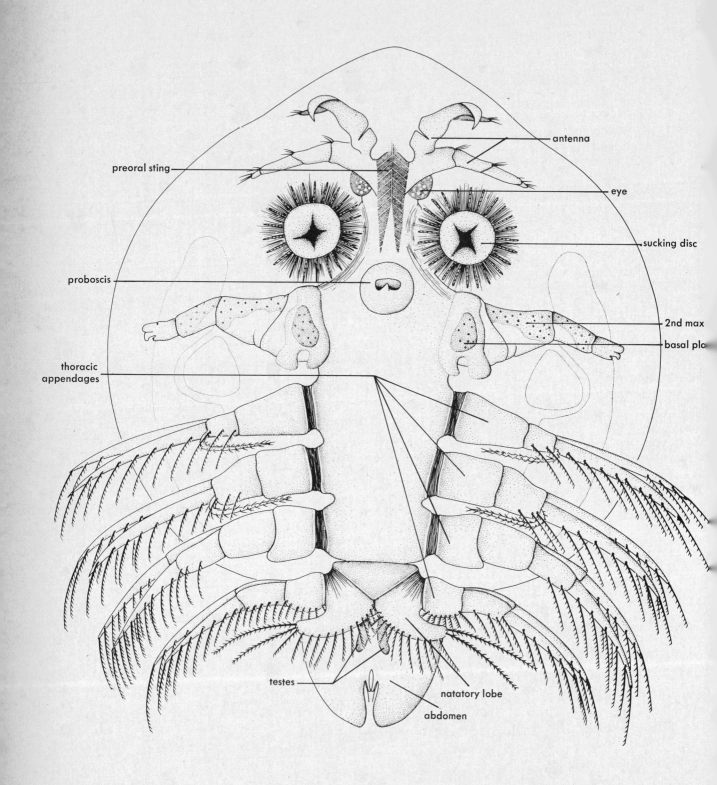

preoral sting

antenna

eye

sucking disc

proboscis

2nd max

basal plo

thoracic
appendages

testes

natatory lobe

abdomen

Figure 9.7. *Argulus*, ventral view.

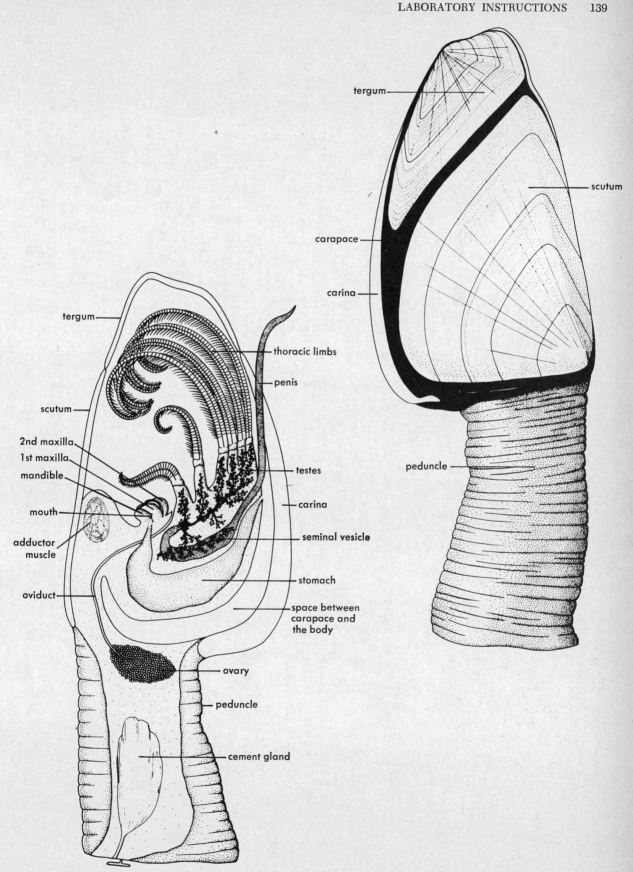

tergum

scutum

carapace

carina

peduncle

tergum

thoracic limbs

penis

scutum

2nd maxilla
1st maxilla
mandible

mouth

adductor
muscle

oviduct

testes

carina

seminal vesicle

stomach

space between
carapace and
the body

ovary

peduncle

cement gland

Figure 9.8. *Lepas anatifera*: (*left*) internal anatomy; (*right*) external anatomy.

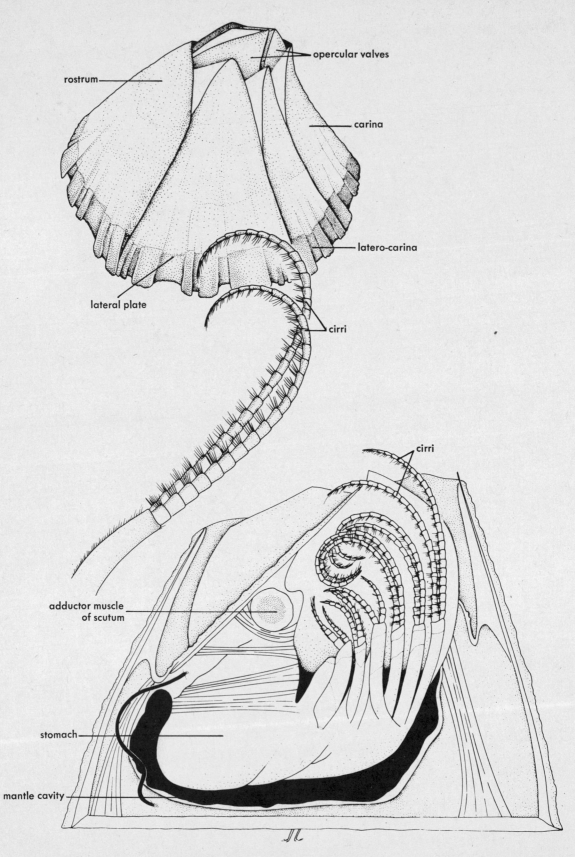

Figure 9.9.   *Balanus*: (*top*) external features; (*bottom*) internal structures.

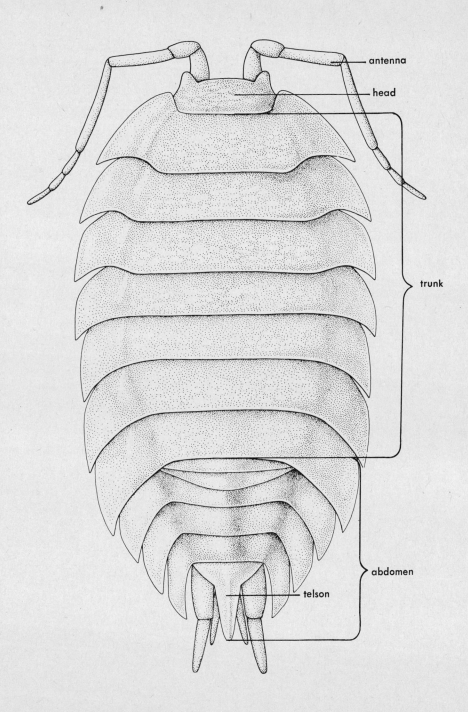

Figure 9.10. *Oniscus asellus,* dorsal view.

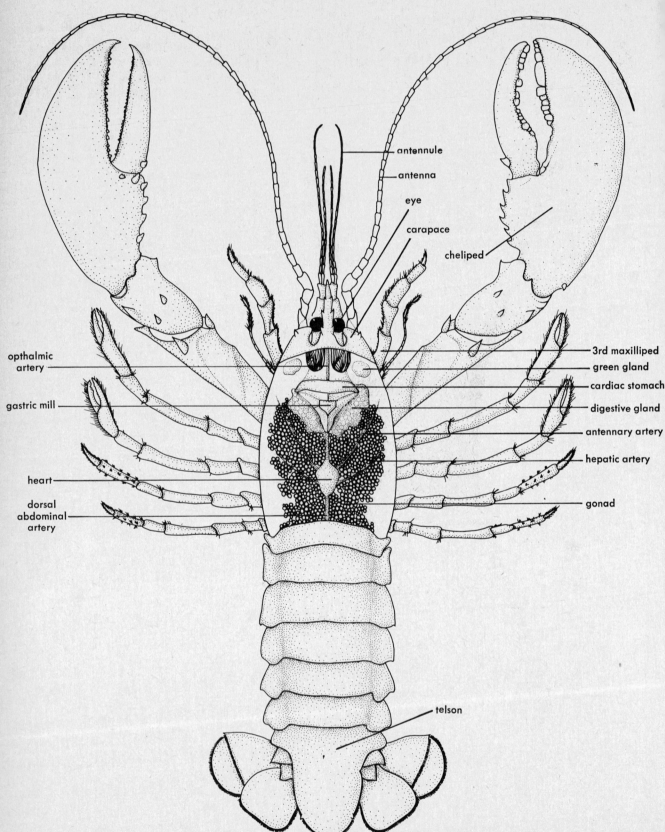

Figure 9.11. *Homarus* (lobster), dorsal view with carapace removed to show internal structures.

joints and a balancing organ or *statocyst* on the flattened dorsal surface of the basal joint. Next are the *antennae*, a pair of long, slender, many-jointed appendages with *excretory openings* at the basal segment of each. Then come the *mandibles* or chewing jaws, followed by two pairs of *maxillae* for food handling. Three pairs of *maxillipeds* follow; these serve for manipulation and sensory perception of food. Then, completing the series, are the *chelipeds* (claw legs) containing the largest claws; four pairs of walking legs (*pereiopods*, a term taken from *pereion*, or thorax, the thoracic legs); five pairs of *swimmerets*; and, finally, the *uropod*, to tail.

Next review the location of the following body openings: *mouth*, *anus* (on ventral surface of *telson*, the central portion of the tail), and *excretory pore* on each antenna.

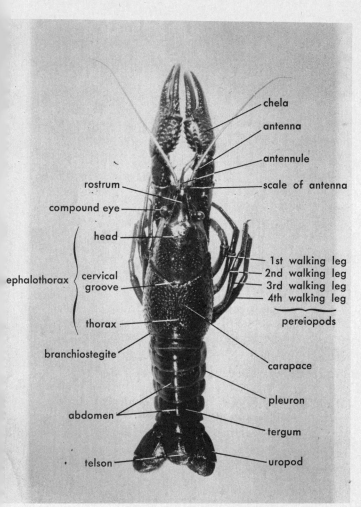

Figure 9.12. *Cambarus*, dorsal aspect.

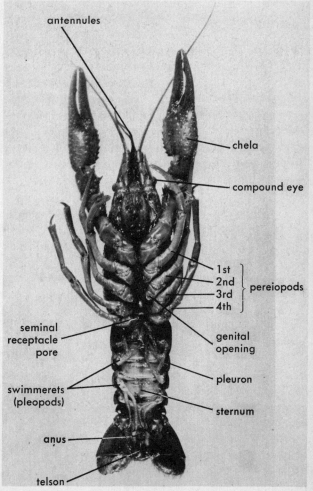

Figure 9.13. *Cambarus*, ventral aspect.

Observe the peg-and-socket joints between abdominal segments. The exoskeleton of each body segment is divided into a dorsal *tergum*, a lateral *pleuron*, and a ventral *sternum*. Find these parts on the abdominal and thoracic somites of your specimen.

b. *Body parts.* Before removing the appendages and studying their relationship, identify the major body parts visible dorsally (Fig. 9.11)—*head, antennule, antenna, cephalothorax, carapace, cervical groove, rostrum, eyes, cheliped pereiopods, uropod, telson.* Then observe the crayfish from the *lateral* view and locate the *gill chambers* under the pleura of the carapace (branchiostegites). Study the animal *ventrally* and observe its 19 paired appendages. First, find the anteriormost sensory *antennules*, biramous (two-branched) structures with many

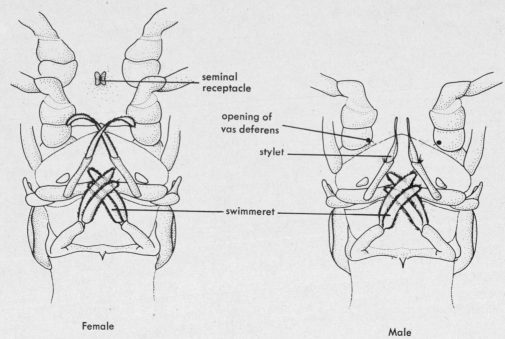

Figure 9.14.   *Homarus* (lobster): ventral views of female (*left*) and male (*right*).

Find the *external sex openings*. The male genital ducts or *gonoducts* open at the base of the last leg. A trough for the transfer of *spermatophores* (sperm capsules) into the female *seminal receptacle* is formed from the fused tubular first and second swimmerets (Fig. 9.14). The *oviducts* open at the base of the second walking legs; another pore at the base between the fourth walking legs serves to receive the spermatophores. The first swimmeret in females is small or absent; the second is a typical swimmeret rather than a specialized organ as in the male. The female telson and filamentous swimmerets hold the egg cluster to form an external *brood pouch* for the eggs and young. Females carrying such a mass of eggs are said to be "in berry."

c.   *Appendages* (Fig. 9.15). After this initial ex-

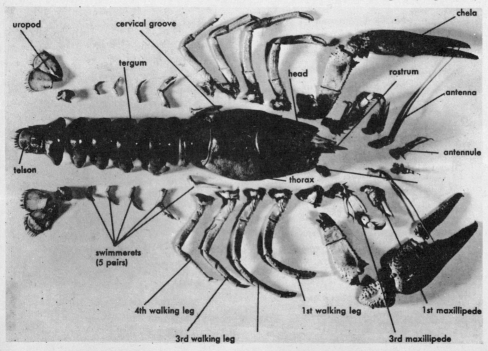

Figure 9.15.
Disarticulated view of crayfish, male.

amination of your specimen, you should be ready for detailed review of the appendages.

Learn not only the functional and structural differences between appendages but also their relationship to a common prototype which is thought to be a simple, biramous (two-branched) appendage such as the swimmeret. Species thought to be primitive show a repetition of parts with little specialization as seen in the brine shrimp. In more highly evolved animals, the specialization of appendages, especially on the head, is apparent. There are striking differences between a brine shrimp with its many similar biramous appendages (homologous metamerism) and a crab, shrimp, or crayfish with highly modified appendages (heteronomous metamerism), apparently derived from more simple appendages. Structural changes in serially homologous parts can often be traced by comparing them with an unspecialized segment or appendage.

Examine a sample dissection of crayfish appendages if one is available. Read the following instructions carefully *before* starting your dissection.

Each limb consists of three basic portions: the stem or attachment, the *protopod*; and two branches, the innermost of which is the *endopod*, the outer the *exopod*, each of which may also be jointed. The protopod usually consists of two joints (*podomeres*), a basal *coxopod* attached to the body, and the distal *basipod* to which the two branches (endopod and exopod) are attached. Each appendage fundamentally follows this structural pattern, though with considerable modification in the head region. In some cases, homology can be determined only by embryological study, owing to the loss of some portions during growth and to the development of more specialized appendages. In certain cases, as in antennules, the precise homology of the segments is still uncertain.

Carefully cut away the lateral extension of the carapace covering the gills, the *branchiostegite*, from one side to expose the internal sheet of gills (Figs. 9.16 and 9.17).

Begin your dissection with the *uropod* and work forward. The simple structure of the abdominal appendages more closely resembles what is presumed to have been the primitive crustacean biramous appendage, and makes a more suitable starting point. Each appendage should be examined critically with a hand lens

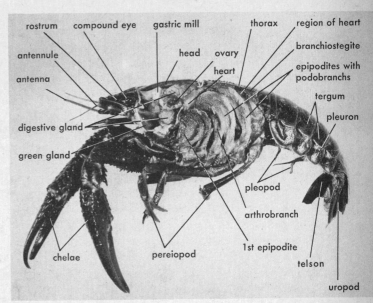

Figure 9.16. Lateral view of female crayfish with lateral surface of cephalothorax removed to show disposition of internal organs.

or under a dissecting microscope. Observe the structure, check it against its stated function, and note the *increased degree of structural specialization as you proceed toward the head*. You may find it more instructive to pin the appendages alongside the animal under water in a wax-bottom dissecting pan, in order to view them from the same orientation. Afterwards the appendages can be dried and glued onto a cardboard sheet and labeled for later reference (Fig. 9.15). Be sure to remove a bit of attachment membrane along with each appendage to ensure having the intact structure.

Beginning with the uropod, remove one appendage at a time from the side on which you have exposed the gills. Identify the three basic parts (if all are present) and pin out the appendage, carefully orienting the protopod to the left, the endopod anteriorly, the exopod posteriorly. Compare the normal plane of movement of each part and of each limb with the others to get an idea of how the appendage functions. From the uropod proceed anteriorly through the simple pleopods, noting again the sexual differences and water-circulating role of these organs. The walking legs, removed next, are of interest for their endopod specialization (*the exopod disappears entirely during development*) and for their respiratory role. Special

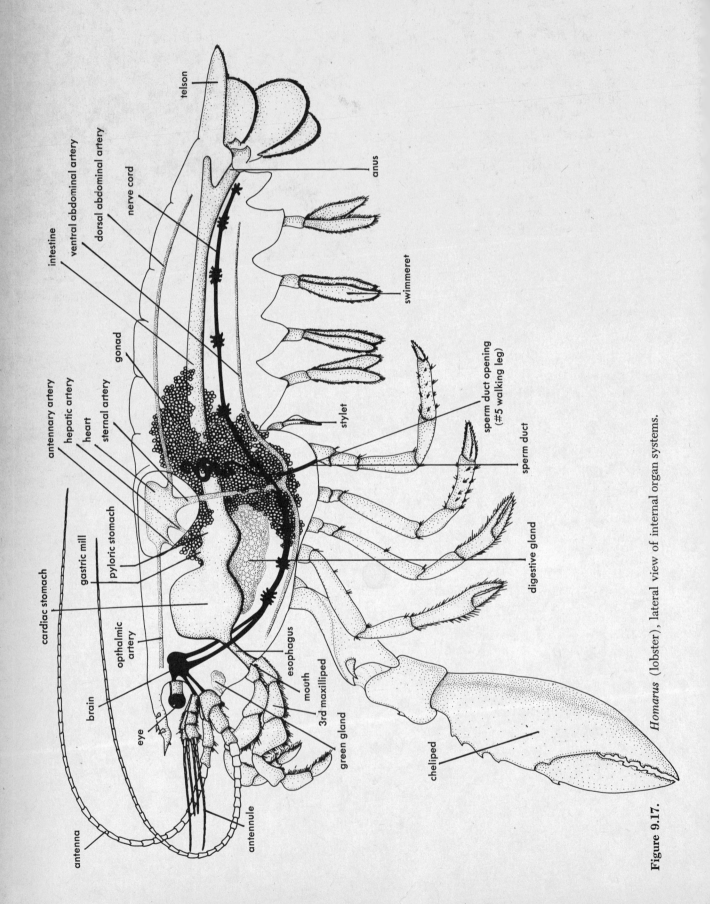

Figure 9.17. *Homarus* (lobster), lateral view of internal organ systems.

gills, *podobranchiae*, are attached to certain legs (*which ones?*). The coxopod has a specialized extension or *epipod* to which the gill filaments attach.

■ **How are these gills moved?**

Anterior to the pincer-bearing chelipeds are three pairs of maxillipeds, part of the complex food handling, grinding, and sensory mechanism. The third maxilliped, the largest of the three and most similar to the walking legs, is a good appendage to observe closely for signs of specialization, evolutionary modification that will become more marked in the anterior mouthparts. The exopod is present in all three maxillipeds, but the endopod becomes progressively smaller (opposite to the modification of the walking legs). The first maxilliped is marked by a large flattened extension of the coxopod. This is the *epipod*, a paddlelike structure that helps to maintain a water current through the gill chamber. The more anterior segments are parts of the head; their appendages form the *mouthparts* and related *sensory organs*. The second maxilla, anterior to the first maxilliped, also maintains a water flow across the gill membranes by *its* specialized paddle, the *scaphognath*. Actually a fusion of epipod and exopod, this paddle may be referred to as the *bailer* because of its constant sweeping action as it draws water from the gill chamber. The extremely small first maxilla or *maxillule* consists of a much reduced endopodite and larger basipodite and coxopodite. The exopodite is entirely absent, as in the walking legs.

■ **Can you tell where the exopod probably was originally attached?**

Next is the *mandible*, a heavy crushing structure (actually the coxopod) with a cutting inner edge and a brushlike sensory projection; its much modified endopod is called the *palp*. The basal segment of the palp is actually the basipod. Again, as in the walking legs, an exopod is lacking. A well-developed tendon attaches the mandible to mandibular muscles on the carapace tergum. The next appendage is the *antenna*. Find the opening of the excretory organ (*green gland*) in the coxopod. Segmentation here is fairly typical, all parts being present. The long filament of the *antenna* is an ex-

tension of the endopodite. Finally, examine the *antennule*, the anterior-most appendage, with its two equal divisions. These, however, are not easily homologized with the endopod and exopodite of more typical appendages.

Draw each appendage, labeling the important segments. Be able to compare each with a typical three-part appendage and to explain its function.

■ **What is meant by *serial homology*?**
■ **How would you compare serial homology with the homology represented by a comparison between the wing of a bat and a human hand?**
■ **How does the serial homology of CRUSTACEA (heteronomous metamerism) compare with that of the ANNELIDA (or primitive arthropods such as the brine shrimp)—homonomous metamerism?**
■ **What is meant by each of these terms?**
■ **How is anteroposterior differentiation in brine shrimp and crayfish related to habits and food-handling capacity in these animals?**
■ **Does this differentiation imply an increased degree of nervous control?**

4. **Respiratory system** Keep your specimen immersed under water in a dissecting pan. The *podobranchiae* have already been removed, as they were attached to the walking legs and to the second and third maxillipeds. The two gill blankets that still remain consist of four pairs, including a single anterior gill. These joint gills or *arthrobranchiae* (can you tell why the term is used?) are attached to the base of the *arthropodial membrane*. They are called side gills or *pleurobranchiae*. In *Cambarus*, side gills are lacking.

■ **Are any of these gills inside the body?**
■ **Is the branchiostegite the true outer wall of the thorax?**

Remove an arthrobranchia and examine it under a lens. Find its central axis and the many fine lateral filaments.

■ **What is the function of these lateral filaments?**

Cut across the gill and observe the *afferent* and *efferent branchial blood vessels* visible in cross section within the central axis. (Afferent means flowing *to*; efferent means flowing *from*. To or from *what*?).

■ **How does walking affect the podobranchiae?**

■ **How is a current of fresh oxygenated water kept constantly moving over the gills?**

■ **Name the structures involved in control of this respiratory stream.**

■ **Where does gas exchange occur?**

5. **Internal anatomy** (Fig. 9.18)

a. Systems to be examined in this section are *digestive*, *muscular*, *circulatory*, and *reproductive*.

It is difficult to separate the individual internal systems of the crayfish clearly, as they form a tightly compacted whole, morphologically intertwined. We shall therefore study them as they appear—as interfunctioning parts of a single physiological complex.

A fresh specimen is best for this exercise. If available, use one that has been injected with colored latex to show the circulatory system. You can still use your original specimen if it has been carefully dissected. Place it in water in your dissecting pan, dorsal side up; pin it if you are using a wax-bottom pan.

Cut under the lateral margins of the cephalothorax tergum. Remove this shield after making crosscuts behind the eyes and the posterior edge of the carapace. Over the exposed surface lies the *epidermis* (part of which may have adhered to the carapace) which secretes the exoskeleton.

Remove the epidermis and examine the exposed organs. You will see the gill surfaces outside the body proper in the branchial chambers. The first internal structure to locate is the *heart* in its *pericardial sinus* along the dorsal midline. The two dorsal pores, or *ostia*, through which blood enters the heart are visible. Later you will observe two other pairs of ostia in the isolated heart. The *stomach*, more descriptively called the *gastric mill* because of the hardened, internal sclerites for grinding food it contains, is a thin translucent sac lying in the thoracic cavity. Moving the sclerites are short, powerful muscle bands that pass from the posterior sclerites to the undersurface of the carapace. Find the cut edges of these muscles on the carapace shield you removed. The anterior set forms the *anterior gastric muscles*; the posterior set, the *posterior gastric muscles*.

If your specimen is a female "in berry," you will notice two masses of ovaries beside the heart and fusing posterior to it. Although male *testes* are less conspicuous they also have the same Y-shaped pattern.

Before continuing your dissection, trace some of major muscles. There are two pairs of *abdominal extensor muscle* bands that pass through the pericardial sinus and diverge to attach anteriorly along the floor and in front of the exoskeletal framework. These muscles continue into the abdomen, where their dorso-ventral position causes the abdominal sclerites to extend when the muscles contract. This extension moves the tail *dorsally*, setting the uropods in position for the stronger *abdominal flexor muscels* to deliver the powerful kick, the escape spurt you have already observed in these ani-

Figure 9.18. Anterior end of crayfish with dorsal surface of cephalothorax removed to show disposition of internal organs.

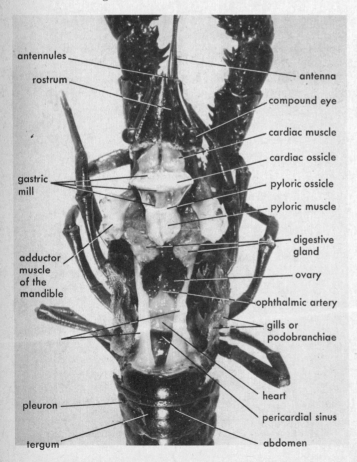

antennules

rostrum

antenna

compound eye

cardiac muscle

cardiac ossicle

gastric mill

pyloric ossicle

pyloric muscle

digestive gland

adductor muscle of the mandible

ovary

ophthalmic artery

gills or podobranchiae

heart

pleuron

pericardial sinus

tergum

abdomen

mals. These flexors will be seen later when you open the abdomen.

Now the specimen can be exposed for further dissection. On the other side with the gills removed, cut away the inner thoracic wall and remove the carapace behind the eye. (Do not damage the green gland.) At this point you will have a lateral view of the internal organs in their normal position.

Notice the distribution of the digestive gland and ovaries (or testes), the heart with its lateral pairs of ostia, the coils of *vas deferens* (sperm ducts) in the male, and the *green gland* (excretory organ). Use forceps (with caution) where necessary to move the organs aside or to remove portions of digestive or gonadal tissue that obscure other organs.

Cut away the dorsal abdominal sclerites along their entire length by making parallel incisions down the lateral margins (Fig. 9.19). Lift back this plate after cutting the extensor muscles of the abdomen behind the heart. Observe them under the dorsal skeleton and review their function. Then note the powerful abdominal flexor muscles forming bilateral masses on either side of the abdomen below the intestine. These are the previously mentioned muscles that give the power kick in swimming (and most of your gastronomic pleasure in these crustaceans). You may also find the *posterior aorta* (dorsal abdominal artery) lying dorsal to the intestine. This important blood vessel is best seen in latex-injected specimens.

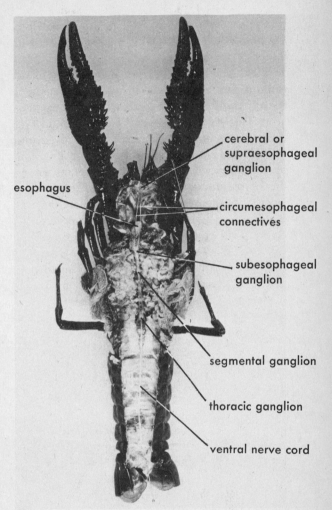

Figure 9.19. *Cambarus*, dorsal view of nervous system.

Labels: esophagus; cerebral or supraesophageal ganglion; circumesophageal connectives; subesophageal ganglion; segmental ganglion; thoracic ganglion; ventral nerve cord

■ **Do any parts of the gonads or digestive gland extend into the abdomen?**

At this point, pause to review what you have already examined: the gill system, major muscles, heart and posterior aorta, gastric mill, intestine, digestive gland, excretory organ, and gonads. Be able to identify each and show its structural relationship to the rest of the internal anatomy.

Make an outline *drawing* of the crayfish from the lateral aspect and include all organs you have been able to locate. Later, add those you find in subsequent dissection.

Starting from the heart, work out as much of the circulatory system as possible, moving the organs aside to find the primary vessels. The anterior artery leaves the heart anteriorly and supplies the head. A pair of diverging *lateral cephalic arteries* (*antennary arteries*) emerges from the heart at the same point and supplies the gastric mill, the anterior muscle groups, and the green gland. Two *lateral visceral arteries* (*hepatic arteries*) pass ventrally from the anterior end of the heart to the digestive gland, anterior intestine, and gonads. The *posterior aorta* and the *sternal artery* leave the posterior end of the heart. The sternal artery passes ventrally and divides into the *ventral thoracic* and *ventral abdominal arteries*.

You have now seen the major distributive pathways for freshly oxygenated arterial blood. Remember that this is an example of an *open* type of circulation. Blood vessels carry blood from the heart as described above, but it then passes into open blood *sinuses* and a *hemocoel* through which it flows prior to movement into the branchial (gill) vessels for oxygenation and return to the heart.

The blood sinuses of *Cambarus* carry blood to the tissues and cells after which it seeps into the *afferent branchial vessels* where it is oxygenated. The oxygenated blood leaves the gills via the *efferent branchial vessels*, and moves into *branchiopericardial canals* that carry it to the *pericardial chamber.* There the blood enters the *heart* through the three pairs of ostia and is ejected by the contraction of the heart through the arteries you have just viewed. Compare this system with the *closed* circulatory system of *Lumbricus.*

Remove the heart, place it in a small dish of water, and study it with a hand lens or dissection microscope. Locate the other ostia, in addition to the dorsal pair that you have already observed.

■ **Can you find any *valves* that prevent backflow through the ostia?**

Now complete your examination of the genital and digestive systems. Find the *oviduct* (*vas deferens* in the male) and note the general shape of the gonads. Trace the ducts to the genital pore on one side, and add these details to your drawing.

With care to avoid injury to the nervous system, clear away some of the gonadal and muscular tissue to locate the complete digestive system. Trace the *intestine* from the *stomach* to the *anus.* Examine the area anterior to the stomach, then find the short *esophagus*, which passes ventrally to the mouth through the *nerve ring* near the green gland. Try not to damage either. You can locate the esophagus by cautiously inserting a rounded probe into the mouth and carefully dissecting intervening tissues from the lateral aspect. Next try to remove the entire digestive tract including the intact digestive gland. Cut the intestine near the anus and tease it free to the stomach. Then work the digestive gland clear, cut the esophagus posterior to the nerve ring, and remove the entire alimentary canal. Let the structure float free in water and observe its pores, especially the connections of the digestive gland ducts to the stomach. Add this system to your drawing,

or place it in a separate drawing if that is more convenient. *Label* completely.

Cut the *gastric mill* from the rest of the gut, slit it ventrally, and examine its parts carefully. Find the gastroliths or grinding stones ingested by the crayfish, which act against the heavy, toothlike projections to masticate the food. Move the sides of the organ to see how the stones and gastric mill operate.

■ **What had your crayfish been eating?**

Pin out the gastric mill, let it dry to show the *ossicles* more clearly. Study it under the dissecting microscope. Determine the pattern of food flow into the cardiac section; find the pressure chamber and valves leading to the intestinal connection, and the opening from the digestive gland (or "liver").

b. *Nervous system*

i. The *sense organs* to be studied include: many *sensory bristles* (hairs), *compound eyes*, and *statocysts.* Sensory bristles over the entire body provide the animal with touch receptors on the otherwise relatively insensitive exoskeleton. Hairs on the mouthparts and antennae probably have *chemotactic* (taste) as well as *tactile* (touch) receptors. Nerve cells lying near tactile hair bases are stimulated by movement of the hair.

The characteristic arthropod eye is a compound structure formed by a number of many equal units or *ommatidia* (like cells of a photographic light meter), covered by a thin, transparent, cuticular *cornea* (Fig. 9.20). A cross section of the eye made with a razor will disclose black radiating lines marking separate ommatidia or visual units, which emanate from a central *optic ganglion. Pigment cells* around each *rhabdome* and the *crystalline cone* of a single ommatidium may be seen by microscopic examination of a portion of the eye that has been placed in a drop of water on a microscope slide.

The statocyst, or gravity receptor, is located in the basal segment of each antennule. Remove the remaining antennule and examine the basal segment under a dissecting microscope or hand lens. Search for a thin sac attached to the dorsal wall of this segment. As the crayfish moves, sand grains placed by the animal in this sac stimulate sensitive tactile hairs.

■ **What function does the organ serve and how might it work?**

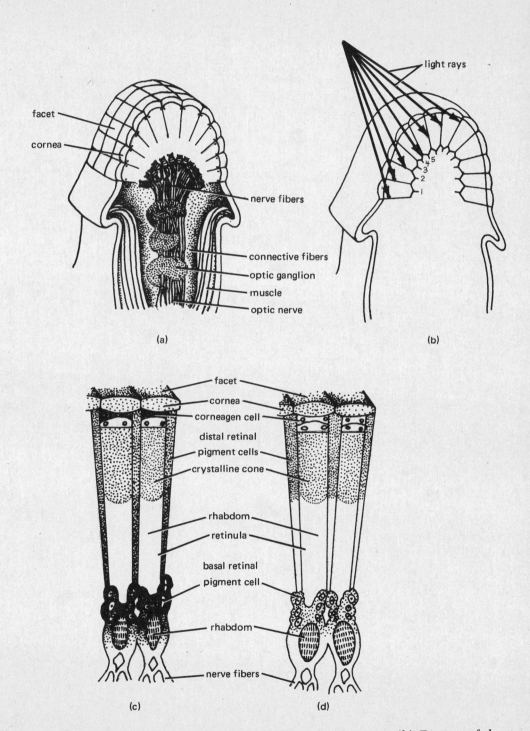

Figure 9.20. Crayfish. (a) Eye in longitudinal section to reveal general structure; (b) Diagram of the same eye illustrating how light rays are absorbed by the pigment surrounding the ommatidia; only those that pass through the center such as (4) reach the nerve fibers. This results in a separate image from each ommatidium; (c) One ommatidium in the light; pigment extended so that it completely surrounds the ommatidium, isolating it from its neighbors; (d) Ommatidium in the dark; pigment in the basal pigment cells is withdrawn.

■ **What equivalent organs have you already studied in other groups?**

An unusual experiment can be performed by placing a newly molted crayfish in a tank of clean, filtered water to which iron filings have been added. The animal will place these filings in its statocysts, since no sand grains are available.

■ **How will the crayfish orient itself if a magnet is held over its head? at the side of its head?**

It would be interesting to see if these reactions eventually disappear and if the animal learns to associate normal positioning of its body with a different stimulus, such as that produced by a magnet held in a dorsal position for a long period. Why not try it?

ii. *Central nervous system* (Fig. 9.19). The *ventral nerve cord*, in the midventral portion of the animal above the ventral abdominal artery, lies below the abdominal flexor muscle and under a shelf of connective tissue and *apodemes* (ingrowths from the ventral exoskeleton serving for muscle attachment).

Expose the nerve cord from the dorsal surface by stripping away the *dorsal flexor muscles* of the abdomen and cutting carefully through the thoracic apodemes on either side of the midline. It is sometimes simpler to make a single cut, spread the opening apart with the fingers to locate the cord, and then cut away projecting tissues. Expose the nerve as completely as possible.

■ **Can you find *lateral nerves* arising from the abdominal portion of the nerve?**
■ **Are they connected to *ganglia*?**
■ **How many *thoracic ganglia* are there per segment?**
■ **What modification has altered the anterior ganglia compared with the posterior ones?**
■ **What is the evolutionary significance of this development?**
■ **Can you trace any thoracic nerves from the cord?**

A special anatomical study would be required to work out the *peripheral* nerve connections. Notice the space for the sternal artery between the sixth and seventh thoracic ganglia. Trace the cord anteriorly and dissect clear the *subesophageal ganglion*, the *esophageal nerve ring* (*circumesopha-*

geal connectives), and the *brain* (*cerebral ganglion*) above the esophagus. The dorsal part of the nerve ring is sometimes called the *supraesophageal ganglion*. Locate the central pair of *optic nerves*, the *antennulary nerves* lateral to them, and the lateral pair of *antennary nerves*. Trace these nerves to their respective organs.

Draw the ventral nerve cord and brain, showing the location of ganglia and lateral nerves.

After completing your study of *Cambarus*, observe other crustaceans available on demonstration.

II. *Romalea* and *Periplaneta* (Either can be used for detailed study.)

1. **Classification**
Phylum **ARTHROPODA**
Class INSECTA
Subclass PTERYGOTA
Division HEMIMETABOLA
Order ORTHOPTERA
Species *Romalea microptera* (lubber grasshopper)
Order DICTYOPTERA
Species *Periplaneta americana* (American cockroach).

2. **Introduction** Representative insects we shall study are the lubber grasshopper, a member of the order ORTHOPTERA (grasshoppers, crickets, and related forms); and the American cockroach, in the related order DICTYOPTERA (cockroaches, praying mantis, and their allies). These are in the division HEMIMETABOLA, insect orders characterized by *gradual* (or *incomplete*) *metamorphosis* in which immature forms are smaller "editions" of their parents. Their wings gradually develop externally, eyes are like those of the parents, and sexual organs remain undeveloped until the last *instar* (stage) is reached. Insect growth is *discontinuous*, as is typical of all arthropods. Because of the size limits imposed by the exoskeleton, growth is most rapid during the brief period after each molt, when the cuticle is soft and can be stretched. Internal pressures, such as from blood (hemolymph) and air, exerted against the soft cuticle at this stage, stretch the soft skin, creating space for internal growth until once again the nymph is literally bulging at the seams.

Other hemimetabolous orders include

DERMAPTERA (earwigs), PLECOPTERA (stone-flies), ISOPTERA (termites), EMBIOPTERA (embiids), ODONATA (dragonflies and damsel flies), EPHEMEROPTERA (mayflies), MALLO-PHAGA (biting lice), HOMOPTERA (cicadas, scale insects, aphids), and THYSANOPTERA (thrips).

Varied and numerous as these insects are, they are far less so than are the HOLO-METABOLA, the higher insects, characterized by *complete metamorphosis*. In holometabolous orders, the life cycle can be divided into four clearly distinct stages: *egg, larva, pupa,* and *adult*. Each stage is an excellent example of morphological specialization. The larva, usually a worm or grub, is a voracious eater and represents the growth stage. The sexually mature adult, the flying stage, is specialized for both reproduction and distribution. The egg and pupa represent periods of protected bodily reorganization, each ending in a dramatic structural and functional change. Growth is confined to larval instars, as adult insects do not molt. The limited adult "growth" that does occur is only a distension with eggs as in queen bees or ants. There is no such thing as a "baby fly" (except the wormlike grub, of course). Small adult flies are simply small species, or dwarfed flies hatched from poorly fed larvae. Specialization of body structure and stage of development to the metabolic and ecological needs of the higher insects represents a unique type of adaptation of the entire life cycle to distinct functions and diverse habitats.

Compare the larval and adult stages of a *beetle, mosquito, honey bee,* and *butterfly*.

The largest of insect orders is the COLE-OPTERA (beetles). LEPIDOPTERA (moths and butterflies), HYMENOPTERA (bees, wasps, and ants), and ever-present DIPTERA (true flies) are also giant groups. The HYMENOPTERA, with their complex social organization, are perhaps the most highly developed of all arthropods. SIPHONAPTERA (fleas) are among the most specialized. Lesser known orders of HOLOMETABOLA include the NEUROPTERA (lacewings), MECOPTERA (scorpionflies), and TRICHOPTERA (caddisflies).[3]

This is quite a collection—about 27 orders and about a million species. Name a terrestrial or freshwater environment and it will have insects. Atom-blast an island and insects will be among the first to repopulate it. Prolific, adaptable, varied, rapidly evolving insects are here to stay. It is increasingly questionable whether man can devise insecticides as rapidly as his insect targets can become resistant to them by mutation and natural selection. Nor are we any longer convinced that this is the best means to achieve a balanced level of insect control. It is a race in which we are far from ahead. Our knowledge of insects and the reasons for the ubiquity of this highest group of arthropods is of more than academic interest.

SUGGESTIONS

Make an insect collection representing at least 10 orders and 50 families. Key out your specimens; label, pin, and mount them on composition board in cigar boxes, or museum insect boxes. Available textbooks and guides (such as H. E. Jacques, "How to Know the Insects"—see references at end of chapter), and numerous federal or state publications on entomology are excellent sources for methods of killing, preserving, pinning, and labeling. Make careful field notes of date, locality, habitat, and other collecting data. Use as many techniques as possible for collecting and searching different habitats. Develop your acuity and sense of observation for the characteristic habits and habitats of different orders. If you do this conscientiously and for fun as well as knowledge, you will be rewarded by a far more satisfying and instructive lesson in entomology and biology than any laboratory can provide.

3. **External anatomy** Both the voracious lubber grasshopper *Romalea microptera*) and the unpopular American cockroach (*Periplaneta americana*) (Figs. 9.21 through 9.23) are useful in these dissections. The cockroach is easier to dissect and may be considered a somewhat more representative

---

[3] A few other insect groups are placed in a separate subclass, the APTERYGOTA, owing to certain primitive or highly distinctive characteristics. Included are wingless forms like the DIPLURA (japygids) and book-loving silverfish or THYSANURA (bristletails). Still more primitive, scarcely even conforming to the strict definition of insects, are the tiny PROTURA and the equally small but very very abundant COLLEMBOLA (springtails).

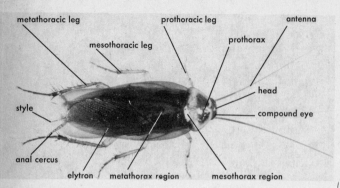

Figure 9.21. Dorsal view of the cockroach, *Periplaneta americana*, male.

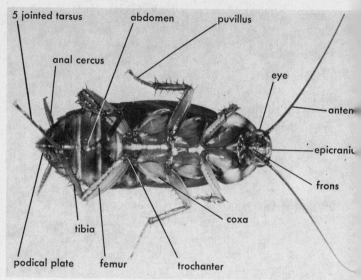

Figure 9.23. Ventral aspect of *Periplaneta americana*, female.

insect than the grasshopper, owing to its lack of specialized jumping adaptations.

a. Identify the major body segments—*head, thorax, and abdomen.*

■ **How do these segments compare with those of the crayfish?**

The insect exoskeleton, a complex, many-layered structure of chitin and other materials, is covered by a thin but critically important outer waxy layer that prevents desiccation. Some "desiccant" insecticides now being developed abrade or dissolve minute pores in this wax and allow rapid evaporative water loss, which quickly kills the insect.

The head consists of six fused segments; the thorax consists of three segments (*pro-, meso-,* and *metathorax*), and the abdomen of

eleven, including the terminal reproductive organs. A tendency toward fusion is as marked in insects as in other arthropods.

Movement is restricted to soft *conjunctivae* between sclerites and segments. Growth, or body expansion, is limited to stretching during the brief period from molting of the old larval exoskeleton to hardening of the new one.

Review how the exoskeleton *limits* as well as *permits* evolutionary change in insects (see Kennedy, 1927, in reference section).

Notice that each thoracic somite consists of several sclerites. On a thoracic or abdominal somite locate the *tergum, pleuron,* and *sternum.*

■ **What, then, would be the metapleuron, the mesosternum, and so on?**
■ **How many sclerites are there per tergum? per pleuron? per sternum?**

b. Closely examine the *head.* Locate the compound eyes (with many small facets or *ommatidia*) and simple eyes (*ocelli*) between them (how many?). Head sclerites, from dorsal surface of the head down the face, include the *epicranium, frons* (below the attachment of the antennae), *clypeus,* and *labrum* (upper lip). The "cheek" areas are termed the *gena.* The other mouthparts will be examined later.

c. The thorax is divided into three body segments, as already noted.

Figure 9.22. Dorsal view of *Periplaneta americana*, male, with wings removed.

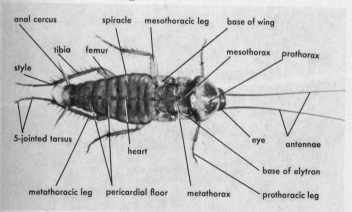

■ **To which segments are the legs at-tached?**

Identify the following leg segments starting from the point of attachment and extending to the terminal claws: *coxa, trochanter, femur* (bearing a few spines), *tibia* (very spiny), *tarsus* with five joints, a number of *pads*, and a terminal pad (*pulvillus*) bearing a pair of hooks or *claws*. Find the corresponding parts in each leg along one side of the body.

■ **How do they differ?**

Spread the *forewing* (*elytron*) and the flying or *hindwing*. Note the segments to which they attach.

■ **How do the two wing pairs differ in structure?**
■ **How do they differ in beetles? in bees? in butterflies? in flies?**
■ **What functional advantages do these differences afford?**

The abdomen has eleven segments, though the terminal segments 9–11 are difficult to discern. Abdominal segment 1 is interrupted in *Romalea* by the *tympanum* (function?) and the metathorax.

Segments 2–8 bear *spiracles* on each side. These are the apertures for gas exchange into and from the tracheal network. (Did you also notice the two pairs of spiracles on the thorax?)

The tip of the abdomen shows sexual dimorphism. *Romalea* females possess two pairs of pointed *ovipositors*, digging structures for egg-laying in soil. Near the upper pair of ovipositors, a pair of small sensory projections—the *cerci* (singular: *cercus*)—marks the border between the tenth and eleventh segments. The male grasshopper has a prolonged genital plate extending backwards and upward from the ninth sternite.

Cockroaches lack specialized digging structures. The female has large cerci with a pair of triangular ventral *podical plates* between them. The anus opens between these plates at the posterior tip of the insect. The male has cerci plus a pair of fine terminal spines, each called a *style*.

Draw a lateral view of your insect showing relationships of body sections and somites and the parts you have identified.

d. *Mouthparts* (Fig. 9.24). Much of the adaptability among insects is reflected in the diversity of their mouthparts, manner of feeding, and types of food. Compare the mouthparts of a *mosquito, butterfly, housefly,* and *squash bug*. All of these extraordinary functional differences can be homologized to the same structural elements. Mouthparts of roaches and grasshoppers are thought to be fairly conservative or non-specialized; hence, they are useful for demonstration of basic anatomy.

As you identify and remove each part, place it on a clean sheet of paper in the same orientation as in the intact head (Fig. 9.24). Remove the clypeus and labrum, already identified. This will expose the powerful mandibles or jaws (usually black or brown). Notice their serrated inner surface.

■ **Can you determine probable food habits on the basis of the mandibles? Explain.**

Observe that the mouthparts are paired, rather like legs. They appear, in fact, to be

Figure 9.24. Isolated mouthparts of the grasshopper, *Romalea.*

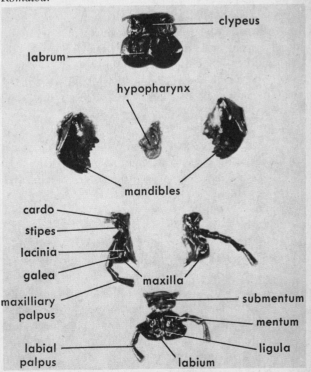

serially homologous with legs (compare with crayfish and annelids) and to be derived from leglike appendages.

Separate the mandibles with a needle and remove them with forceps. Next are the *maxillae*, rather complex sensory food handlers (and chemoreceptors) to which *maxillary palps* are attached. In addition to the palps, find two other projections—the rounded *galea* and the sharp, pointed, hairy *lacinia*. Remove maxillae and add to the array of mouthparts.

The *labium*, or lower lip, underlies the maxillae and bears a second pair of sensory palps or "feeders." Though the labium appears to be single, it actually is a pair of maxillalike structures fused along the midline to form a plate, or *mentum*. Removal of the labium reveals a narrow, tonguelike *hypopharynx*. Arrange all the mouthparts in their correct relative position and review their names and functions in the food-handling process. Glue the mouthparts in this position; label them and their principal parts.

4. Internal anatomy

a. *Introduction.* Few animals are as well packaged as insects. Miniaturization is carried to a degree that would astonish even our satellite designers. There is a high survival premium for efficient packing of internal organs into the limited available space. Consequently dissection of insects requires dexterity and patience to avoid destruction of one organ system while another is being disentangled from its neighbors. These dissections can well be considered special projects owing to their demanding technical skill.

Several dissection techniques can be followed. If the specimen is small and difficult to handle, it is advantageous to embed it in wax before dissection. Melt some clear paraffin and pour it into a glycerine-coated watch glass. (Glycerine allows easy removal of wax.) Trim off the legs as close to the body as possible. Holding the specimen by the wings, carefully immerse it in the wax halfway up the pleurites, deeply enough to allow the wax a firm grip on the body. Keep the insect in place until the paraffin hardens on the surface, then speed the hardening process by placing the watch glass in cold water. After it has hardened, remove the entire wax piece by slipping it from the glass. If the wax is still malleable, it can

be pinned to the wax bottom of your dissecting tray without cracking. You now have a firmly held insect ready for dissection under a binocular microscope or under water.

A somewhat simpler method is to pin the insect directly to the dissection tray through the base of the metathoracic legs and forewings after trimming off the other legs and hind wings close to the body. The wax embedding method permits a somewhat more controlled dissection and allows microscopic examination of the specimen during work. Either technique can be satisfactory or discouraging, depending upon one's interest and patience. The result, however, is your introduction to a remarkable, complex animal as successful in its realm as man is in his, yet constructed on a markedly different body plan.

b. *Dissection.* Open the dorsal surface of the abdomen to disclose the heart, aorta, and general disposition of internal organs (clearly seen only in fresh specimens). Loosen one of the terga near the end of the abdomen; then make a lateral cut from this edge of the loosened dorsal plate to the first abdominal segment. Raise this loosened flap and work it free, being careful not to damage the heart and aorta directly under the plate. Continue to cut forward toward the head, then across the top of the prothorax behind the head, and slowly work free the entire exoskeletal shield. Check the midline of the inner surface of this shield to see whether the heart was stripped from the abdomen. You will find it to be a delicate tube, either dorsal to the thin mass of *alary muscles* adhering to the removed portion of the exoskeleton, or still intact on your specimen. Examine the heart closely and try to locate the *ostia*.

■ How do the ostia function in circulation?

■ What sort of circulation is typical of insects?

■ How does insect circulation compare with that of annelids? the crayfish?

The *skeletal* system consists essentially of the *exoskeleton*, with its internal projection (apodemes) that serve as additional surfaces for muscle attachment. The extremely complex *muscular system* is too difficult to work out in a brief laboratory period. In fact, the cockroach is said to have about 10 times as many separate muscles as man. These separate muscles are not grouped and tied to *tendons* as in vertebrates. Grouping the ends of muscles to a tendon attaches them to a single point on a bone, a necessary development in vertebrate structure. The insect has more muscles, yet each has a separate point of attachment and insertion, allowing for considerable diversity and flexibility of movement. All of this of course is due to the fact that the insect exoskeletal tube offers about 10 times more surface area for attachment relative to the animal's bulk than does the human endoskeleton.

Closely connected to the nervous system are various recently discovered hormonal centers that trigger the various steps of maturation and metamorphosis. These structures are best reviewed in your textbook rather than on the specimen. Details of the sensory apparatus are other important structural and functional aspects for interesting study in more advanced courses.

Remove the heart, with its adjacent pericardial floor and alary muscles, by carefully stripping it from the exposed surface of the insect without damaging the underlying organs. Pick off the covering layer, largely *fat body*. Examine the glistening white tubes spreading through the fat body and covering the crop and other organs visible below it, the trachea.

c. *Tracheal system.* The air-tube or tracheal system is another anatomical feature related to the exoskeleton. It supplies oxygen from the outside air directly into the tissues and cells without a blood carrier, unlike most other animals. In this rapid gas exchange by diffusion air flows through major tubes and numerous smaller branches that penetrate tissues and reach each cell. If all but the trachea of the insect were to disappear, you would see remaining a tight network of fine interlacing branches reaching around the cells, just as capillaries do in our own bodies. Air passes from the *spiracles* to the major tracheal *ducts*, then to *air sacs* and a multiplicity of smaller tubes lined with spiral *taenidia* that keep the channels open. Remove a bit of muscle tissue, place it in a drop of water on a slide, tease it apart with needles, and then press a coverslip down carefully. Examine the preparation under a compound microscope; observe numerous tracheal branches among the muscle fibers.

Draw a group of fibers and accompanying tracheae.

If your specimen is freshly killed, air sacs and large tracheae will be glistening white and filled with air. If it is preserved, the ducts will be harder to see but their location should nonetheless be sought. Trace one duct from a spiracle to its primary divisions, the air sacs, and into smaller divisions. Use a blunt probe to push the intervening organs aside without damaging them.

d. *Digestive system.* Remove the overlying mantle of tracheae and remaining fat body tissue to expose the *crop*, largest portion of the alimentary canal. Carefully work the crop loose without breaking attachments at either end and lift it out holding it to one side with a pin placed against the inner margin of the crop. Avoid damaging the reproductive organs in the posterior portion of the body. As you lift the crop free, the *stomach* and coiled *intestines* will appear. These are marked anteriorly by fingerlike *gastric caeca*, and at the start of the hindgut by numerous hairlike *Malpighian tubules*, which form the excretory organ. Work the coils loose and try to expose the complete digestive system. In the thoracic region it will be necessary to cut away most of the heavy limb and wing musculature. As the anterior end of the crop is lifted from the thorax, look for the paired *salivary glands* that lie ventral to the *esophagus* but extend halfway into the thoracic cavity.

■ **Where do these glands empty?**

The anterior part of the alimentary canal consists of an anterior *foregut* (*stomadeum*) including the *mouth, esophagus, crop*, and *proventriculus* (gizzard). This latter structure grinds like the gastric mill of the crayfish. In some insects, a muscular sucking organ—the *pharynx*—is found at the anterior end of the esophagus.

■ **Name several examples of insects with a pharynx.**

The middle section of the alimentary canal, the *midgut* (*mesenteron*), consists of *gastric caeca* and *stomach* (*ventriculus*). Most digestion takes place in this region.

The beginning of the third region, the hindgut (*proctodeum*), is marked by Malpighian tubules, as already noted. The hindgut consists of an enlarged anterior and a slender posterior portion of intestine and the heavy-walled *rectum* ending at the *anus*. The foregut and hindgut are both derived from ectodermal infoldings, whereas the midgut is formed from endodermal tissue. (Foregut and hindgut linings are shed, along with the rest of the exoskeleton, during molting.)

Disentangle these structures and lay out the intact digestive system, leaving it attached at mouth and anus. Remove interfering bits of fat body and tracheae, but do not damage the reproductive system.

e. *Excretory system*. The thin, brownish Malpighian tubules between the stomach and intestine have previously been identified as excretory in function. These blind ducts in the hemocoel around the gut are washed by blood-flow emptying from the arteries. Their function is apparently to absorb metabolic waste products from the blood and to pass them into the intestine. Products not later reabsorbed in the hindgut are passed out the anus with undigested food wastes.

f. *Reproductive system*. Expose the reproductive organs in the posterior end of the abdomen. This highly complex system is similar in general organization and symmetry in the two sexes. The male grasshopper or roach, however, has a single, mushroom-shaped *utricular gland* (lacking in other insect groups), which receives reproductive products from the small testes lying under the terga in the fourth and fifth abdominal segments. Sperm pass through a fine duct (*vas deferens*) from each testis and fill the paired sperm chambers (*seminal vesicles*). In some roaches, testes and sperm ducts atrophy in the adult after the seminal vesicles are filled. These vesicles and a surrounding mass of fingerlike glands form the utricular gland, in which sperm are packaged in *spermatophores*, which then pass out the mushroom "stem," a muscular *ejaculatory duct*, and out the *penis* or copulatory organ. The latter is associated with a complex external *intromittent organ* consisting of an asymmetrical arrangement of lobes, plates, hooks, spines, and bristles with controlling internal musculature.

The female system is also paired and, as in the male, the gonads (*ovaries*) are really two clusters of smaller units. Each ovary consists of a group of tapering *ovarioles*. Eggs are produced in the narrow tips and passed posteriorly into enlarged *follicles* (egg chambers), which feed into an oviduct, one per ovary. The two oviducts then unite into a common oviduct, which passes into an enlarged, egg-holding *vagina*. The vagina in turn leads into the external *ovipositor*, or egg-laying structure. The dorsal wall of the vagina is connected to two other important structures. One is the *spermatheca* (seminal receptacle), a small bulb for storing sperm received at copulation. It also allows sperm to pass a few at a time through a small duct into the vagina for insemination of newly produced eggs. A pair of *accessory glands*, also attached to the vagina, secrete the special gluelike material that surrounds the eggs at laying.

Few of these smaller structures can be discerned in your specimens. However, in female specimens you should find the two ovaries with their ovarioles, the paired oviducts leading to the common oviduct, and the many-branched accessory glands that, with the spermatheca, empty into the vagina. Find the central, composite gland in the male, with the medial ejaculatory duct into which the accessory glands empty. Between the testis and ejaculatory duct are the coiled vas deferens and the enlarged utricular gland. Lay these organs out, free from fat body and tracheae, so that they can be seen clearly.

Draw a full-page outline of the dorsal view of your dissection. Sketch the alimentary canal along one side, and label all parts. Add the excretory system and the reproductive system, spread out as in your own dissection, and label as far as visible. Leave space in the midline for adding the ventral nerve cord.

g. *Nervous system* (Fig. 9.25). Expose the *ventral nerve cord* by carefully picking away overlying muscle, tracheae, and fat body tissues. Work from the posterior end, exposing each *ganglion* in turn. Remove the rectum and reproductive organs to expose the terminal ganglia. To show the anterior ganglia, cut the esophagus, place it to one side, and remove the salivary glands. Cut the exposed dorsal sclerites of the head away from the neck area to a point between the eyes. Use fine pointed scissors or forceps to avoid damaging the brain. This

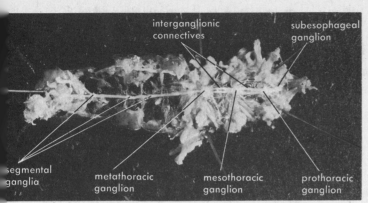

interganglionic connectives

subesophageal ganglion

segmental ganglia

metathoracic ganglion

mesothoracic ganglion

prothoracic ganglion

Figure 9.25. Ventral nerve cord of *Periplaneta americana*, dorsal ganglion removed.

should expose the entire nerve cord, showing the *abdominal*, *thoracic*, and *cranial* ganglia.

■ **How many of each ganglion type are there?**

■ **In what segments are they located?**

Find the *subesophageal ganglion* with its *circumesophageal connectives* joining the *cerebral ganglion* or *brain* (also called *supraesophageal ganglion*).

■ **Can you find nerves stemming from each ganglion?**

Note that all ganglia are paired, joined transversely by *commissures*, and to the posterior and anterior ganglia by *connectives*. Even apparently single ganglia consist of fused pairs of ganglia. Considerable modification of the nervous system with a marked tendency toward fusion of anterior ganglia is found among insects. This is particularly evident in the so-called higher orders, especially in some DIPTERA and HYMENOPTERA. Remember, however, that throughout the phylum **ARTHROPODA** the pattern of the nervous system shows a marked uniformity, a similarity that can be extended back to include even the **ANNELIDA.**

■ **What phylogenetic relations does this similarity suggest?**

Clear away the tissue around the brain to show the precise location of the subesophageal ganglion and its connection to the brain and to the next posterior, or *first thoracic* ganglion. A group of apodemes from a special supporting structure in the head (*tentorium*) serves as an area for attachment of numerous muscles controlling movements of the mouthparts and antennae. This structure must be cut on each side and removed to expose the subesophageal ganglion lying beneath.

Add the nervous system to the drawing of your dissected insect, and have your instructor check your work. Be prepared for an oral review of all these systems, including names and functions of important parts.

■ **How do these organ systems compare in general with those of CRUSTACEA?**

■ **How would you compare their general degree of specialization?**

■ **How has the exoskeleton affected size, growth *pattern, type of flight, feeding habits,* and *general habitats* of insects and provided them with both biological advantages and limitations?**

Describe the arthropod basic body type, then modifications as seen in the classes CRUSTACEA and INSECTA. Finally, explain how these major structural adaptations are related to the way of life and distribution of these two large, important classes of animals.

In sum, do not despise the lowly insects. They appear able to take over any time we radiate or explode ourselves out of existence.

Suggested References

CRUSTACEA

Barnes, R. D., 1974. *Invertebrate Zoology*, 3rd ed. Philadelphia, Pa.: W. B. Saunders.
Brown, F. A., Jr., 1950. *Selected Invertebrate Types*. New York: J. Wiley & Sons.
———, 1961. "Physiological rhythms," in *The Physiology of Crustaceans*. New York: Cambridge University Press.
Buchsbaum, R., 1956. *Animals Without Backbones*. Chicago, Ill.: University of Chicago Press.
Bullough, W. S., 1962. *Practical Invertebrate Anatomy*. New York: St. Martin's Press.

Carlisle, D. B., and F. Knowles, 1959. *Endocrine Control in Crustaceans.* New York: Cambridge University Press.

Cohen, M. J., 1955. "The function of receptors in the statocyst of the lobster *Homarus americanus.*" *J. Physiol.,* 130:9.

Costlow, J. D., Jr., and C. B. Bookhout, 1953. "Moulting and growth in *Balanus improvisus.*" *Biol. Bull.,* 105: 420.

——, 1956. "Shell development in *Balanus improvisus.*" *J. Morph.,* 99:359.

Cushing, D. J., 1951. "The vertical migration of planktonic Crustacea." *Biol. Rev.,* 26:158.

Dahl, E., 1963. "Main evolutionary lines among recent Crustacea," in *Phylogeny and Evolution of Crustacea,* H. B. Whitington and W. D. I. Rolfe (eds.). Cambridge: Mus. Comp. Zool.

Darwin, C., 1851–1854. *A Monograph on the Subclass Cirripedia* (2 vols.). London: Ray Society.

Glaessner, M. F., 1957. "Evolutionary trends in the Crustacea." *Evol.,* 11:178.

Green, J., 1961. *Biology of Crustacea.* London: H. F. & G. Witherby.

Hansen, H. G., 1925. *On the Comparative Morphology of the Appendages in the Arthropoda. A. Crustacea.* Copenhagen: Glydenda.

Jennings, R. H., and D. M. Whitaker, 1941. "The effect of salinity upon encystment of *Artemia.*" *Biol. Bull.,* 80: 194.

Lockwood, A. P. M., 1960. "Some effect of temperature and concentration of the medium on ionic regulation of the isopod *Asellus aquatica.*" *J. Exp. Biol.,* 37:614.

Marshall, S. M., and A. P. Orr, 1955. *The Biology of Calanus Finmarchicus.* London: Oliver & Boyd.

Passano, L. M., 1961. "The regulation of crustacean metamorphia." *Am. Zool.,* 1:89.

Richard, A. G., 1951. *The Integument of Arthropods.* Minneapolis, Minn.: University of Minneapolis Press.

Richman, S., 1958. "The transformation of energy by *Daphnia pulex.*" *Ecol. Monogr.,* 28:273.

Scudmore, H. H., 1948. "Factors influencing molting and the sexual cycles in the crayfish." *Biol. Bull.,* 95:229.

Southward, A. J., 1955. "Feeding of barnacles." *Nature,* 175:1124.

Tiegs, O. W., and S. M. Manton, 1938. "The evolution of the Arthropoda." *Biol. Rev.,* 33:255–337.

Waterman, T. H. (ed.), 1960–1961. *The Physiology of Crustacea,* vols. 1 and 2. New York: Academic Press.

Wilder, J., 1940. "The effects of population density upon growth, reproduction and survival of *Hyalella azteca.*" *Physiol. Zool.,* 13:439.

INSECTA

Borror, D. J., and D. M. DeLong, 1971. *An Introduction to the Study of Insects.* New York: Holt, Rinehart and Winston.

Brian, M. V., 1952. "The structure of a dense natural ant population." *J. An. Ecol.,* 21:12.

Brown, F. A., Jr., 1950. *Selected Invertebrate Types.* New York: J. Wiley & Sons.

Chu, H. F., 1949. *How to Know the Immature Insects.* Dubuque, Iowa: William C. Brown.

Corbet, P. S., 1963. *A Biology of Dragonflies.* New York: Quadrangle Books.

Crampton, G. C., 1925. "The external anatomy of the head and abdomen of the roach, *Periplaneta americana.*" *Psychd.,* 32:197.

Davey, K. G., 1965. *Reproduction in the Insects.* San Francisco, Calif.: W. H. Freeman.

Gersh, M., 1961. "Insect metamorphosis and the activation hormone." *Am. Zool.,* 1:53.

Goetsch, W., 1957. *The Ants.* Ann Arbor, Mich.: University of Michigan Press.

Gutpa, P. D., 1947. "On copulation and insemination in the cockroach *Periplaneta americana* (Linn.)." *Proc. Nat. Inst. Sci.,* 13:65.

Helfer, J. R., 1953. *How to Know the Grasshoppers, Cockroaches and Their Allies.* Dubuque, Iowa: William C. Brown.

Jaques, H. E., 1947. *How to Know the Insects.* Dubuque, Iowa: William C. Brown.

——, 1951. *How to Know the Beetles.* Dubuque, Iowa: William C. Brown.

Kennedy, C. H., 1927. "The exoskeleton as a factor in limiting and directing evolution of insects." *J. Morph. Physiol.,* 44:267.

Pierce, G. Q., 1949. *The Songs of Insects.* Cambridge, Mass.: Harvard University Press.

Pringle, J. W. S., 1957. *Insect Flight.* London: Cambridge University Press.

Richards, O. W., 1953. *The Social Insects.* London: MacDonald.

Snodgrass, R. E., 1928. "Morphology and evolution of the insect head and its appendages." *Smithson. Misc. Coll.,* 81:1.

——, 1933. "Morphology of the insect abdomen. Part II, The genital ducts and the ovipositor." *Smithson. Misc. Coll.,* 96:1.

——, 1944. "The feeding apparatus of biting and sucking insects affecting man and animals." *Smithson. Misc. Coll.*, *104*:1.

Van der Kloot, W. G., 1961. "Insect metamorphosis and its endocrine control." *Am. Zool.*, *1*:3.

Von Frisch, K., 1950. *Bees, Their Vision, Chemical Senses, and Language*. Ithaca, N.Y.: Cornell University Press.

Wigglesworth, V. B., 1953. *The Principles of Insect Physiology*. London: Methuen.

CHELICERATA

Baerg, W. J., 1958. *The Tarantula*. Lawrence, Kans.: University of Kansas Press.

Baker, E. W., and G. W. Wharton, 1952. *An Introduction to Acarology*. New York: Macmillan.

Brown, R. A., Jr., 1950. *Selected Invertebrate Types*. New York: J. Wiley & Sons.

Bullock, T. H., and G. A. Horridge, 1965. *Structure and Function in the Nervous System of Invertebrates*. San Francisco, Calif.: W. H. Freeman.

Carthy, J. D., 1965. *The Behavior of Arthropods*. San Francisco, Calif.: W. H. Freeman.

Cloudsley-Thompson, J. L., 1958. *Spiders, Scorpions, Centipedes, and Mites*. New York: Pergamon Press.

Comstock, J. H., 1940. *The Spider Book*, 2nd ed. Ithaca, N.Y.: Comstock Publishing.

Gertsch, W. J., 1949. *American Spiders*. Princeton, N.J.: Van Nostrand.

Kaston, B. J., and E. Kaston, 1953. *How to Know the Spiders*. Dubuque, Iowa: William C. Brown.

Savory, T., 1964. *Arachnida*. New York: Academic Press.

Snodgrass, R. E., 1948. "The feeding organs of Arachnid, including mites and ticks." *Smithson. Misc. Coll.*, *110*:1.

——, 1952. *A Textbook of Arthropod Anatomy*. Ithaca, N.Y.: Cornell University Press.

Vachon, M., 1953. "The biology of scorpions." *Endeavour*, *12*:80.

CHAPTER 10

PHYLUM MOLLUSCA

1. Introduction

How can such a varied assemblage including chitons (Figs. 10.1–10.4), clams (Figs. 10.5–10.9), snails (Figs. 10.10–10.14), and octopuses (Figs. 10.15–10.21) be placed in the same phylum? Like frogs and man, which share a single phylum, these animals all have the same essential body organization. The molluscan basic body type is typically *unsegmented and soft* with an *epithelial mantle* that secretes a *calcareous shell*, an *anterior head* and *ventral foot* (the head-foot), *dorsal visceral body mass*, and a *mantle cavity* containing the *ctenidia* (gills). Special aspects of the nervous system, mouth, body cavity, and gut further characterize the phylum **MOLLUSCA.**

Of particular biological interest is the manner in which molluscan body organization has undergone variation in the six classes within this phylum, especially with regard to feeding and locomotion. For example, in the octopus the body is organized with structural and functional adaptations for speed and predation, in the clam for filtering fine food material that rains upon it, in the snail for gliding and protection, and in the chiton for algal grazing and adherence to wave-beaten rocks.

Each evolutionary pattern is sufficiently distinct and important to characterize an entire class of mollusca.

2. Classification

I. Class AMPHINEURA, Chitons

Chitons (Figs. 10.1–10.4) are mollusks with eight calcareous linked plates surrounded by a variable fleshy girdle (*mantle*). Their elongate, flattened body bears a well-developed foot for clinging to rock surfaces. Chitons have a small head without tentacles, numerous gills on either side of the foot between foot and mantle, a single gonad, and separate sexes. They are algal feeders and occur chiefly in marine intertidal areas.

II. Class PELECYPODA (LAMELLIBRANCHIATA or BIVALVIA)

The PELECYPODA are bivalves—clams, scallops, oysters, mussels, and so on (Figs. 10.5–10.9)—which, as the name implies, have two symmetrical valves forming a shell. The dorsally hinged valves are tightly closed by well-developed adductor muscles. The key to understanding this specialized but large mollusk group is its manner of feeding. Bivalves are essentially water pumping and filtering organisms enclosed in a protective shell. Aided by one pair of enlarged, curtainlike gills covered with cilia and mucus-secreting cells, oral palps, and excurrent and incurrent siphons, bivalves feed by filtering food from an internally controlled water flow. Other characteristics of the group are secondarily related to the mode of feeding: loss of both head and rasping radula, loss of motility (except in scallops, which swim by clapping their shells together). Four orders of pelecypods are distinguished by structural details of the gill system.

III. Class GASTROPODA

This class includes snails, welks, limpets, slugs, and nudibranchs (Figs. 10.10–10.14), characterized by a single, well-developed spiral shell (in snails, not in slugs or nudibranchs) and an embryological phenomenon called *torsion*. During development, the body of the embryo twists 180° with reference to the head and foot. Differential rates of growth cause the visceral mass to turn counterclockwise so that the anus is moved to an anterior position and the gills to a posterior one within the mantle cavity. This growth pattern in effect allows withdrawal of the head into the shell mantle cavity *prior* to withdrawal of the foot. The latter structure is often equipped with a tough doorlike *operculum* that blocks the shell opening. Thus the delicate head with its exposed sense organs is well protected by the shell and by the covering foot.

■ **Had torsion not occurred, where would foot, gills, and head be located with respect to the shell opening, upon withdrawal of the snail into its shell?**

The head has tentacles and a well-developed rasping radula. Marine, freshwater, and terrestrial gastropods[1] are mostly slow-moving plant feeders and scavengers, although some are active swimmers and others are predators or even internal parasites.

IV. Class CEPHALOPODA, Squids and Octopuses

Cephalopods (Figs. 10.15–10.21) are highly modified for motility and active predation. They possess a large head and eyes, a highly developed brain and central nervous system, eight or ten arms (or more in some cephalopods), with rows of sucking discs encircling the head, a mouth with a horny beak and radula, and a large siphon for controlled, rapid movement. The group includes the nautili (order TETRABRANCHIA), and cuttlefishes, squids, paper nautilus, and octopus (order DIBRANCHIA).

V. Class SCAPHOPODA, Tooth Shells or Elephant-Tusk Shells (Example: *Dentalium*)

These animals live in a slender, tusk-shaped tubular shell. Adapted to life in mud or sand, often at great depth, they are seldom seen alive. The mantle is fused along its midventral line, and the shell is open at both ends. Small ciliated tentacles and a knob-shaped foot that protrudes from the larger opening carry food particles into the mouth. The radula and other alimentary organs are present, but gills have been secondarily lost.

VI. Class MONOPLACOPHORA, primitive, Segmented Mollusks (Example: *Neopilina galathae*)

These animals (Fig. 10.22) were discovered off the Central American coast by Danish Professor H. Lemche in 1952. They are *segmented molluscs*, suggesting an evolutionary relationship between the mollusks and the segmented annelids, but now considered an interesting but aberrant form. This collection of living specimens was one of the most unexpected biological finds in recent times. Previously, internal structure of these animals has only been surmised from impressions of muscle scars and other hard parts on the 300-million-year-old

---

[1] Gastropods are divided into three subclasses, each defined by adaptive changes in general structure. In the subclass PROSOBRANCHIA—primitive snails with characteristic torsion, well-developed shell, and unspecialized gill(s)—are limpets, slipper shells, abalones, periwinkles, cowries, rock shells, and oyster drills. Gastropods of the subclass OPISTHOBRANCHIA have undergone reverse rotation or *detorsion* in development. They have a reduced shall (or none at all) and a modified single gill and nephridium. Bizarre and often brilliantly colored nudibranchs are in this group along with sea hares and other tectibranchs. Often called sea slugs, opisthobranchs play an ecological role in the ocean similar to that of familiar slugs on land (the latter are in the subclass PULMONATA, considered next). Members of this third subclass, PULMONATA, characteristic terrestrial and fresh-water snails and slugs, have a lunglike mantle cavity modified as a respiratory sac, instead of gills; hence, the name PULMONATA. This important evolutionary modification enables the pulmonates to become air breathers and to occupy terrestrial habitats. *Helix*, the European garden snail, is an example of a typical pulmonate gastropod. (This snail is, as well, the epicure's delight).

fossils. It is a mollusk with *segmentation*—five pairs of retractor muscles, five pairs of gills in the mantle cavity, six pairs of nephridia, one or two pairs of gonads. The heart has two pairs of auricles, and a single ventricle. The digestive tract is very much like that of certain primitive gastropods. A crystalline style is also present, as is the case for primitive gastropods and pelecypods. *Neopilina* externally resembles gastropods and chitons (limpet-shaped shell) and possesses such typical molluscan characteristics as mantle, shell, anterior mouth, and radula. As more *Neopilina* are collected and studied, it may be necessary to modify even further our thinking in regard to their relationships and proper taxonomic position.

We shall review several evolutionary patterns by studying representatives of four major molluscan adaptive types—chiton, clam, snail, and squid. As is often necessary, these organisms are samples selected from their respective classes for availability and convenience. Each should be considered as one of many, and not as definitive prototypes.

3. Laboratory Instructions

I. Class AMPHINEURA, chitons (*Ischnochiton, Chiton, Stenoplax, Cryptochiton, Chaetopleura, Tonicella*)

Most of these intertidal rocky-coastline animals show essentially the same anatomical structure. Different genera vary chiefly in size, shape,

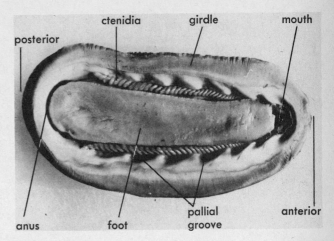

Figure 10.2. *Stenoplax*, external anatomy, ventral view.

form of the girdle, and other external characteristics. Chitons are among the simplest mollusks, rather like a presumed ancestral type, although clearly not as primitive as *Neopilina*.

1. **External anatomy** (Fig. 10.1) Carefully observe your specimen. Notice the eight overlapping dorsal plates or *valves* surrounded by the fleshy *girdle* of the *mantle* (Figs. 10.1 and 10.4). These plates do not indicate segmentation, they are secondary adaptations permitting the animal to bend and cling to rocky surfaces with its large ventral foot. The chiton can also coil up in a defensive posture like the familiar garden sow bug or pill bug. Note linkage and mark-

Figure 10.1. *Stenoplax*, external anatomy, dorsal view.

Figure 10.3. *Stenoplax*, internal anatomy, ventral view.

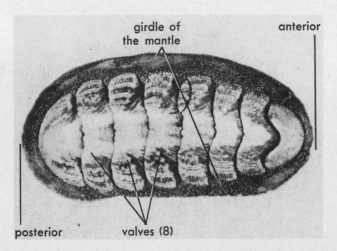

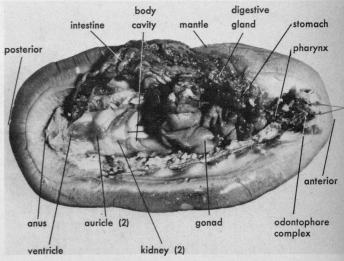

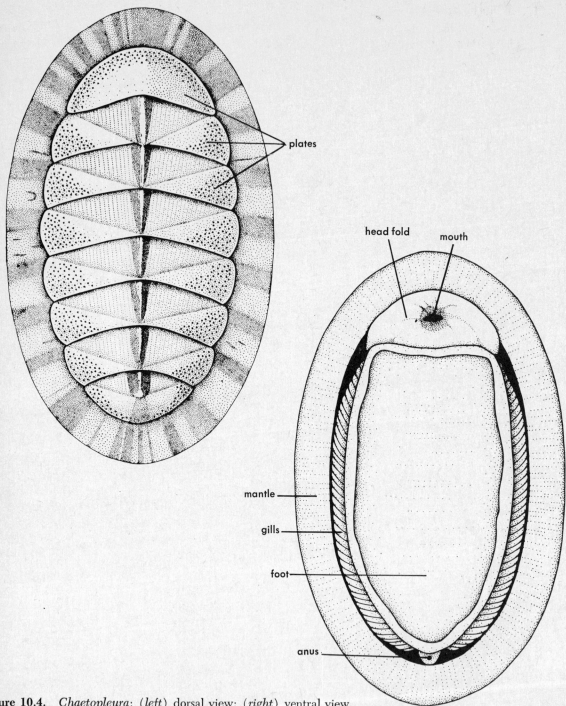

Figure 10.4. *Chaetopleura*: (*left*) dorsal view; (*right*) ventral view.

ings of the plates, and extent, thickness, and type of girdle. *Draw* a dorsal view, showing the relationships of the parts you see.

Study the animal ventrally and observe the large flat muscular foot, usually flanked by protruding portions of many *gills* that lie in a *pallial groove* between foot and mantle (Figs. 10.2 and 10.4). Examine this groove and find the gills if they are withdrawn. The *mouth*, surrounded by a circular mass of tissue, is a small opening in the head. An anus is visible at the posterior end. Try to locate the paired *genital openings*, each with a *nephridiopore* behind it, in the pallial groove on each side about one fifth the body length from the anus.

2. Internal anatomy

a. Remove the foot by inserting the sharp point of scissors or the blade of a scalpel into the foot alongside the pallial groove near the mouth. Extend this cut posteriorly, but not deeply, around the entire foot. However, be sure to leave both the mouth and anus attached to the remaining portion of the animal. Carefully remove the large plate cut from the foot and expose the internal organs (Fig. 10.3). Observe the *gills* in the pallial groove (several to many depending upon the genus). Much of the exposed tissue between loops of coiled, elongated *intestine* is the glandular filamentous *digestive gland* or "liver."

■ **What does the length of the intestinal tract indicate about the food habits of this animal?**

Expose the *stomach* and trace the intestine to the anus. Note the *pharynx*, a portion of the gut connected to the mouth, with an enlarged *buccal capsule*, containing the many-toothed, rasping *radula*. Cautiously cut the tissue around the mouth to expose this capsule and observe the red muscular attachments that control radular movement. A *subradular organ* anterior to the buccal capsule secretes digestive enzymes.

b. Probe carefully around the pharynx and locate the encircling *nerve ring* just anterior to the buccal swelling. This ring consists of *cerebral* and *pleural ganglia*, and their *connectives*. Two pairs of *nerve cords* arise from this ring, one passing to the foot, the other out to the girdle. Fine cross connectives join the *ventral nerves*.

c. Dorsal to the intestine lies the large single gonad, either a *testis* or an *ovary*. Sever the connection between pharynx and buccal capsule and remove the intestine to expose the gonad. Try to locate two ducts that carry gametes laterally and ventrally out the two *gonopores* in the pallial grooves near the anus (already observed in the intact animal). If possible, trace these gonadal ducts.

d. The *aorta*, dorsal to the gonad and just beneath the mantle, carries blood pumped from the three-chambered *heart* (two *auricles* and a posterior *ventricle*). Locate the heart in a special *pericardium* between the gonad and anus. Paired vascularized *nephridia* that drain this cavity carry waste liquids from the pericardium to the nephridiopores.

■ **Where are the nephridiopores with respect to the gonopores?**
■ **What is the function of the blood vessels associated with the nephridial ducts?**

e. Cut out the pharynx, split the tube, and remove the *radula*. Lay the organ out flat on a slide in a drop or two of water, cover with a coverslip, press flat, and examine the structure microscopically. Observe the neatly arranged rows of fine teeth that characterize the radula.

■ **List important chiton characteristics that help them adapt to their environment.**

II. Class PELECYPODA, Clams and Relatives

Freshwater forms characteristic of this class are *Anodonta, Margaritana, Lampsilis*, and *Unio*; marine forms are *Venus, Pecten, Ostrea, Crassostrea, Mytilus*, and *Mya*.

1. **General aspects** In a casual appraisal of **MOLLUSCA** we might think of bivalves as typical of the phylum. Actually, from tiny fingernail clams to the 500-pound *Tridacna* of the Pacific reefs, they are among the most specialized of animals, and far from typical mollusks. The name PELECYPODA, meaning "hatchet foot," reflects the laterally compressed body. As already mentioned, this class is modified for a passive filter-feeding existence, having lost the head and having developed a complex sheet of gill-derived tissues for screening out microorganisms from an induced water current.

Many fresh-water pelecypods possess a distinct specialization: a parasitic larva, called a *glochidium*, which hooks or clamps onto the flesh or gill filaments of fish. The host provides a safe, nourishing location for larval development and serves as an agency for their dispersal into an otherwise harsh fresh-water environment with extreme temperature and water-level fluctuations. The young clams later escape from the cysts in which they are embedded, and, if they fall into appropriate mud or sand bottoms, develop to familiar adult clams.[2]

2. **External anatomy**   Examine the empty shell of a clam. The paired valves are joined by a tough *hinge ligament*, which holds the toothlike projection of the shell together at a swollen *umbo* junction. Orient a pair of valves so that the *umbo* is up and the open edge is below. The shorter, rounded end is *anterior*; the opposite end, through which *siphons* protrude, is *posterior*; the umbo end is *dorsal*; the open or gaping edge is *ventral*. Usually the clam digs into mud by shoveling with its foot. The anterior end remains in the substrate with the posterior siphon-bearing end pointed upward. Note the rough shell with concentric *lines of growth* and characteristic grooves and markings. Break off a piece of shell and observe the layers under a hand lens: first is the thin, horny outer layer (*periostracum*), which protects the rest of the shell from dissolved carbon dioxide in water.

■   **What reaction occurs between carbonic acid and the shell?**
■   **How would strong alkali affect the shell?**
■   **Test your answers experimentally and record the results.**

The next layer is the *prismatic* layer, the crystalline calcium carbonate potion of the shell. Below it is the irridescent mother-of-pearl inner *nacre*, or nacreous layer.

■   **Which is the oldest part of the intact shell?**
■   **How old was the specimen you are examining?**
■   **Why are shells from limestone-rich waters thicker than those in other areas?**
■   **Study the microscopic structure of these**

---

[2] Generally, the freshwater clams *Anodonta, Unio,* or *Margaritana* are used in laboratories near major rivers, but marine clams such as the quahog, *Mercenaria* (*Venus*); the soft-shelled clam, *Mya*; the mussel, *Mytilus*; or the edible oysters, *Ostrea* or *Crassostrea* are equally useful. Each has similar structures but is modified in body shape, shell form, and organ arrangement. Illustrations and descriptions in the ensuing discussion are based on *Margaritana* and *Mercenaria* (or *Venus*).

layers in a prepared slide and draw a small portion of such a cross section.

Observe the inner surface of a valve. Notice the smooth rounded areas at the anterior and posterior ends. These are *muscle scars*, points of attachment of valve-closing and foot-moving muscles. They include the large scar where the *anterior valve adductor* was attached and the small scar marks where the *anterior foot protractor* (behind the anterior adductor) originated. Muscles attached to these scars close the valves and withdraw and extend the foot, respectively. The posterior end is marked by a large *posterior adductor muscle* scar (another strong muscle that helps close the valves) and by the small *posterior retractor* of the foot.

■   **Why are the adductors so much larger than the other muscles?**
■   **How is the shell *opened* in the absence of specific muscles for this purpose?**

Next, locate the *pallial line* that marks the outer margin of mantle attachment. It is a long ridge following the curve of the shell, extending between the two adductor scars. It also marks the margin of the retractor muscle of the mantle.

3.   **Internal anatomy**   (Figs. 10.5–10.9)

a.   *General aspects.* Formalin-preserved specimens will be available for dissection. Orient the animal, identifying right, left, dorsal, ventral, anterior, and posterior surfaces. Separate the valves slightly, insert a scalpel, and cut across the adductor muscle as close as possible to the muscle scars on the *right* valve to prevent injury to other organs. Then twist off the right valve and cut the dorsal hinge ligament attaching it to the other member.

■   **What other structure helps attach valves at this point?**
■   **Identify each visible muscle on the left valve.**

You can now study the exposed organs of the *right* side of the animal as it lies in the *left* valve. Find the cut muscles and see how they operate to close the valves and to control move-

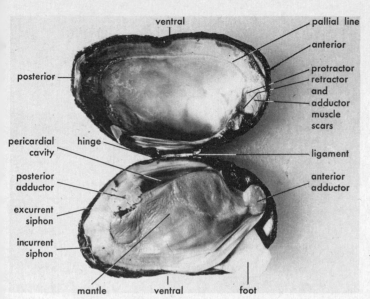

Figure 10.5. *Margaritana*, internal anatomy.

ment of the foot. The organs form a central *visceral mass* and are covered with a sheet of bounding mantle epithelium (Fig. 10.5). The mantle both encloses the body cavity and secretes the entire shell. Organs visible through the mantle include a fleshy *foot* protruding through the anteroventral margin; the greenish brown *digestive gland* in the dorso-anterior region; the *pericardial cavity* (reduced coelom) posterior to the digestive gland; and the *nephridium*, a dark triangular excretory organ ventrolateral to the digestive gland. The *pericardial gland*, sometimes called *Keber's organ*, is anterolateral to the digestive gland. This brief examination provides you with a general orientation to pelecypod anatomy.

Float the specimen in water to facilitate viewing detailed structures in a more natural position, each in proper relation to surrounding organs. Locate the siphons. The larger *incurrent siphon* has *sensory papillae* inside the margin; the *excurrent siphon* lies just above it.

■ **What is the function of these sensory papillae?**

Both siphons are actually tubes formed by fusion of posterior ends of the mantle. Probe carefully into the incurrent siphon.

■ **How does this siphon connect to the mantle cavity?**

Remove the exposed half of the mantle, carefully cutting it out with scissors. Below it lies the *visceral mass*, *foot*, and sheetlike *gills*, two on each side of the *visceral mass*. Trace the *alimentary canal* which coils back on itself in the visceral mass and then passes posteriorly through the *pericardial cavity* to its terminus at the anus, above and slightly behind the posterior adductor muscle. The inconspicuous mouth between the anterior adductor muscle and the foot is bordered by a pair of small triangular *oral palps*. Food particles are enclosed in small mucus balls and passed "hand-over-hand" by ciliary action to the oral palps where edible portions are selected by size.

b. *Filter feeding*, *water conduction*, and *respiratory system* (Figs. 10.6 and 10.7). The water current entering the claim via the incurrent siphon carries (1) food particles into the mantle cavity, and (2) dissolved oxygen to the mantle and into gill water tubes where respiratory exchange occurs. Dissolved carbon dioxide and wastes are carried out via the excurrent siphon. Microorganisms or bits of organic matter are trapped from the passing water by cilia and mucus on the gill surfaces and on the foot and mantle. Acceptable food is transferred into the mouth and rejected items are passed by other cilia down the foot and through the gap between valves to the outside. Water from which food has been selected flows into the

Figure 10.6. *Margaritana*, internal anatomy: mantle tissue removed to show oral palps and ctenidia.

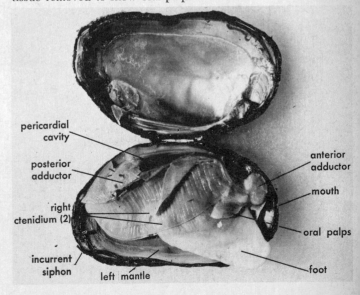

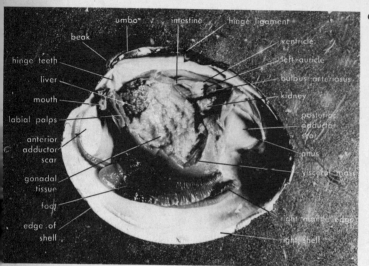

Figure 10.7. *Mercinaria*, internal anatomy: digestive, circulatory, and reproductive systems.

mantle cavity, through numerous pores (*ostia*) into the gills, and into many vertical cilia-lined *water tubes* bordered by vertical *gill bars*. Fine capillaries that parallel the water tubes permit rapid respiratory exchange by diffusion through the thin intervening membranes. The water stream, with dissolved carbon dioxide, passes dorsally up the water vessels and posteriorly through water-collecting *suprabranchial chambers* (one above each pair of gill tissue sheets, four in all). Each collecting chamber then drains the thoroughly used water out through the excurrent siphon.

Cut a piece of gill tissue, float it in water, and examine it under a hand lens or dissecting microscope. Find the ostia and vertical gill bars on each side of the rows of ostia. Each gill is composed of two sheets. Each sheet in turn is formed by two *lamellae*. Water tubes are formed when the two lamellae of each sheet partially fuse. Therefore a single, W-shaped gill consists of two gill sheets (four lamellae) on each side of the visceral mass. Pelecypod gills have evidently undergone considerable modification and enlargement in conjunction with the filter-feeding mode of nutrition.

■ What is meant by "ciliary-mucus-filter-feeding"?

■ Can the structure, distribution, and biological success of these headless mollusks be attributed to their specialized mode of feeding?

c. *Excretory and reproductive systems.* Carefully remove the exposed mantle and gill tissue, leaving the inner right suprabranchial chamber intact.

Fluid wastes from the excretory organs ("kidneys") also flow into the suprabranchial chambers via *nephridial ducts*. These nitrogenous and other wastes enter the ducts after exchange from nephridial capillaries. The freshly cleansed blood flows from excretory capillaries directly into the gill circulation where respiration occurs. It then passes into veins that conduct it back to the heart.

Trace the excretory system from the *pericardial pores* in the pericardial wall to *nephridial ducts* and *nephridia* (dark organs ventral and lateral to the pericardial cavity), and then to the *excretory ducts* leading from the nephridia to the inner (medial) suprabranchial chambers. You will have to cut the intestine and move it away from the pericardial cavity in order to search the walls for pericardial pores and nephridial ducts leading to the nephridia. Cut open a nephridium and try to trace out the loop of the excretory duct. You may even find the *excretory pores* draining the nephridia into the anterior end of the inner suprabranchial chambers. A fine wire probe is useful for exploring these ducts.

Nephridial ducts form a flattened U-loop within the nephridia, the upper arm being quite thin-walled, the lower arm spongy and vascular, surrounded by numerous fine blood vessels. Diffusion of metabolic wastes and resorption of usable salts and other materials occur across the thin capillary and excretory walls. *Notice that the nephridia drain the coelom* (or what is left of it in these animals—the *pericardial cavity*).

■ How does this compare with the activity of nephridia in annelids?

■ Could this functional similarity constitute evidence that the pericardial cavity is a reduced coelom?

■ How is the circulatory system employed in disposal of excretory fluids in the nephridia?

Two openings are found in each inner suprabranchial chamber. The upper one is the excretory pore, the lower one the *genital pore*.

Genital ducts lead from the highly branched *gonad*, enmeshed in the intestinal loop in the basal portion of the foot, to the suprabranchial chambers via the genital pores.

The inner suprabranchial chambers also form *brood pouches* for glochidial larvae in the case of freshwater clams. Study microscopically several preserved glochidia or some taken from the brood pouch of a ripe living female.

■ **Why are the adductor muscles of these larval clams so well developed?**

■ **Review the life cycle of a freshwater clam.**

■ **What is the probable adaptive advantage of parasitism at the larval stage in the life cycle of freshwater clams?**

d. *Digestive tract* (Figs. 10.7 and 10.8). Trace the alimentary canal from mouth and oral palps, along intestinal windings through the base of the foot, through the pericardial cavity, and out the anus at the margin of the posterior adductor muscle.

Remove the body from its remaining valve. Cut the mantle and gill tissue from both sides; note the previously identified foot retractor muscles and their function. Carefully dissect away muscle and tissue of the visceral mass on the right side of the medial intestine. The pasty material surrounding the intestine is largely gonadal (Fig. 10.7). The *digestive gland* (liver)

lies anterior to the intestinal loops, surrounding the stomach or swollen portion of the gut (Figs. 10.8 and 10.9).

Pin out the specimen in your dissecting pan. Keep it clean with water squirted from a medicine dropper. Relate the following previously examined parts both structurally and functionally, including ducts or tubes by which they are connected: *mouth, oral palps, esophagus* (short tube posterior to the mouth), *pericardial cavity* (with *rectum* passing through it), *anus excurrent siphon,* and *posterior adductor muscle.* Inside the coiled intestine you may find the gelatinous, rodlike *crystalline style* in a special intestinal outpocket. This rod apparently releases digestive enzymes supplied by a gradual wearing away of the style at the end entering the gut near the stomach.

Draw, in diagrammatic outline, the relationship of these parts.

■ **How can you account for the lack of a radula and well-developed head in this class of mollusks, in comparison to gastropods and cephalopods?**

e. *Circulatory system.* Examine your specimen from the dorsal aspect. Find the pericardial cavity and open it anterior to the posterior adductor muscle. Identify parts of the heart (previously seen in your search for excretory pores). Locate the muscular *ventricle* enclosing the intestine (rectum), two delicate anterior *auricles* that connect the ventricle to the lateral margins of the pericardial cavity, and the *bulbus arteriosus* anterior to the ventricle. These organs deliver blood from the mantle and gills to the ventricle. The blood is then pumped anteriorly via the bulbus arteriosus and *anterior aorta* to the alimentary canal and mantle, where it empties into the large blood *sinuses* supplying the various organs. It returns via a vein and capillaries to the nephridia and then by capillaries into the gills before it returns to the heart.

■ **What is meant by "open system" of circulation?**

■ **How can the blood flow be controlled?**

■ **What are some of its advantages? disadvantages?**

Figure 10.8. *Mercinaria,* internal anatomy: digestive, circulatory, and reproductive systems.

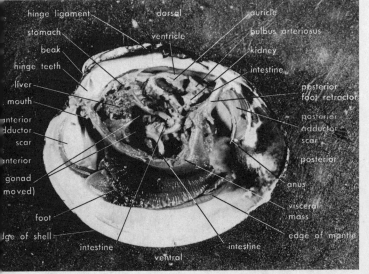

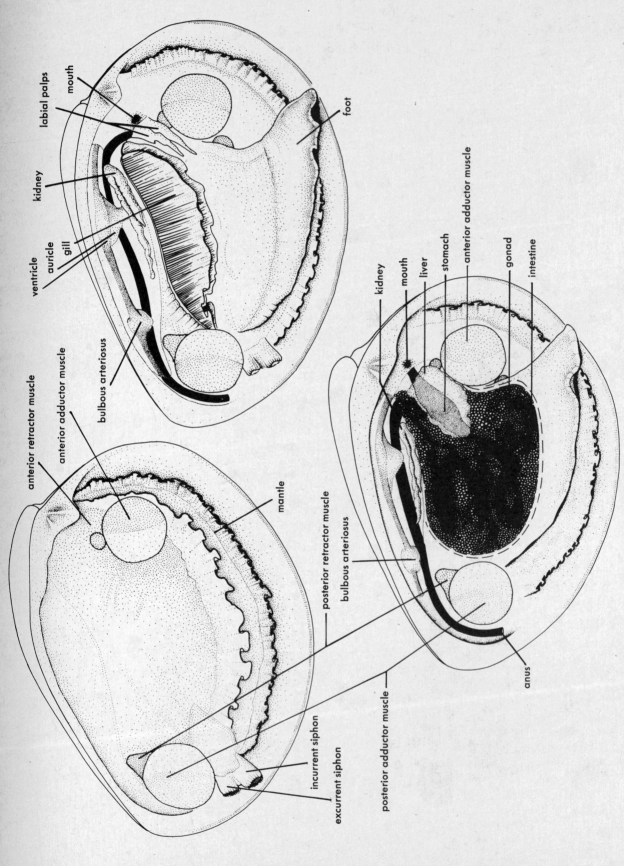

mouth

labial palps

kidney

ventricle

auricle

gill

anterior retractor muscle

anterior adductor muscle

bulbous arteriosus

mantle

posterior retractor muscle

bulbous arteriosus

incurrent siphon

excurrent siphon

posterior adductor muscle

foot

anus

kidney

mouth

liver

stomach

anterior adductor muscle

gonad

intestine

Figure 10.9. *Mercinaria,* semidiagrammatic illustrations: *(upper left)* right valve removed; *(upper right)* right mantle removed; *(bottom)* right ctenidia and visceral body wall removed.

SUGGESTION FOR A SPECIAL EXPERIMENT

Set up a kymograph[3] to show the normal pumping action of a living clam heart, and then the effects of minute amounts of acetylcholine, epinephrin, and other chemical agents known to affect heart rate.

f. *Nervous system.* In the clam, the principal nerves are connected to three ganglia of the *central nervous system.* These small yellowish ganglia with their white *connectives* are all located near major muscle areas. Included are: (1) paired *cerebropleural ganglia* at the posterior margin of the anterior adductor mucle, below the base of the palps; (2) *pedal ganglia* between the foot and visceral mass; and (3) a single *visceral ganglion*, on the ventral surface of the posterior adductor muscle.[4]

■ **Is the nervous system as highly developed in pelecypods as in the other classes of mollusks?**

■ **Does your answer fit the way of life characteristic of these groups?**

g. *Recapitulation.* Review the primary organs examined with respect to their function in the digestive, excretory, circulatory, nervous, muscular, respiratory, water-conducting, and reproductive systems. Try to trace water current, blood flow, food passage, nerve control of major muscle groups, digging activity, shell closure, and sperm or egg passage in the clam. Use your own specimen as much as possible for this review.

III. Class GASTROPODA, *Helix*

1. External anatomy

a. *Observations. Helix*, the common European garden snail and the escargot of gourmets, is one of a vast array of mollusks included in the class GASTROPODA. Many common freshwater snails, such as *Physa, Planorbis, Helisoma,* or *Lymnaea,* make excellent study specimens.

Observe young snails (often quite transpar-

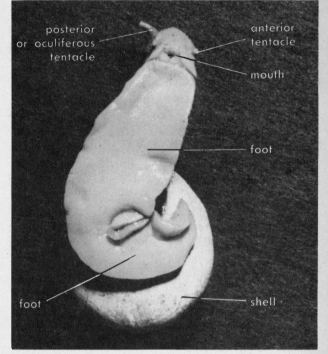

Figure 10.10.   *Helix*, ventral view.

ent) in a watch glass (Fig. 10.10). Study movement and action of head and tentacles; reactions to a bit of meat, to strong light, to moderate probing with a needle, to a few crystals of menthol dropped in the water, and to 1 percent sodium chloride, mild acid, or alkali. Watch for the occasional movement of the *pneumatopore*, the respiratory opening.

■ **How many eyes does the snail have?**
■ **Where are the eyes located?**
■ **How does the mouth function?**

To study the normal feeding pattern, observe aquatic snails feeding on algae-coated glass surfaces of an aquarium. Watch for a wedge-shaped object protruding briefly from the mouth opening. This is the top of the buccal mass in which the radula is embedded. The rasping, tooth-covered surface of this organ scrapes off bits of algae or other plant materials, which are then engulfed. Garden snails and slugs feed on vegetation; many others feed on decayed matter. Most marine gastropods are scavengers. Others, such as the oyster drill, adapted for predation, use the radula to bore

---

[3] Appendix IV contains information on the construction and use of a kymograph.

[4] By careful dissection, the following nerves around these ganglia can also be exposed: *cerebropedal connectives* between ganglia (1) and (2), and *cerebrovisceral connectives* between (1) and (3).

through shells of other mollusks. Still others are highly modified, internal parasites.

It is interesting and useful to observe aquatic snails in an aquarium. You can see feeding and locomotory habits, chemical stimulation from various foods, mucus threads the snail uses for movement in water or to help it support itself at the water surface, responses to light, breathing rates at various temperatures, and the effect of desiccation. Life cycle, patterns of reproduction, and the stages of development can also be studied in aquarium specimens. Your own observations may suggest experiments under better control by which the snail's responses to various stimuli can be more critically tested.

A convenient way to observe the muscular action of the foot and the appearance of the snail's ventral surface is to place the animal in a drop or two of water on a slide, invert the slide over a small dish, and watch the foot under a dissecting microscope (Fig. 10.10).

Young *Physa* or other small freshwater snails are excellent subjects for observation of blood circulation and heartbeat, as well as movement of the intestine and its contents. Your observations can be made with a dissecting microscope, using the snail in shallow water in a watch glass or small dish. The shell can then be cracked and carefully removed and the intact animal studied with startling clarity.

b. *Shell.* Examine an empty shell for *whorls, growth lines* (why are they irregular?), *spire, aperture*, and thickened edge or *collar*. In marine snails, the collar is often drawn out into a lip to enclose the respiratory siphonal canal.

■ **Which is the oldest portion of the shell?**
■ **How does the shell grow?**

Study a bisected or broken shell and note the axis (*columella*) about which the shell twists.

■ **How can you tell a right from a left twist in the shell (right or left with respect to the opening)?**
■ **Do snails have the same three-layered shell seen in the clam?**
■ **Why are nudibranchs and tectibranchs considered gastropods, even though the shell and the body torsion may be lost?**
■ **Cite a few examples to show how vari-** ous shell types adapt the snail to its particular habitat.

2.   **Internal anatomy**

a.   *Dissection*—a painstaking task, but well worth the effort. Study an extended, preserved *Helix* and identify external characters already discussed: shell, whorls, spire, collar, aperture, anterior tentacles, posterior tentacles with terminal eye, mouth with one ventral and two lateral lips, *genital pore* (right side of head), *pedal gland opening, foot, mantle, pulmonary pore*, and *anus* (located on right side of mantle under shell collar).

Remove the shell (Fig. 10.11) by cutting around the spiral to the terminal coil or by carefully cracking the shell and removing the chips. Be careful not to damage the tissues, which may have been made hard and brittle by the fixative. The central *columella* may have to be freed from the body by cutting the *columella muscle.*

Observe the *visceral hump.*

■ **How many coils does it form?**

Notice the thickened *collar* (for shell secretion) at the end of the mantle and the thin, vascular *respiratory mantle cover* elsewhere. Locate the *nephridium* (kidney) (in the second half-turn of the visceral mass), anterior to which is the *heart.* The dark, lobed *digestive gland* (liver) dominates the rest of the visceral hump. Final whorls show the whitish *albumen gland* and *hermaphroditic gonad*, the *ovotestis.*

Next, lay back the intact mantle (both roof and floor) by the following procedure. Cut

Figure 10.11.   *Helix*, lateral view with shell removed.

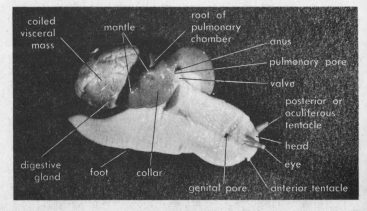

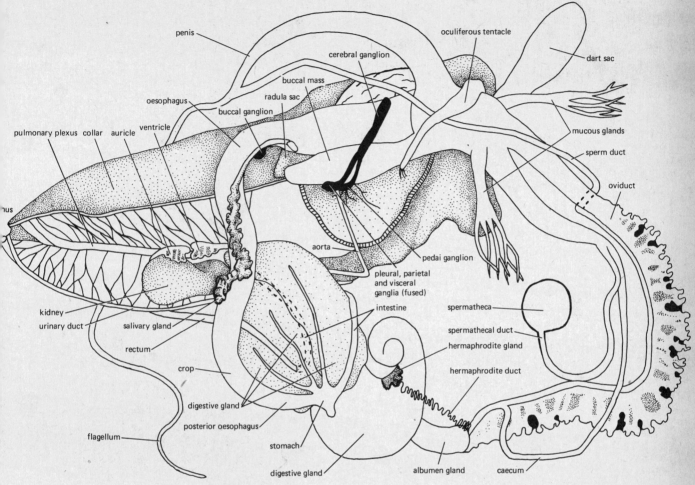

Figure 10.12. Semidiagrammatic representation of the internal anatomy of *Helix*. Also refer to Figures 10.13 and 10.14, which show anatomical structures as revealed in dissection.

from the genital pore back to the base of the collar; make a parallel cut on the other side of the head; then connect these two cuts behind the tentacles and lay back this tissue flap. Then continue the initial cut (on the right side of the snail) around the collar and follow the whorls of the body. This will expose the internal organs (Fig. 10.12) and permit you to fold back the entire roof and floor of mantle tissue. The rectum, vascularized lung system, and heart are contained in this tissue flap, which should be laid back and pinned out (Fig. 10.13).

This exposure of internal organs essentially completes your basic dissection. All that remains is to spread the organs so that the *digestive, reproductive*, and *nervous* complex can

be identified. Be sure that the specimen is covered with water so that the organs float freely. Place the reproductive organs to one side, preferably the right. Separate them carefully and place pins against (not through) them at appropriate points (Fig. 10.14). Cut through connective tissue that may prevent laying out the parts, such as tissue connecting the spermatheca and the albumen gland to the digestive organ. Then move the crop somewhat to the left to allow the sexual organs the fullest degree of display possible. Have your laboratory instructor check your dissection before you proceed to dismember your beast further.

b. *Excretory system.* Each organ system should be traced out as completely as possible. Locate the following in the excretory system:

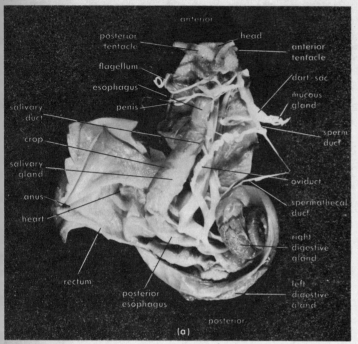

Figure 10.13.   *Helix*, internal anatomy.

i. *Nephridium* (kidney): large structure in roof of mantle cavity, in the second half of the first coil of visceral mass.

ii. *Renal duct*: passing parallel to the rectum, out to the anterior end of the mantle cavity, and to an *excretory pore* at the edge of the *respiratory pore* (pulmonary opening). These relationships have been disturbed by dissection.

c. *Digestive system.* Locate the following:

i. *Perivisceral cavity*: the blood space or *hemocoel*, large blood sinuses in which internal organs are bathed.

ii. *Buccal mass* (with nerve ring): esophagus (passing through the nerve ring), *radular retractor muscle, crop* (thin-walled tube from esophagus to base of visceral hump), and *salivary glands* (alongside crop) with ducts passing through the nerve to the top of the buccal mass.

iii. *Stomach*: wider than crop, forming second coil of visceral hump. Near the stomach–intestine junction are two small ducts leading to the large *digestive glands*. The *right* digestive gland forms the terminal coil of the spire; the left forms the bulk of tissue embedding the intestine.

iv. *Intestine*: a thick, S-shaped tube (eventually returning to the main body whorl as the rectum) which passes through the mantle cavity and exists via the *anus* next to the excretory and respiratory pores at the right collar margin. These pores, you will note, were moved with the mantle during your dissection.

d. *Reproductive system* (Fig. 10.13). Starting with the tip of the coiled visceral hump, find the following:

i. *Ovotestis*: protandrous (the male developing first), it is found on the inner surface of the coiled digestive gland.

■ **What is the biological importance of protandry?**

ii. *Hermaphroditic duct*: tight coils running from gonad to base of the large whitish *albumen gland* where the fertilization chamber is located. Here eggs are fertilized and then coated with *albumen*.

iii. *Common duct*: the prominent duct with separate male and female portions leading from the hermaphroditic duct.

The common duct consists of a thin male half for sperm passage and a larger convoluted female half, where calcareous shells are secreted around eggs moving down the duct. Note the incomplete septum dividing the ducts.

iv. *Oviduct and sperm duct*: produced by separation of the common duct into distinct units. Joining the sperm duct between the origin of this duct and the *penis* is the *flagellum*, a long narrow tube for compressing sperm bundles into *spermatophore* groups.

v. *Penis*: protrusible terminus of the sperm duct, which can be extruded through the *common genital pore* and withdrawn by the *retractor muscle* (previously cut when the mantle floor was removed).

vi. *Oviduct*: a thick-walled, separate tube continuing from the female portion of the common duct to the vagina and joined by the *spermathecal duct,* the blind end of which is a swollen *spermatheca*. In this sac are stored sperm from the male organ of another snail, received during hermaphroditic copulation. The spermatheca is found in the upper coils near the junction of hermaphroditic and common ducts. The long spermathecal duct runs parallel to the common duct and joins the oviduct near the vagina. Another small diverticulum joins the spermathecal duct near the anterior end of the common duct. (It is, in fact, as complicated as it sounds.) Work it out on your specimen and check with the photographs. Appreciate the fact that structural complexity is not an exclusively human or even vertebrate characteristic.

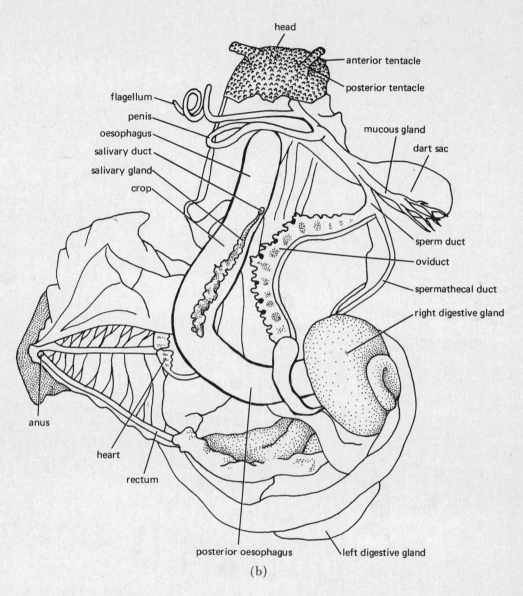

head

anterior tentacle

posterior tentacle

flagellum

penis

oesophagus

salivary duct

salivary gland

crop

mucous gland

dart sac

sperm duct

oviduct

spermathecal duct

right digestive gland

anus

heart

rectum

posterior oesophagus

left digestive gland

(b)

The final portion of the reproductive system includes the *vagina*, joined by 2 branched *mucous glands*, and a *dart sac* (a muscular organ producing a calcareous spine for stabbing into the side of the other snail as a precopulatory stimulus.) The entire complex, joined at the vagina, opens externally via the common genital pore near the base of the right posterior tentacle.

e. *Nervous system* (Fig. 10.14). Cut the esophagus anterior to the *nerve ring*, then cut the adjoining salivary gland ducts and radular retractor. Move the crop aside, pull the anterior stump of the esophagus out from the nerve ring, and pin it aside. Then dissect the *ganglia*

and nerves of the ring free of enclosing connective tissue. Observe the concentration of ganglia in this area.

■ **What is the significance of this concentration of neural elements?**

Find the *cerebral ganglia* above the esophagus, with a small pair of nerves leading to the *buccal ganglia* near the bases of the salivary gland ducts on the buccal capsule. Other nerves from the cerebral ganglia lead to the body wall, mouth, and eyes. The fused posterior ganglia, *cerebropedal*, *cerebropleural*, and *visceral ganglia*, form a ring around the *cephalic aorta*.

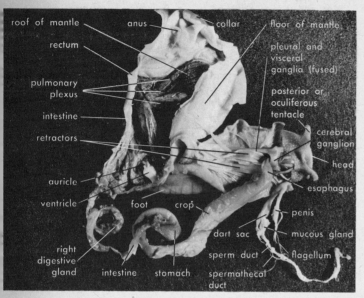

roof of mantle
anus
collar
floor of mantle
rectum
pleural and visceral ganglia (fused)
pulmonary plexus
posterior or oculiferous tentacle
intestine
cerebral ganglion
retractors
head
auricle
esophagus
ventricle
foot
crop
penis
dart sac
mucous gland
right digestive gland
sperm duct
flagellum
intestine
stomach
spermathecal duct

Figure 10.14. *Helix*, internal anatomy.

Return now to the pinned-out mantle tissue. The exposed thin inner lining is the floor of the mantle overlying the foot and viscera. Dorsal to it (actually pinned down below it in your preparation) is the vascularized roof of the mantle cavity embedding the lunglike *pulmonary plexus, pericardium, heart,* and connecting major blood vessels, *rectum,* and *anus.*

f. *Circulatory system.* Locate the large branchial "lung" vein or pulmonary plexus just identified. It returns aerated blood to the heart. Follow the plexus to the pericardial cavity into the thin-walled large *auricle* and small muscular *ventricle.* The latter pumps blood out the *aorta.* This major vessel almost immediately divides into a *visceral branch* (along the ventral surface of the liver, from which it branches to the entire visceral mass) and a *cephalic branch* to the head and foot, encircled by the fused posterior ganglia already observed. In the snail's "open" circulation, blood bathes the organs directly, flowing through large sinuses or blood spaces. Blood returns to the heart through sinuses (forming a hemocoel) rather than by closed veins (*as in what previously studied animal?*). En route back to the heart, blood is collected in a large *afferent branchial vessel* located near the collar, and then passes through the lung capillary bed (*plexus*) *or* to the kidney capillaries. After aeration or excretory exchange, blood returns to the heart from the hemocoel and is again pumped into the general circulation.

A final examination of whatever is left of your snail can include dissection of the *buccal mass* to expose the anterior "jaw" (a rubbing surface for the *radula*); the ventral radula in its *radular sac*; and the *odontophore* or ventral muscle mass, which moves the radula in its beltlike grinding action. Dissect out the radula and examine it microscopically to view the complex array of teeth, arranged in a fashion characteristic for each species.

■ How does the *Helix* radula compare with that of the chiton?

SUGGESTION

Compare radulae of several snail species to see their differences, mount each in balsam on a slide, and identify them by their radular tooth patterns using appropriate taxonomic references on gastropods.

■ Could a taxonomic key be made based on these differences?
■ What general features adapt *Helix* to terrestrial life?

Review essential organs of each system and be able to tell how they are adapted to this animal's mode of life. (All this in an ordinary garden snail!)

IV. Class CEPHALOPODA, *Loligo,* **the Common Squid** (Fig. 10.15)

Figure 10.15. *Loligo opalescens,* adults.

1. **General aspects** Rapid movement, rapid sensory responses—these are characteristic of the evolutionary pattern demonstrated by the squid, octopus, and other cephalopods. The structure of squids and octopuses is adapted for *active predation, rapid swimming, complex behavior*, and a degree of *intelligence* probably unique among invertebrates. Observe how CEPHALOPODA represent a highly specialized but still a recognizable example of the molluscan basic body type. Though squids and octopuses differ widely in their habits and activity patterns, both are characterized by high development of neural control and remarkable eyesight.

■ Where is the shell or the remnant of it? the mantle?

■ What are the other typically molluscan features?

■ What structures are unique to the CEPHALOPODA? Does this imply great age of this group?

■ Name several fossil representatives of the class. How do these fossil forms differ from modern counterparts?

■ What then has been the general evolutionary trend within the cephalopods?

2. **External anatomy** (Figs. 10.16 and 10.17)

a. *Symmetry and orientation.* Formalin-preserved, frozen, or freshly captured specimens may be used. Examine the general body form and gross features, color pattern, mouth, head and eyes, mantle, movable funnel, and mantle cavity (Fig. 10.16).

Figure 10.16. *Loligo opalescens,* dorsal view, male.

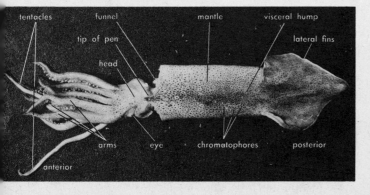

■ **What is the general pattern of symmetry?**

Morphological orientation and *directional* orientation of the squid are distinctly different. Your first requirement, therefore, is to decide which end is "up." The tip of the cone-shaped body is morphologically dorsal, the arms and tentacles are morphologically ventral, and the funnel is the posterior surface. In swimming, however, the animal moves with arms leading and the funnel held below. Therefore, from the standpoint of *directional* as opposed to *morphological* orientation, the morphologically ventral arms become functionally anterior, and the morphologically posterior funnel becomes functionally ventral.

■ **What then would be the *functional* orientation of the morphological *dorsal* surface? Review these changes of orientation. Dissection directions will utilize *functional* orientation.**

b. *Arms, tentacles, mouth, and surrounding membranes.* In the squid, the molluscan *head* is modified into five pairs of *arms* surrounding the mouth. Four pairs, equally long, are nonretractile and have two rows of suckers. One pair (the *tentacles*) is longer, retractile, and has specialized suckers only at the tips. To capture swimming prey, the tentacles dart out rapidly, grab a fish or crustacean with the specialized suckers, then draw it back to the other eight arms. Suckers on the eight arms anchor and move the victim to the horny jaws that bite and poison it. Poison glands open into the buccal cavity; the poison enters with saliva.

Observe how the suckers vary in size. Note the *cup* and attachment *pedicle* or stalk of each.

Remove a sucker and examine it microscopically. Note the chitinous toothed ring supporting the edge of the cup with the basal portion forming a small piston.

■ **How does the sucker function?**

Locate the lower left arm—the *fourth* left—and see if it shows a modification in which size of suckers decreases and length of pedicles increases. This asymmetry marks the *male* and is called *heterocotyly* (different arm). The male fourth left arm is modified for transfer of the sperm-bearing *spermatophore* to the *horseshoe*

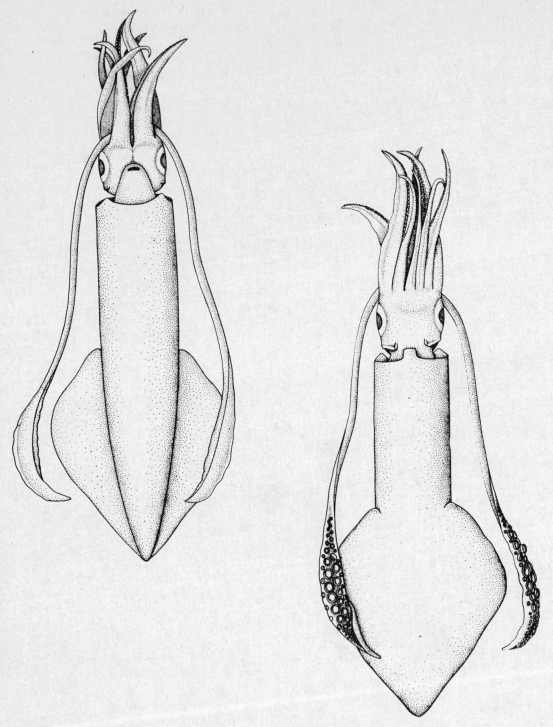

Figure 10.17. *Loligo*: (*left*) ventral view; (*right*) dorsal view.

organ on the female's buccal membrane. In some species this arm breaks off at the time of sperm transfer.

Expose the mouth of your specimen to observe this membrane connecting to the bases of the arms. The outer portion is the *buccal membrane*; the inner, the *peristomial membrane*. The buccal membrane has seven projections, each with suckers, whereas the peristomial membrane lacks suckers.

■ **Are the dark, chitinous, beaklike jaws visible through the mouth opening?**

The ventral jaw usually overlaps the dorsal. If your specimen is a female, look for the horseshoe organ (sperm receptacle) on the buccal membrane below the mouth.

Around the dorsal surface of the head, note the crescentic opening which receives water currents passing into the mantle cavity.

On the dorsal surface of the skin, observe the numerous tiny dark spots, the pigment cells (*chromatophores*). Contraction or expansion of each cell by special enclosing muscle fibers restricts or spreads out the pigment within the cell, accounting for the extremely rapid color changes that sweep over the animal, often seen as rhythmic pulsations.

c. *Eyes.* Remove and examine an eye under the dissecting microscope. Identify *cornea, iris, pupil, lens,* and *aquiferous pore* (small opening in front of each eye) that leads into the eye chamber and presumably functions for pressure equalization. The *olfactory crest,* a fold of tissue behind each eye, partly covered by the mantle, is thought to have an olfactory function (for example, it may serve as a water-tasting organ).

d. *Supporting and locomotory structures.* The rest of the squid's body is covered by the *mantle,* the free anterior edge of which is called the *collar. Pallial cartilages* on either side of the funnel can be seen as lateral projections. A dorsal projection is produced by the anterior end of the *internal shell* or *pen.* The mantle forms a graceful posterior *cone,* with two lateral *fins* for locomotion and guidance. Three sets of muscles (*longitudinal, transverse,* and *vertical*) activate the fins to make slow rhythmic movements or powerful strokes for rapid forward locomotion.

The funnel, projecting from beneath the mantle, is *not* homologous with the siphons of clams. The latter are derived from the fusion of posterior portions of the mantle. The funnel of cephalopods, however, is thought to be derived from the foot of an ancestral mollusk.

■ **How does the function of the cephalopod funnel compare with that of the clam's siphon?**

■ **How does the funnel action control direction and speed of the animal's movement?**

■ **Locate the transverse muscle fibers clearly grouped on the central surface of the fins.**

3. **Internal anatomy** (Figs. 10.18–10.21)

a. *Dissection.* Expose the internal organs of your specimen by making a longitudinal incision on the ventral surface of the visceral hump along one side of the midventral line. Then make a longitudinal cut along the funnel and pin back this thick wall of mantle tissue. Keep the specimen under water to facilitate observation. Observe the set of *valves* and *interlocking cartilages* by which the incurrent water flow is restricted to the dorsal openings and the excurrent flow is confined to the funnel. Cartilaginous grooves fit into cartilaginous ridges dorsally and ventrolaterally where the inner surface of the mantle joins the body anteriorly. Entry of water is restricted to the dorsal openings by two large saclike valves at the base of the siphon (preventing inflow). Incoming currents therefore must first pass over the two large gills (*ctenidia*). Two *anal valves* at the base of the siphon mark the terminus of the *rectum.* Note the functional relationship of these valve openings. Water movement out of the mantle cavity is controlled by *circular* muscles, which squeeze the mantle cavity, draw the collar down tight on the head, and expel water out of the funnel. Contraction of radial muscles expands the cavity, drawing water into the cavity through the openings between collar and head.

Identify the major organs of your specimen (Fig. 10.18). The black or silvery *ink sac duct* lies dorsal to the rectum. The basal swelling of this duct forms the sac. Ejected ink provides an escape screen for the octopus or squid. It probably is less a screen than a false prey, as the ejected ink blob forms a sticky mass about the size of the animal causing a predator to check it momentarily—all the time needed for the octopus to escape.

In the male (Figs. 10.20 and 10.21) the *testis* occupies the posterior end of the mantle cavity dorsal to the *cecum* in the cone-shaped body tip. The *penis* is a muscular duct on the left side of the *rectum.* The paired *nephridia,* elongate or triangular structures between the gill bases, are clearly visible in the male. In the female (Figs. 10.18 and 10.19) these structures are covered ventrally by a pair of large, white, elongate *nidamental glands.* Each has a small

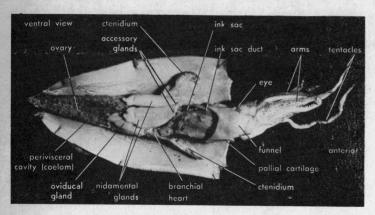

ventral view   ctenidium   ink sac
           accessory
ovary   glands   ink sac duct   arms   tentacles
                                        eye
                                                          anterior
periviceral   funnel
cavity (coelom)   pallial cartilage
oviducal   nidamental   branchial   ctenidium
gland   glands   heart

Figure 10.18. *Loligo opalescens*, internal anatomy, ventral view, female.

accessory nidamental gland at its anterior tip. These glands may be red or orange-speckled before egg laying. The entire conical tip of the mantle cavity may be filled with eggs, as the large ovary in this region sheds eggs directly into the *perivisceral* (coelomic) *cavity*.

b. *Reproductive system*. Not only is the sexual apparatus complex, but intricate behavior patterns have evolved that ensure proper fertilization and placement of eggs. The male's large, single testis opens into the coelom through a slit in the wall. Sperm are picked up by a ciliated funnel (*sperm bulb*) and passed into the posterior *vas deferens* or sperm duct. The duct, at first a slender coiled tube white with sperm, may become considerably enlarged by its contents. Two glands straddle the vas deferens. Anteriormost is the *spermatophoric gland* (seminal vesicle), which receives the sperm and packs them into complex *spermatophores*. The second, the *spermatophoric organ* (*Needham's sac*), receives finished spermatophores and stores them before ejection through the enlarged *anterior vas deferens* that passes near the dorsal surface of the left kidney. The vas deferens ends at the muscular ejaculatory *penis*. The sperm are therefore bundled, packaged, and stored before being used in the complex copulatory act. Examine spermatophores from the spermatophoric sac. Each spermatophore consists of an elongate outer *tunic* enclosing the *sperm mass*. This rather incredible structure is completed by a *cement body* and coiled *ejaculatory organ* at the anterior end of the spermatophore (near the end bearing an elongate cap thread). When the cap is broken, the coiled

ejaculatory organ below it springs out, dragging the cement gland and sperm mass with it. The cement gland sticks to whatever it is thrown against and holds the sperm masses on or near the point—usually near unfertilized eggs. Spermatophores are generally transferred by the male's specialized arm (*which one?*) to the female horseshoe organ, or they may be thrust by the male directly into the female mantle cavity.

The female genital system (Fig. 10.18) consists of the *ovary*, which empties numerous eggs into the coelom, and an *oviduct*, which picks up the eggs with a ciliated funnel. The funnel and adjacent portion of the oviduct are embedded in the egg mass and are thus difficult to locate and more easily traced from the anterior portion. Observe the flared opening of the oviduct through which eggs pass from the coelom into the mantle cavity. (Look in the area halfway down the left gill.) The thick-walled glandular, anterior portion of the oviduct forms the *oviducal gland*. Trace the oviduct back to the ovary through its several loops, removing the left gill and its branchial heart if necessary. Notice the nidamental glands and the accessory nidamental glands, which secrete the outer capsules of the egg masses. The oviducal gland (in the oviduct) forms spherical capsules around individual eggs. Encased in their individual packets and surrounded by a fluid, jellylike matrix, the eggs are then extruded through the funnel, fertilized, and placed by the female on an appropriate rock or substrate below the low-tide line. The female octopus guards and constantly aerates her eggs by controlled water movements. The jelly covering the eggs hardens to enclose a cluster of several dozen eggs in a finger-shaped protective envelope, which, in the case of squid eggs, adheres by its basal cementing material. In an aquarium with running sea water, the development of young squids can be observed through these protective membranes.

c. *Respiratory and circulatory systems*. The cephalopod circulatory system is of the *closed* type. (How does this compare with that of the clam and the earthworm?) The major flow pattern of the blood is best seen in injected specimens, but can also be worked out in noninjected material. *Branchial hearts* receive venous blood returning from passage throughout the body. The principal veins are the large pair of *post-*

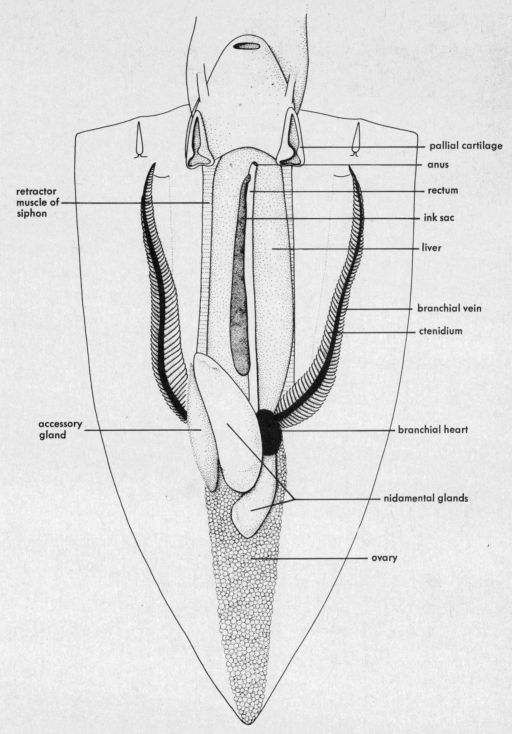

Figure 10.19. *Loligo*, internal anatomy, ventral view, female; semidiagrammatic.

cavae (singular: *postcava*) or *posterior mantle veins*. Blood from the head returns through the *anterior vena cava* (or *cephalic vein*), which receives blood from two large vessels, the *right* and *left precavae*. Find where the precavae pass through the kidney tissues and join the postcavae as they enter the branchial hearts. From the branchial hearts, blood passes out the *branchial arteries* into a fine capillary bed in the gills, where O_2–CO_2 exchange occurs, then flows

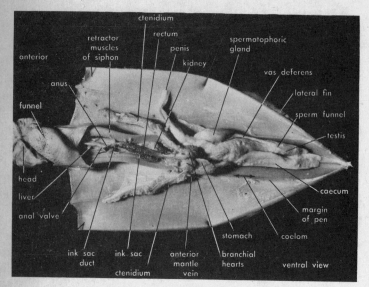

Figure 10.20. *Loligo opalescens*, internal anatomy, ventral view, male.

back through the *branchial veins* into a large *systemic heart* between and somewhat anterior to the branchial hearts. Remove the *pericardium* to examine the heart. Oxygenated blood is pumped out of the heart and passes forward to the head through the *anterior aorta* (to the right), or through the *posterior aorta* that then divides into a medium and two lateral *mantle arteries*. Near its point of origin, the posterior aorta also gives off branches to the ink sac, rectum, branchial hearts, and gonoduct. Both the veins entering the heart and the arteries leaving it are protected from backflow by *semilunar* valves.

d. *Excretory system.* Expose the paired nephridia or kidneys. A pair of papillae, each enclosing a *nephridiopore*, is seen as a small projection on the anteroventral surface of each kidney. The large, bilobed *urinary gland* is enclosed by kidney tissue.

■ Are relationships of kidney, gills, and heart (direction of blood flow) the same in the classes CEPHALOPODA, PELECYPODA, and GASTROPODA?

e. *Digestive system.* Expose the muscular, bulbous *buccal mass* in the head by cutting through the base of the funnel and tissues overlying the *mouth.* Lay these parts to either side, and dissect through the muscles, cutting open the head

along the median line. The *radula* and *radular muscle* are enclosed in tissues between *mandibles* of the *beak.* Try to dissect out the radula and examine it microscopically as you did with the snail and chiton. Be careful, however, not to damage ganglia of the head region. A pair of buccal *salivary glands* lie in muscular tissue posterior to the buccal cavity. Trace the thin *esophagus* from the buccal mass to the *stomach* along its loop through the *digestive organ* (liver). Posterior to the muscular stomach is the *cecum*, a long sac (considerably shorter in starved specimens; compare Figs. 10.18 and 10.20) extending out to the end of the cone. Try to locate the U-shaped *pancreas* anterior to the stomach. A single *hepatopancreatic duct* carries fluids from digestive gland and pancreas into the cecum. Near the junction of the esophagus and stomach, find the *intestine*, which passes forward between lobes of the pancreas, narrows into the rectum, and ends at the *anus* near the funnel base.

f. *Nervous system.* Cephalopods are said to be the most intelligent of invertebrates. We mentioned how this can be correlated with the predatory habits and activity of these animals. In addition to massing of nerve ganglia in the head region, a definite skull-like structure—the *cephalic cartilages*—has developed.

■ What is the advantage of this structure?

Extremely rapid nerve conduction is possible through *giant nerve fibers*, which originate in the *cephalic ganglia*, pass through the *stellate ganglia*, and end in the circular muscles of the mantle. Find the stellate ganglia embedded in the dorsal surface of mantle tissue, near the tip of the gills. Trace the giant nerve fibers—some of the largest nerves of this type known, and a favorite research tool of nerve physiologists the world over. The largest fibers innervate the most distant muscles.

■ How does the difference in fiber size make possible the *nearly simultaneous contraction* of the mantle?
■ What does this function imply about *integration* of movement and degree of swimming control?

Dissection of the central nervous system, a rewarding challenge, requires care and skill, but

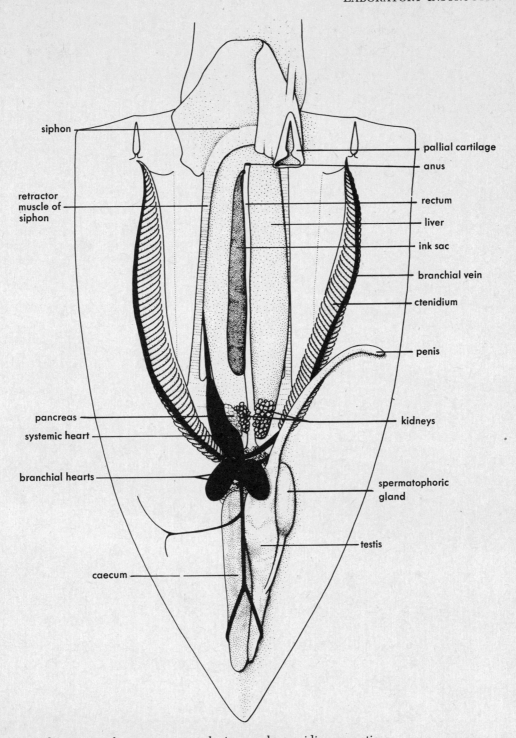

siphon

retractor
muscle of
siphon

pancreas
systemic heart

branchial hearts

caecum

pallial cartilage
anus
rectum
liver
ink sac
branchial vein
ctenidium
penis
kidneys
spermatophoric
gland
testis

Figure 10.21. *Loligo*, internal anatomy, ventral view, male; semidiagrammatic.

promises the interested observer a view into a brain that shows striking developmental parallels between cephalopod and vertebrate nervous systems. Expose as much of the squid brain as possible without destroying primary nerves;

then identify the various ganglia and their connectives.

Perhaps the simplest method is to cut a sagittal section through the head with a razor. This will expose the *dorsal ganglia* and the *cerebral* and small *suprabuccal ganglion* behind the buccal mass. Below the esophagus the main brain elements can be identified: the *visceral ganglion*, the anterior *pedal ganglion,* and the *propedal ganglion* closer to the mouth.

A demanding but more illuminating dissection, a special project, exposes the brain from the dorsal surface so that the entire brain and its contributory nerves can be isolated. Remove the skin, dorsal muscles, and upper portion of the cartilaginous skull, being careful not to destroy the *postorbital nerves* passing through the dorsal cartilaginous shield. Dissect laterally to clear tissues from around the optic ganglia and major connectives. The "*white bodies,*" or white blood cell centers, should be removed to expose the ganglia more fully. Work out the relationships of the different ganglia and their connectives. Contrast this view with that obtained in the sagittal section. Then dissect anteriorly to expose the *buccal ganglia* and nerves, and the *fused nerve ring* of ganglia about the esophagus. Continue posteriorly to expose the *medial visceral nerve* and giant nerves from the stellate ganglia.

■ Which ganglia are paired?
■ Does the cerebral ganglion appear to be derived from a pair?
■ What other evidence of *fusion* can you find?
■ How does the squid brain compare with that of the clam? the snail?
■ What correlation can be made between *muscular activity, brain size,* and *ganglion concentration?*

Cut off one eye and optic ganglion to view the brain both dorsally and laterally; place the eye in a separate dish of water for later examination. Review the parts of the newly exposed brain.

■ Does the cerebral ganglion give off nerves? What might this imply?

Observe the relatively large proportion of the brain that seems to be associated with the eyes (*optic nerves, optic ganglia, oculomotor nerve* to the *eye muscles*).

Now check the structure of the dissected eye. Remove it from enclosing tissues. Observe the short optic nerve that connects it to the optic ganglion. Locate the *anterior* and *posterior chambers* between which is the lens. (See pp. 179–181 for external structures.)

■ How does the eye focus?
■ How does this compare with methods of focusing among vertebrates?

Find the *cornea, iris,* and *pupil.*
Draw the squid eye; name its principal parts.
Compare the function and position of these parts with those of the vertebrate eye. This similarity of eye structure in two such divergent groups represents one of the most remarkable examples of *evolutionary convergence* in nature.

g. *Skeletal system.* The *pen,* a chitinous, translucent, middorsal internal shell, has already been identified. Remove it and note its relationship with the *nuchal cartilage* at its anterior end in muscle tissue near the collar. *Infundibular cartilages* lie laterally along the base of the funnel, and these in turn articulate with *posterior mantle cartilages* (*pallial cartilages*) in mantle tissues.

■ Is the skeletal system of the squid a basic molluscan feature?
■ What is the chief function of the cartilages?

Some authorities say that the pen is a nonfunctional remnant of the heavier and more typical external shell seen in the related pearly nautilus or in the extinct ammonites. Others feel it might be a new and useful structure, rather than a vestigial relic.

■ What information would be required to choose properly between these alternative views?

Review the names and chief characteristics of the four mollusk classes and their representatives studied here.

■ What is the basic body type of mollusks?
■ Which would you consider a "typical mollusk?"

Correlate feeding habits and degree of motility in each class with the pattern of evolution and structural adaptation exemplified by that class.

V. Other Special Projects

1. Kymograph studies of selected molluscan hearts.

2. Studies of functional morphology, such as the following:

a. Orientation and response of selected gastropods to various physical and chemical stimuli.

b. Comparison of radular structure and action in a variety of gastropods.

c. Gross structure and ciliation of the ctenidia and palps in a selected series of simple to more complex bivalves to determine evolutionary changes.

3. Embryology of *Physa* (or other fresh-water snails). Often clusters of transparent eggs are carried adhered to the shell of aquatic snails. These eggs are easily studied with the aid of a dissecting microscope without disturbing the developing snails. Periodic observation will allow you to observe stages of embryo development, torsion, distribution of cilia, growth of organs and of shell. *Rate* response to temperature change and to a variety of other variable is also easily tested.

4. Growth-rate analysis of freshwater snails in aquaria kept at different temperatures, with different degrees of crowding and different levels of nutrition.

5. Study of snail parasitism in different host species from different habitats (see Chapter 6).

6. Study the variety of shell types in your

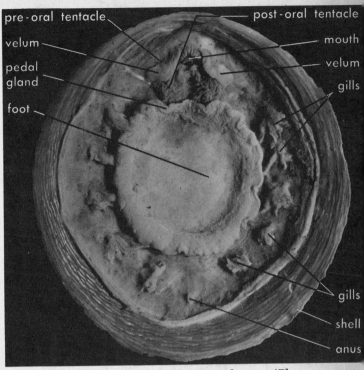

Figure 10.22. *Neopilina galathae*, ventral view. (Photograph courtesy Dr. Henning Lemche.)

local collection or museum display. Compare *Helix* with *Lymnaea*, *Physa*, a limpet, *Neopilina* (Fig. 10.22), abalone, and moon shell. An interesting and biologically enlightening hobby is to collect these abundant, varied, and often beautifully colored marine gastropods and pelecypods. Shell collecting has initiated many lifetime hobbies and successful careers in biology.

Suggested References

Abbott, R. T., 1954. *American Sea Shells*. Princeton, N.J.: Van Nostrand.

Atkins, D., 1936. "On the ciliary mechanisms and interrelationships of lamellibranchs. pts. I, II, and III." *Quart. J. Microsc. Sci., 314*:181, *315*:339, *315*:375.

Barnes, R. D., 1974. *Invertebrate Zoology*, 3rd ed. Philadelphia, Pa.: W. B. Saunders.

Barnes, H. F., and J. W. Weil, 1944. "Slugs in gardens: Their numbers, activities, and distribution, pt. I." *J. Anim. Ecol., 13*:140.

Baylor, E. R., 1959. "The response of snails to polarized light." *J. Exp. Biol., 36*:369.

Brown, F. A., Jr., 1960. *Selected Invertebrate Types*. New York: J. Wiley & Sons.

Bullock, T. H., and G. A. Horridge, 1965. *Structure and Function in the Nervous System of Invertebrates*. San Francisco, Calif.: W. H. Freeman.

Bullough, W. S., 1962. *Practical Invertebrate Anatomy*. New York: St. Martin's Press.

Coe, W. R., 1935. "Sexual phases in *Crepidula*." *J. Exp. Zool., 72*:455.

Drew, G. A., 1900. "The anatomy, habits, and embryology of *Yoldia limatula*." *Mem. Johns Hopkins Univ. Biol. Lab.*, no. 4.

Duncan, C. J., 1959. "The life cycle and ecology of the fresh-water snail *Physa fontinalis*." *J. Anim. Ecol.*, 28:97.

——, 1960. "The evolution of the pulmonate genital system." *Proc. Zool. Soc. (London)*, *134*:601.

Hyman, L. H., 1968. *The Invertebrates*, vol. 4, *Mollusca I*. New York: McGraw-Hill.

Kline, G., 1956. "Notes on the stinging operation of *Conus*." *Nautilus*, 69:76.

Lemeche, H., 1957. "A new living deep-sea mollusk of the Cambro-Devonian class Monoplacophora." *Nature*, 179:413.

MacGinitie, G. E., and Y. MacGinitie, 1949. *Natural History of Marine Animals*. New York: McGraw-Hill.

Meglitsch, P., 1967. *Invertebrate Zoology*. New York: Oxford University Press.

Morton, J. E., 1958. *Molluscs. An Introduction to Their Form and Function*. London: Hutchinson.

Mozley, A., 1954. *An Introduction of Molluscan Ecology*. London: H. K. Lewin.

Orton, J. H., 1913. "The mode of feeding of *Crepidula*." *J. Mar. Biol. Assoc. (U.K.)*, 9:444.

Peterson, R. P., 1959. "The anatomy and histology of the reproductive system of *Octopus bimaculoides*." *J. Morph.*, *104*:61.

Prosser, C. L., and F. A. Brown, Jr., 1961. *Comparative Animal Physiology*. Philadelphia, Pa.: W. B. Saunders.

Raven, C., 1958. *Morphogenesis: An Analysis of Molluscan Development*. Oxford: Pergamon Press.

Ricketts, E. G., and J. Calvin, 1968. *Between Pacific Tides*. (rev. ed. by J. Hedgepeth). Stanford, Calif.: Stanford University Press.

Rowett, H. G. Q., 1957. *Dissection Guides*, vol. 5, *Invertebrates*. New York: Holt, Rinehart and Winston.

Sherman, I. W., and V. G. Sherman, 1970. *The Invertebrates: Function and Form. A Laboratory Guide*. New York: Macmillan.

Webb, W. F., 1951. *Handbook for Shell Collectors*, 9th ed. St. Petersburg, Fla.

Wilbur, K. M., and C. M. Yonge, 1964. *Physiology of Mollusca*, vol. 1. New York: Academic Press.

——, 1966. *Physiology of Mollusca*, vol. 2. New York: Academic Press.

Yonge, C. M., 1932. "The crystalline style of the Mollusca." *Sci. Progr.*, *26*:643.

Young, J. Z., 1961. "Learning and discrimination in the Octopus." *Biol. Rev.*, *36*:32.

CHAPTER 11

PHYLUM ECHINODERMATA

1. Introduction

Sea stars, sea urchins, brittle stars, sea cucumbers, and feather stars are all echinoderms (a word meaning "spiny skin"). They comprise a large phylum of specialized marine organisms that are interesting for both their striking body form and their possible relationship to the precursors of the vertebrate line. Though adults of most echinoderms show *radial symmetry*, this does not ally the **ECHINODERMATA** with the **CNIDARIA.** Even a hundred years after the time of Linnaeus, the superficial similarity of echinoderm and coelenterate symmetry obscured the basic differences between these two phyla. Further research made it clear, however, that the radially symmetrical pattern of coelenterates is present in *each stage* of their life cycle. This is a basic phylogenetic characteristic called *primary* radial symmetry. In echinoderms, the larval stages are actively swimming *bilaterally symmetrical* animals. Radial symmetry in this phylum is an evolutionary modification found only in the *adults* and is associated with a sessile or creeping existence. This is considered *secondary* radial symmetry, a subsequent modification or specialization added later in the evolutionary history of the group. Though this interpretation would hardly interest the sea star, it tells us that adult symmetry is not a reliable reading of the phylogenetic history of **ECHINODERMATA.** It also helps clarify our thinking about the importance of considering larval stages in working out evolutionary antecedents, as we shall soon see, in the question of vertebrate origins. Finally, it permits classification of echinoderms with other eucoelomate animals to which they have close structural and embryological ties, and releases them from an artificial alliance with coelenterates to which they have no basic similarity.

We will begin our study of **ECHINODERMATA** with a description of its basic body type and then study examples of the five classes and their structural and functional characteristics. The embryological and phylogenetic importance of the phylum is discussed in a subsequent section.

Approach your specimens questioningly. Keep the same curiosity you felt as a child for the strange and wonderful beasts that washed up on the beach. Opportunities to acquire knowledge of *living* animals in their *natural* environment are offered by several university marine biological stations, where summer courses in marine biology stress living activities and ecological interrelationships. A lifetime of interest in marine life usually results from such a study.

2. Classification

The fundamental adult morphological pattern of this ancient phylum is *pentamerous* (five-armed) and radially symmetrical around an oral–aboral axis (aboral indicates the side *away* from or opposite the mouth). The adult develops from a bilaterally symmetrical larva that undergoes a striking metamorphosis, changing to a body form adapted to slow-moving life on the ocean floor.

Perhaps the most unique characteristic of echinoderms is the *water-vascular system*, which consists of numerous water-filled tubes ending in a large number of *tube feet*. The latter control motility and, to a varying degree, respiration. There is a spacious coelomic (perivisceral) cavity lined with flagellated peritoneum developed in the embryo by the *enterocoelous* (out-pocketing) method. An *endoskeleton* of calcareous *plates*, *ossicles*, or *spines* appears to be external but is actually covered by a thin epidermis. Additional features are ciliated organs; lack of cephalization or segmentation; minute respiratory structures (*dermal papillae*) protruding through fine pores in the endoskeleton and epidermis; and a simple nervous system of *circumoral ring* and *radial nerves* to the arms. Many extinct groups and an extensive fossil record are known. Living representatives are divided into the following five classes.

I. Class ASTEROIDEA, Sea Stars (Figs. 11.1–11.7)

This class includes the predaceous star-shaped or pentagonal sea stars. In some species, however, the number of arms may be as high as 50. Ossicles are separate, permitting movement; short spines and *pedicellariae* are present; two or four rows of tube feet line the open *ambulacral grooves* in each arm. Oral surface is ventral; *madreporite* is aboral. *Asterias, Pisaster, Patiria, Henricia, Solaster*, and *Pycnopodia* are common genera. Sea stars are often called starfish, though this obviously is a misnomer.

II. Class OPHIUROIDEA, Brittle Stars (Figs. 11.8–11.10)

Similar to sea stars, brittle stars have a central *disk* to which highly flexible, jointed limbs are attached. Tube feet, confined to five rows and lacking suckers, have a sensory function. Pedicellariae and anus are lacking; madreporite is aboral. Typical genera are *Gorgonocephalus, Ophiothrix, Ophioderma*, and *Ophiocoma*.

III. Class ECHINOIDEA, Sea Urchins and Sand Dollars (Figs. 11.11–11.17)

These spiny, herbivorous echinoderms are constructed as though their arms were folded back into a ball and fixed into a calcareous skeleton (*test*), then covered with long, sharp, movable spines and three-jawed pedicellariae. The test is globular in sea urchins, and disk or heart-shaped in sand dollars. Tube feet are long, slender, and equipped with suckers. Mouth and anus are central or lateral. The large gut fills much of the test cavity (except during spawning periods). Typical genera are *Arbacia, Strongylocentrotus, Echinorachnius*, and *Dendraster*.

IV. Class HOLOTHUROIDEA, Sea Cucumbers (Figs. 11.18–11.23)

Sausage-shaped garbage collectors, sea cucumbers are among the chief scavenging organisms of the ocean bottom. They represent a different evolutionary direction among echinoderms—the adult phase retains the larval bilateral symmetry. With their sausage shape and warty papillated skin, sea cucumbers are well named. They vary in length from an inch to several feet, with body wall consistency from leathery to papery. Arms, spines, pedicellariae, and endoskeleton (except for scattered tiny plates in body wall) are all absent. Tube feet are present. The mouth with *tentacles* is at one end of the body and the anus is at the other, the latter often bearing a complex called the *respiratory tree* (which may be ejected with other organs in a sticky mass when the sea cucumber is attacked or disturbed). Examples are *Thyone, Leptosynapta*, and *Parastichopus*.

V. Class CRINOIDEA, Sea Lilies and Feather Stars (Fig. 11.24)

These stalked, flowerlike echinoderms have five arms that branch to ten or more, each bearing five branchlets or *pinnules* to form a cuplike central disk (*calyx*). No spines, pedicellariae, or suckers arise from the tube feet lining the open ambulacral grooves. In sea lilies a long jointed *stalk* with rootlike projections may attach the animal to the substrate. In feather stars the adult may lack a stalk and be free-swimming with motile, gripping *cirri* and a mouth and anus on the upper (oral) surface. Examples are *Metacrinus*, *Antedon*, and *Heliometra*.

■ Review the general definition of the phylum. Also describe a generalized *echinoderm body type.*

■ Define the five classes and state in what way the general adaptive pattern—feeding habits and locomotion—of each is related to its structure (for example, position of the mouth).

■ Explain how sea stars, sea urchins, sea cucumbers, brittle stars, and crinoids still can be combined in the same phylum in spite of their remarkable variation.

3. Laboratory Instructions

I. Class ASTEROIDEA; *Pisaster*[1]

1. **Observations**

a. Seeing living sea stars (or other echinoderms) in their natural habitat is important for an appreciation of their movements, feeding methods, and natural appearance. A field trip to an intertidal zone is particularly helpful, but much information can be obtained by observing them in a marine aquarium. With low temperatures and proper aeration, these animals remain alive and active for several weeks. They can then be studied at leisure for righting reactions, response to stimuli (such as a molluscan prey), and feeding methods.

Watch their locomotion. Observe coordination of the tube feet, sucking action on the glass wall of the aquarium, and righting reactions.

■ Which arm leads in general movement?
■ Is there a functional anterior end?

If possible, observe normal feeding reactions, including seeking, finding, opening, and digestion of prey. Digestion involves eversion

of the stomach and *external* digestion. To see the reaction to food, place a bit of meat near the mouth of an inverted sea star. The feeding posture may be preserved for more detailed observation by collecting and quick-freezing sea stars in the act of opening clams. The tube feet will remain in position and the stomach everted for digestion of the molluscan victim.

■ How are the tube feet of a sea star able to pull open even the largest mussel?
■ How does the sea star sense the presence of its prey?

External structures best seen in the living specimen include the soft respiratory papillae or *dermal branchiae*, which may be withdrawn or everted, and tiny pincers or *pedicellariae*. The latter go quickly into action when the surface of a living specimen is slowly stroked with a camel's hair brush. Under a dissecting microscope, pinching tips of the pedicellariae can be seen adhering to the brush hairs.

■ Could these pedicellariae explain the freedom of the sea star from surface-encrusting hydroids and other organisms?
■ Almost any surface in the intertidal zone is soon encrusted with organisms, yet sea stars are nearly always clean. Why is this advantageous?

b. The following are suggestions for experiments on asteriods.

i. Test the effect of water drippings from a living sea star on oysters or clams when the latter are open and "pumping."

ii. Test the effect of various types of meat or clam tissue on the feeding reaction of a sea star.

iii. Quick-freeze a sea star in the act of opening a mussel; study the position of the tube feet pulling open the mollusk valves and note the everted baglike stomach.

iv. Test the ability of sea stars to adhere to a silica or grease-coated aquarium glass surface.

v. Observe the action of *cilia*, both inside the body cavity and on the outer surface. To do this, inject powdered carmine or India ink into the body cavity and watch the dermal branchiae under a binocular dissecting microscope.

■ What causes the particles to move?
■ How is this movement related to respiration and to the internal circulation of fluids?

---

[1] Our principal organism for study is the sea star *Asterias* (from the Atlantic coast) or *Pisaster* (from the Pacific coast). Though numerous distinct species are found, each in a specific habitat and locality, their structure is quite similar. Photographs used here are of the giant Pacific sea star *Pisaster gigantea*.

2. **External anatomy** (Figs. 11.1–11.3)
a. Using a preserved or freshly killed specimen, find the disk and the aboral *madreporite* (small, colored plate, markedly off center on the disk). A pair of arms, the *bivium*, borders the madreporite. The other arms form the *trivium*. The *anus* is a fine pore in the center of the aboral surface. Turn your specimen over and observe the *mouth* on the ventral or oral surface. Around the mouth is a soft membranous zone, the *peristome*, and a protective circle of movable *oral spines*. Examine the other spines.

■ **Can you find an epithelial covering on these spines?**

■ **Are the spines a portion of an *ectoskeleton* or of an *endoskeleton*?**

■ **What forms the remainder of the skeleton?**

Study the ventral *ambulacral grooves* in the arms. Two or four rows of *tube feet* (*podia*) are found, depending on the species. Observe their arrangement and position.

■ **What is the function of tube feet?**
■ **Are they interconnected? how?**

Figure 11.1. *Pisaster gigantea*, aboral view.

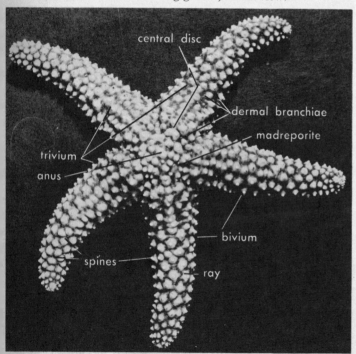

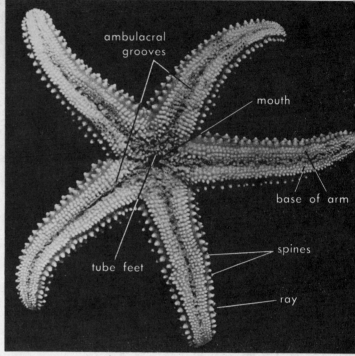

Figure 11.2. *Pisaster gigantea*, oral view.

b. Examine the animal's aboral surface more closely, using a hand lens or dissecting microscope. Keep the specimen submerged in order to float the smaller structures. The thin-walled *dermal branchiae*, representing the major surface in contact with the aquatic environment, have internal cilia that maintain a moving current of blood cells and body fluids. They therefore offer an excellent means for respiratory exchange. Search for the still smaller pedicellariae, some of which are stalked, others attached basally. They can be seen as tiny white specks among the papillae and spines. Scrape the aboral surface and examine the scrapings microscopically in a drop of water covered by a coverslip. Try to locate some pedicellariae jaws. Rock or press gently on the coverslip while you watch the specimen. This motion will sometimes cause the pedicellariae to roll over or even to open. Attempt to work out the scissorlike action of the tiny snap jaws.

c. *Drawings*
    i. Make a large outline drawing of a sea star, showing the external dorsal structure on one arm and the external ventral structure on another arm. Later add internal structures in the disk and arms.

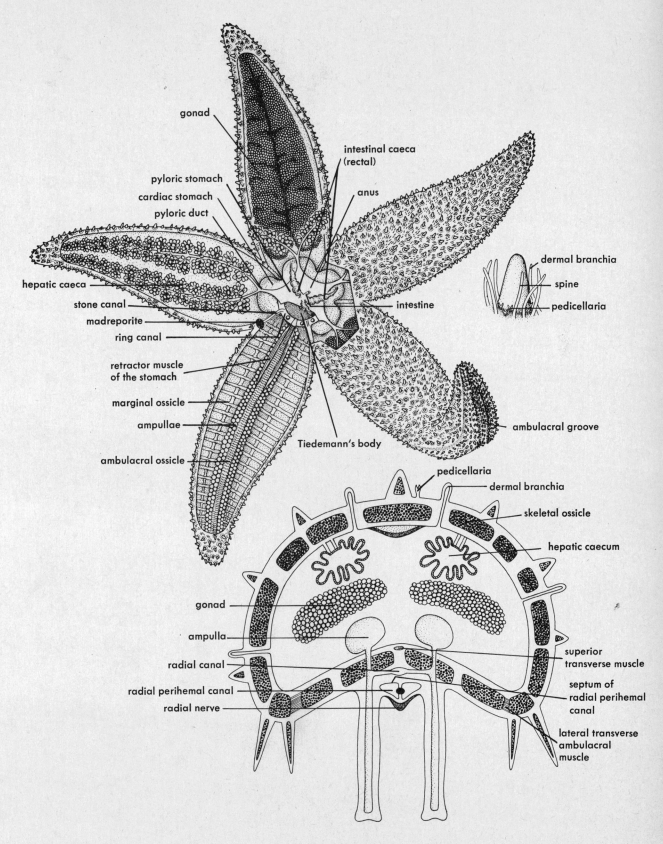

Figure 11.3. *Asterias*: (*top*) semidiagrammatic illustration of internal anatomy, aboral view; (*bottom*) cross section of ray.

ii. Sketch a group of spines, papillae, and pedicellariae to show their normal arrangement. Select a pedicellaria from your microscopic preparation and draw it in lateral view to show the four articulations and attached muscles.

3. **Internal anatomy** (preserved specimen) (Fig. 11.3) As an example of echinoderm organization, you will be provided with a sea star, *Pisaster*, for study of skeletal, digestive, reproductive, water-vascular, and nervous systems.

a. *Skeletal system.* This can be examined on a dried or preserved specimen. Dry sea stars should be soaked in potassium hydroxide to dissolve fleshy portions and to demonstrate the pattern of plates or ossicles forming the internal framework, with openings for feet and mouth. However, many plates are embedded individually and will fall free in the KOH. Therefore compare dried and macerated specimens. Portions of the body wall can be removed and studied to show structure and pattern of the inner skeletal framework.

Cut one arm off your specimen and study the cross section of the stump. Make several more cuts if needed to observe the arrangement of *ossicles* with their fixed *dorsal spines* and movable *ventral spines* along the ambulacral groove.

b. *Digestive system.* Digestive organs are best seen by removing the dorsal (aboral) body wall (Fig. 11.4). *Hepatic ceca*, large paired organs are suspended by *mesenteries* from the roof of each arm.

Make all dissections under water to expose other organs with greater facility and clarity. Cut the tip from one arm, then make two lateral cuts extending toward the disk. Lift the endoskeleton by trimming the mesenteries attached to it so as not to disturb the *pyloric ceca*, large paired organs just under the dorsal endoskeleton. Then cut across the shield where it joins the disk.

Next remove the aboral disk by making a circular cut, loosening adherent mesenteries as before. Avoid cutting out the madreporite; allow it to remain attached to the specimen. For additional views, make a cut from the exposed aboral area out to the tip of one or more remaining arms, and then remove the excised body wall from the disk outward (Fig. 11.5).

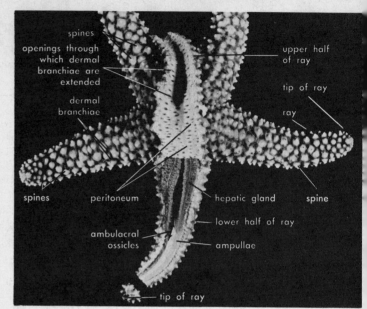

Figure 11.4. *Pisaster gigantea*, internal anatomy, aboral view.

You have exposed the large *coelom* and mesenteries supporting the organs.

■ **How are these mesenteries related to the coelom?**

Carefully probe around the digestive glands.

Figure 11.5. *Pisaster gigantea*, internal anatomy with aboral region of disc removed to show disposition of internal organs.

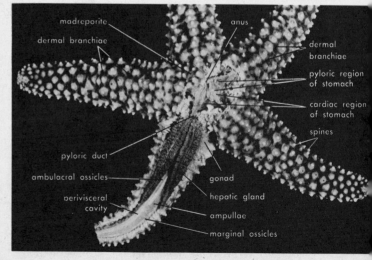

■ How are the paired pyloric ceca attached?

■ How many combined glands are there in each arm?

Trace the connection between the digestive glands and disk. Note where the two *hepatic ducts* of each arm join and enter the *stomach*. Observe that the stomach has two portions, the small aboral *pyloric stomach* connecting with the hepatic ducts, and the large, whitish, baglike *cardiac stomach* that lies ventrally and is the portion everted through the mouth while feeding. Five pairs of *retractor muscles* (one from each arm) attach to the cardiac stomach walls.

■ What is the functional importance of such a large, distensible sac?
■ Why is external digestion advantageous to the sea star?

Dorsal to the pyloric stomach, near the center of the stomach, is the single pair of small, lobed *rectal ceca*. These are attached to a fine *intestine* passing dorsally to the aboral disk and opening externally to a small *anal pore*.

c. *Reproductive system.* A pair of buff-colored glandular *gonads* lies under the hepatic ceca in each arm, usually ventrolaterally, alongside the ambulacral groove (Fig. 11.6). Gonads vary in size from the insignificantly small to those filling most of the coelomic space, depending upon the breeding cycle phase at time of capture. *Gonoducts* between the arms send sexual products out via extremely small *genital pores* in the disk, one per gonad (two per arm), around the periphery of the aboral disk.

Make a microscopic examination of the gonadal tissues and products to determine the sex of your sea star.

d. *Water-vascular system.* The water-vascular system, a unique internal water-pressure system, consists of a *madreporite* leading ventrally through a *stone canal* (Fig. 11.7) to a *ring canal* circling the mouth and branching out into each arm via *radial canals*. Nine small swellings, *Tiedemann bodies*, on the inner margin of the ring canal are probably related to the formation of certain cells in water-vascular fluid. Between each radial canal is the *Polian vesicle*, also of uncertain function. Tube feet,

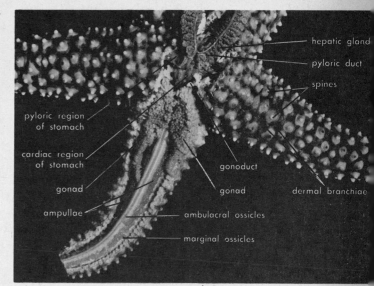

Figure 11.6.   *Pisaster gigantea*, internal anatomy; hepatic gland displaced aborally to show disposition of underlying organs.

basic units of the water-vascular system, lie alongside the radial canals, each tube foot connected to the canal by a short *lateral canal*.

A tube foot consists of a tubular portion protruding into the ambulacral groove, and an internal swollen *ampulla*. This system, which serves as a fluid-pressure mechanism operating the tube feet, provides echinoderms with slow motility, adherence to the substrate, and prolonged sucking action for opening the strongest

Figure 11.7.   *Pisaster gigantea*, internal anatomy; gonads, removed.

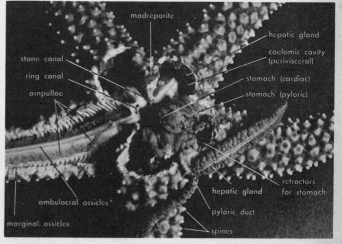

bivalve (by alternately holding with one group of tube feet, then another).

Apply pressure to one ampulla and note the effect on its tube feet. (In preserved specimens there is no such effect.)

■ **How do the ampullae control liquid pressure in the external tube feet?**

Next find the radial canal in the center of the ambulacral groove of the cut arms. Dissect further in this cut portion until you can discern the lateral canal between the radial canal and a tube foot. Trace the radial canal toward the disk and find its connection to the ring canal. Expose the ring canal in turn and find its inner swellings, the Tiedemann bodies. (Note that two are found between each pair of arms except on the bivium, where the tenth space is occupied by the stone canal.) The ossicle-reinforced stone canal rises from the ring canal and emerges to the outside via the madreporite. This structural complex, operated by valves and reservoir chambers to control and maintain internal water pressure in the tube feet and in their suction disks, permits the system's remarkable operation.

Drawing. Complete your outline with details visible in *dorsal view* of the disk. Include both portions of the *stomach, rectal ceca, hepatic ducts*, portions of *digestive* and *reproductive glands*, and visible parts of the *water-vascular system.*

■ **How is respiration accomplished in the ASTEROIDEA?**
■ **How are wastes excreted?**

e. *Nervous system.* Nervous coordination in sea stars is difficult to study morphologically, hence the nervous system will be described only briefly. *Radial nerves* pass from the ambulacral groove of each arm (near the radial canal) to an *oral nerve ring* (near the ring canal, at the outer margin of the peristomial membrane). Coordination of arm movement is centered in *ganglion cells* at the junction of each radial nerve and the nerve ring. The nerve ring serves to coordinate arm movements for directed activity. Local nerve centers are found in the stomach wall, near various muscles and movable spines, and near the tube feet and ampullae. These complete the readily visible branches of the nervous system. In addition, a *subepider-*

mal nerve net serves as a more independent network for control of local reactions.

f. *Circulatory system.* Circulation in sea stars is accomplished by coelomic fluids that move by action of the ciliated peritoneum lining the coelom. This fluid carries a large number of specialized *coelomocytes*, many of which are amoeboid wandering cells. The cilia-directed currents of coelomic fluid appear to have replaced the blood system of other organisms, as evidenced by the presence of a nonfunctional vascular system in sea stars. The *hemal system* consists of an *axial sinus* (outer surface of the stone canal), a *hemal ring*, remnants of branches into the arms (*radial hemal canals*), and branches along the hepatic duct.

g. *Suggested experiments*

i. *Circulation*

a. Open an aboral flap in a living sea star placed in a pan of sea water and trace coelomic currents with a few drops of India ink or carmine. Observe particularly ciliated cells of the peritoneum.

b. Inject these dyes into the pyloric ceca and trace the particles under a dissecting microscope.

c. Observe the passage of particles in distended dermal branchiae.

ii. *Reproduction*

a. Fertilization of sea star ova in sea water (in *clean* glassware previously washed and preferably soaked overnight in sea water) can easily be observed and the developmental stages followed. This is now a standard procedure for embryological study. Consult Chapter 17 and Costello et al. (1957) listed at the end of this chapter for details.

b. Stained preparations of *zygotes, segmentation stages, blastula, gastrula*, and early *bipinnaria* larvae can be studied, identified, and drawn. Be sure to observe the sequence through the blastula and gastrula stages.

i. *Field observation.* Visit a rocky intertidal zone and study the habits and distribution of as many echinoderms as you can discover. List the species encountered and their habitats. Record observations on food, numbers, and any particular activities observed or tested.

II. Class OPHIUROIDEA, Brittle Stars, Serpent Stars, Basket Stars

Study dried or bottled specimens of brittle stars or other ophiuroids (Figs. 11.8–11.10).

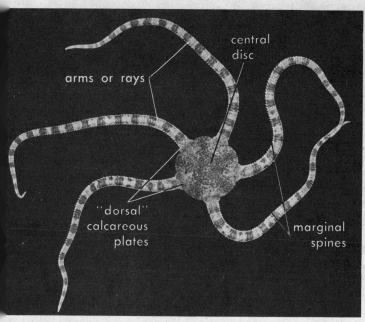

Figure 11.8. *Ophioderma panamensis*, aboral view.

■ **How do they compare with asteroid sea stars?**

■ **What is the chief external characteristic of the class OPHIUROIDEA?**

Notice the absence of pedicellariae and position of the *madreporite*. Observe *ambulacral grooves* and *tube feet* emerging from lateral margins of these grooves.

■ **Can you locate an *anus*?**
■ **What permits the unusual motility of brittle stars arms?**

III. Class ECHINOIDEA, Sea Urchins; *Strongylocentrotus* (Figs. 11.11–11.14)

1. **Introduction and observations** Sea urchins (*Arbacia* on the East Coast, *Strongylocentrotus* on the West Coast are striking and intriguing echinoderms (as many skindivers can painfully attest). As noted previously, they structurally resemble a sea star with its arms folded back and with a dense covering mat of spines. Their way of life, however, is totally different. Sea urchins feed on all types of organic material, but are chiefly algae grazers rather than predators. Sea urchin structure, characterized by a spined protective test, a large coelom, and a greatly en-

larged digestive system has evolved along defensive and herbivorous lines.

■ **How are these structures adapted to the habitat and distribution of sea urchins?**

Examine living sea urchins. Compare their appearance and habits with those of sea stars. Locate the elongated *tube feet* and *pedicellariae* among the numerous spines. While observing your live urchin under the dissection scope use a glass rod to poke at the tube feet, pedicellariae, and spines. What is the urchin's reaction? If possible, observe a specimen crawling on the glass of a marine aquarium.

2. **Internal anatomy** (Fig. 11.14) Study either living or preserved specimens for internal structure. Cut the test along its equatorial axis using heavy scissors or bone cutters. Notice the extremely large *intestine* and digestive apparatus, adapted for herbivorous life. Observe the *pentamerous* pattern of the *gonads, ampullae,* and *test.* Cut around the *peristomial membrane* and remove the mouth and entire feeding apparatus. Study the unusual complex jaw apparatus, "*Aristotle's lantern,*" a chuck-and-bit arrangement remarkably well-adapted for rock scraping. Observe its complex array of muscles for movement of the five parts—

Figure 11.9. *Ophioderma panamensis*, oral view.

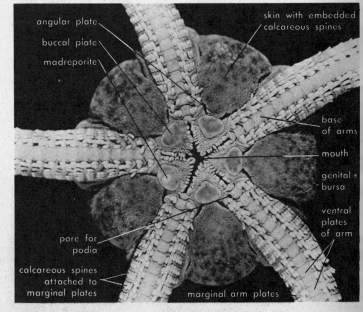

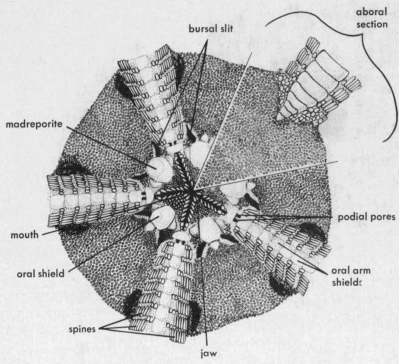

Figure 11.10.   *Ophioderma*, semidiagrammatic illustration of oral and aboral surfaces.

some 92 separate muscles have been identified.

■ What characteristics mark echinoids as a distinct class?

■ What characteristics mark echinoids as echinoderms?

■ What does the name ECHINOIDEA mean?

■ How do other animals in this class

Figure 11.11. *Strongylocentrotus purpuratus*, aboral view.

Figure 11.12. *Strongylocentrotus purpuratus*, oral view.

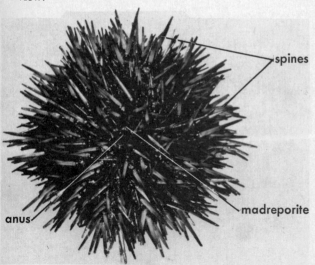

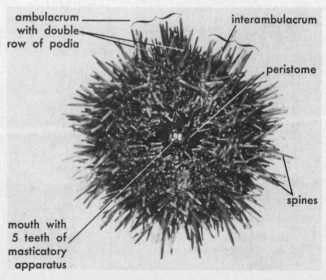

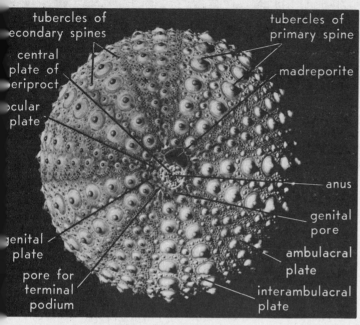

tubercles of
secondary spines

central
plate of
periproct

ocular
plate

genital
plate

pore for
terminal
podium

tubercles of
primary spine

madreporite

anus

genital
pore

ambulacral
plate

interambulacral
plate

Figure 11.13. *Strongylocentrotus franciscanus*, aboral view of test.

differ from sea urchins with respect to gross morphology? (See Figs. 11.15–11.17.)

IV. Class HOLOTHUROIDEA, Sea Cucumbers; *Cucumaria, Parastichopus, Thyone*

1. **Observations**   Among the most specialized echinoderms, sea cucumbers have slender elongate bodies that are bilaterally symmetrical along the oral–aboral axis. Presumably this is an adaptation for directed, crawling movement, derived from an ancestral pentameral pattern that initially was adapted to sessile or slow-moving life. (As noted earlier, the pentameral pattern is, in turn, probably derived from an even earlier bilaterally symmetrical free-swimming form.)

Sea cucumbers are scavengers, feeding at the ocean bottom. Their flowerlike anterior tentacles (Fig. 11.18) entrap material in mucus or scrape together debris or other available food particles.

The holothuroids are of added interest because of their capacity to cast out their internal organs, to fragment, and to eject the entangling viscous tubes of the respiratory mechanism. By these defensive mechanisms, sea cucumbers trap a potential predator or satisfy it with a partial meal. The loss of

organs is not fatal to the former owner and they are rapidly regenerated.

2. **External anatomy**   Examine the external surface of a sea cucumber. Note its leathery, sometimes warty appearance with many tube feet and numerous embedded *ossicles* that can be felt under the skin (Figs. 11.19 and 11.20). Specialized sensory and respiratory tube feet on the dorsal surface are usually in two zones or tracts, and locomotory tube feet in three ventral zones. Notice how this repeats the basic *pentamerous pattern*.

Find the retractile *oral tentacles* (Fig. 11.18).

■   **How many are there? Describe their function.**

3. **Internal anatomy**
a.   Dissection of a large sea cucumber like *Parastichopus* or *Thyone* demonstrates very well the fundamental echinoderm pattern (Figs. 11.21–11.23).

■   **Do any internal systems show evidence of pentamerism?**

Work out the long digestive tract, including *tentacles*; *mouth*; short *esophagus*; enlarged *stomach*; long, looped *intestine*; muscular *cloaca* (attached by numerous mesenteries to the body wall); and *anus*. Extending forward into the *coelom* from the cloaca are two *respiratory trees* with many small branches. The cloaca pumps water through these trees; hence it serves for egestion, excretion, and respiration.
b.   The water-vascular system consists of a *madreporite* that usually opens directly from the coelom, a *ring canal* around the esophagus, and five longitudinal *radial canals* that connect along the body wall to the feet. The tube feet follow five longitudinal internal *muscle bands*. Blood vessels line the intestine. The brushlike *gonad* has fine tubules joining a *gonoduct* opening near the dorsal tentacles.

Make a pinned-out dissection of your specimen to show these organs.

Draw the specimen's internal anatomy.

■   **What features adapt this class to its mud or sandy-bottom marine environment?**
■   **What features make sea cucumbers a distinct class?**

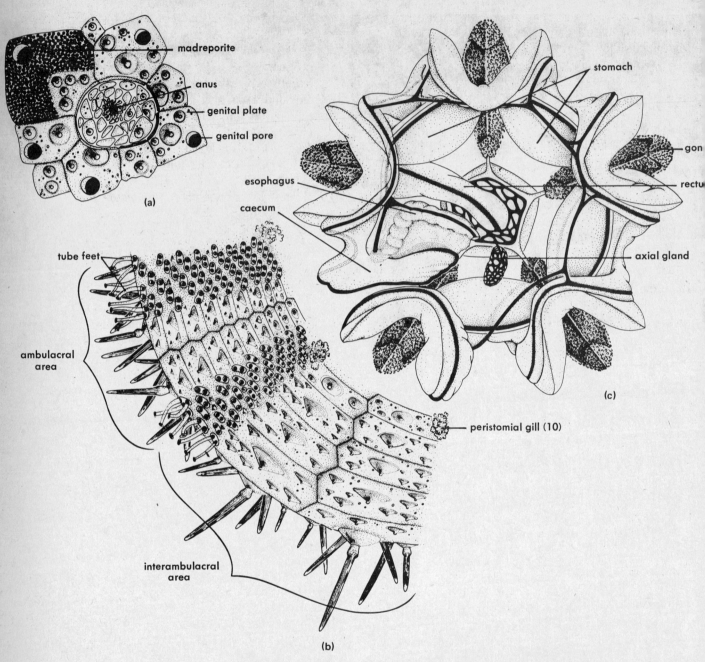

Figure 11.14. *Arbacia,* external anatomy: (a) details of periproct (aboral surface); (b) spines, test, and tube feet; (c) internal anatomy: digestive, reproductive, and hemal (black vessels) systems.

■ **What features ensure its inclusion within the phylum ECHINODERMATA?**

V. Class CRINOIDEA, Sea Lilies

If specimens are available, study these strange sessile or crawling echinoderms (Fig. 11.24).

Observe the *ringed stalk* and the five petal-like arms, often branched.

■ **To what type of habitat are these organisms best adapted?**

The ringed stalk is attached to the substrate by rootlike *cirri*, and topped by a *calyx* from

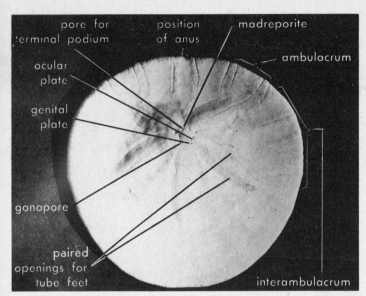

Figure 11.15. Sand dollar, *Dendraster excentricus*, aboral view.

which the five arms emerge. Each arm divides into two branches and again into small branchlets or *pinnules*. A large *anal tube* also protrudes up from the calyx. Some crinoids lack the stalk and swim by arm movement.

Relatives of this class include numerous fossils in the classes CARPOIDEA, CYSTOIDEA, BLASTOIDEA, and EDRIOASTEROIDEA, all of which are placed with CRINOIDEA in the subphylum **PELMATAZOA**. These groups are combined on the basis of the skeleton, especially the calyx, as this is the fossilized portion that is most distinctive.

Echinoderms in which the mouth is *ventral* and which lack a stalk are in the subphylum **ELEUTHEROZOA**, which includes all classes *except* crinoids and their related extinct classes.

VI. Special Projects

If several living specimens of echinoderms are available, a number of simple, yet illuminating experimental studies can be undertaken.

1. **Locomotion and righting** Carefully observe the tube feet of a sea star to show how they control movement. Isolate several tube feet by cutting them off and placing them in a small vessel of sea water. Do they contract when touched? Observe the animal

when turned onto its aboral surface. Are the rays coordinated in the righting response? Cut the ring nerve once, observe twice, and cut again. Are any differences noted in the rate of righting?

2. **Ciliary currents** Using any echinoderm, study the ciliary currents on oral and aboral surfaces by dropping particles of carmine or India ink on various areas. Observe and diagram the direction and pattern disclosed by movement of the particles.

3. **Sensory physiology** Study the responses of a sea star, sea urchin, or brittle star to a graded series of tactile stimulations. *Phototactic studies* can provide another illuminating project. Use a narrow beam of strong white light and direct it on a specific body region. Record your observations. Do this experiment both before and after dark adaptation.

■ **What additional experiments can you suggest for the study of *phototaxes*?**

■ **What other stimuli would be worthwhile testing to give a better idea of the animal's adaptive responses in nature?**

4. **Coelomocytes and clotting** Remove a small volume (10 ml) of coelomic fluid from a sea star, sea urchin, or holothuroid. Note the color and turbidity of the fluid. Prepare

Figure 11.16. *Dendraster excentricus*, oral view.

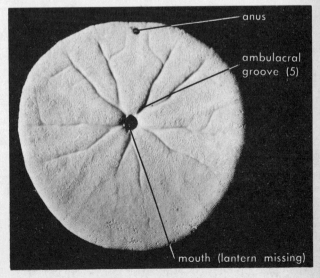

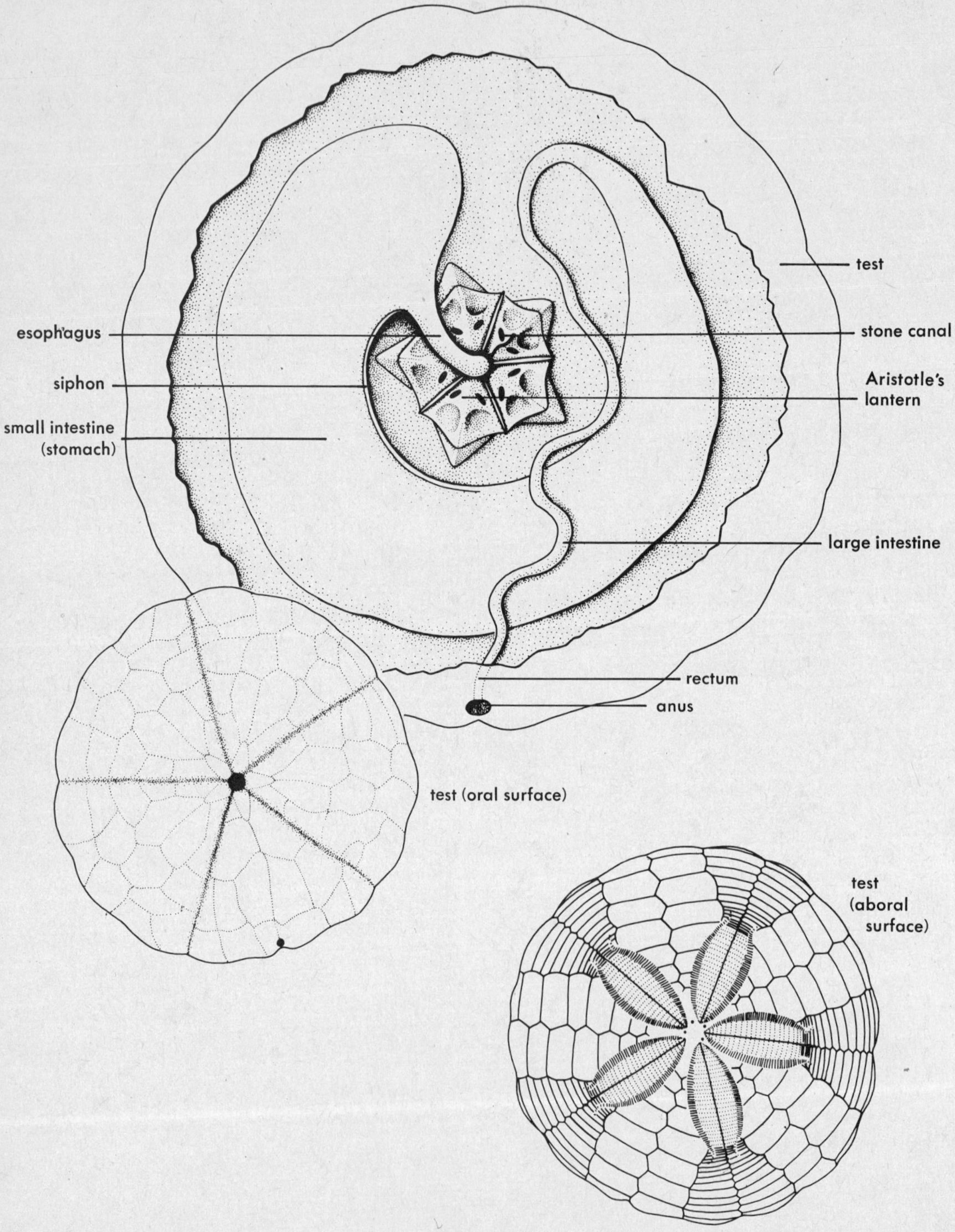

esoph'agus

siphon

small intestine
(stomach)

test

stone canal

Aristotle's
lantern

large intestine

rectum

anus

test (oral surface)

test
(aboral
surface)

Figure 11.17. Sand dollar, *Echinarachinus*: (top) internal anatomy of digestive system; (*lower left*) oral surface of test; (*lower right*) aboral surface of test.

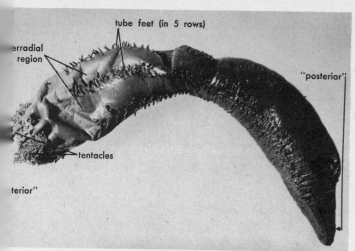

Figure 11.18. *Cucumaria*, external view.

Figure 11.19. *Cucumaria*, external aspects of anterior end.

Figure 11.20. *Thyone*, external view.

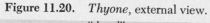
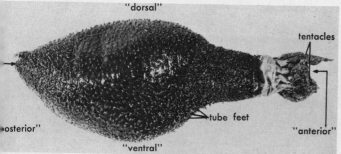

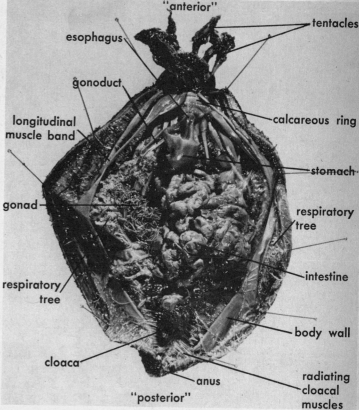

Figure 11.21. *Thyone*, internal anatomy.

Figure 11.22. *Parasitichopus californicus*, internal anatomy.

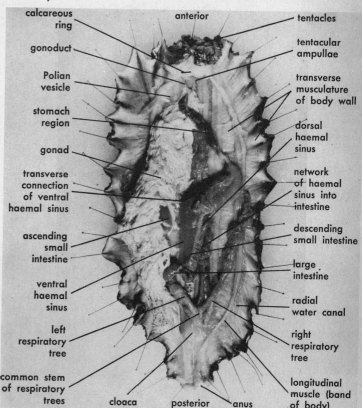

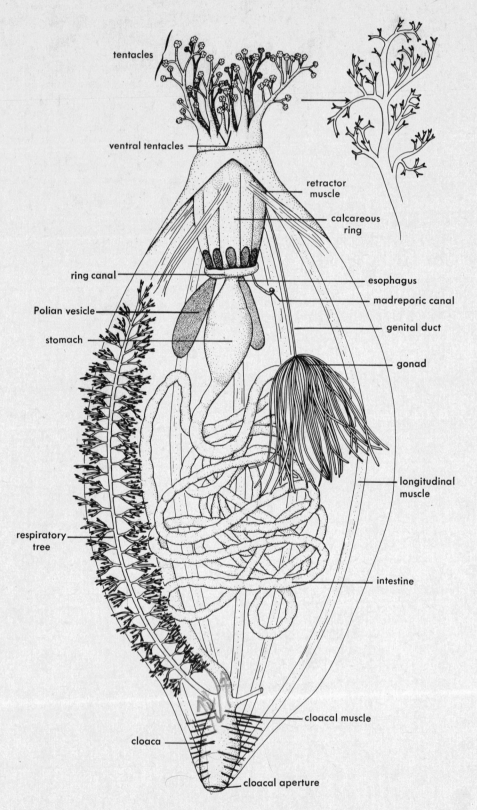

Figure 11.23. *Thyone*, internal anatomy, semidiagrammatic.

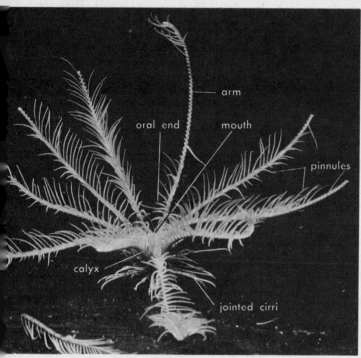

Figure 11.24. The crinoid, *Heliometra glacialis*.

a microscope slide for each organism tested and examine under the high-dry lens. Look at several fields and draw the cell types you observe. Place a rim of vaseline around the four edges of the slide to keep it from drying out. Look at the slides once every 5 min for a 30-min period. What has occurred? Look at the test tube in which the coelomic fluid was collected. Do you note any change? Inject 1 ml of carmine suspension into the coelomic cavity. After 10 min remove a drop of coelomic fluid and examine it under the high-dry objective lens of the compound microscope.

■ **Where do you find the carmine particles? Compare with fluid from another specimen.**
■ **What conclusions can you draw as to the function of the coelomocytes you observe?**

Suggested References

Barnes, R. D., 1974. *Invertebrate Zoology*, 3rd ed. Philadelphia, Pa.: W. B. Saunders.

Boolootian, R. A. (ed.), 1966. *Physiology of Echinodermata*. New York: Interscience Publishers.

——, and A. C. Giese, 1959. "Clotting of echinoderm coelomic fluid." *J. Exp. Zool.*, 140:207.

Brown, F. A., Jr., 1950. *Selected Invertebrate Types*. New York: J. Wiley & Sons.

Buchsbaum, R., and L. J. Milne, 1960. *The Lower Animals; Living Invertebrates of the World*. New York: Doubleday.

Bullough, W. S., 1966. *Practical Invertebrate Anatomy*. New York: St. Martin's Press.

Burnett, A. L., 1955. "A demonstration of the efficacy of muscular force in the opening of clams by the starfish, *Asterias forbesi*." *Biol. Bull.*, 108:355.

Bury, H., 1895. "The metamorphosis of echinoderms." *Quart. J. Microscop. Sci.*, 38:45.

Chadwick, H. C., 1923. "Asterias." *Liverpool Marine Biol. Comm. Mem.* 25.

Clark, A. M., 1962. *Starfishes and Their Relations*. London: British Museum (Natural History).

Costello, D. P., M. E. Davidson, A. Eggers, M. H. Fox, and C. Henley, 1957. *Methods for Obtaining and Handling Marine Eggs and Embryos*. Woods Hole, Mass.: Marine Biological Laboratory.

Crozier, W. J., 1918. "The amount of bottom material ingested by holothurians." *J. Exp. Zool.*, 26:379.

Feder, H. M., 1955. "On the methods used by the starfish, *Pisaster ochcaeus*, in opening three types of bivalved molluscs." *Ecology*, 36:764.

Fell, H. B., 1948. "Echinoderm embryology and the origin of the chordates." *Biol. Rev.*, 23:81.

Fisher, W. K., 1911–1930. "Asteroidea of the North Pacific and adjacent waters." *U.S. National Museum Bull.*, 73, 3 parts.

Gislen, T., 1930. "Affinities between echinoderms, *Enteropneusta* and chordates." *Zool. Bidrag.*, 12:197.

Goodbody, I., 1960. "The feeding mechanism in the sand dollar, *Mellita sexiesperofrata*." *Biol. Bull.*, 119:80.

Harvey, E. B., 1956. *The American Arbacia and Other Sea Urchins*. Princeton, N.J.: Princeton University Press.

Hyman, L. H., 1955. *The Invertebrates*, vol. 4, *Echinodermata*. New York: McGraw-Hill.

Kerkut, G. A., 1954–1955. "The mechanisms of co-ordination of the starfish tube feet." *Behavior*, 6:206; 8:112.

MacGinitie, G. E., and N. MacGinitie, 1949. *Natural History of Marine Animals*. New York: McGraw-Hill.

Meglitsch, P., 1967. *Invertebrate Zoology*. New York: Oxford University Press.

Millott, N., 1953. "Color pattern and the definition of the species." *Experentia*, 9:9.

——, and M. Yoshida, 1960. "The shadow reaction of *Diadema antillarium*." *J. Exp. Biol.*, 37:363.

Nichols, D., 1962. *Echinoderms*. London: Hutchinson University Press.

Pearse, A. S., 1908. "Behavior of *Thyone*." *Biol. Bull.*, 15:259.

Ricketts, E. G., and J. Calvin, 1968. *Between Pacific Tides*. (Ref. ed. by J. Hedgepeth.) Stanford, Calif.: Stanford University Press.

Sherman, I. W., and V. G. Sherman, 1970. *The Invertebrates: Function and Form; A Laboratory Guide*. New York: Macmillan.

CHAPTER 12

PHYLUM CHORDATA

1. Introduction

A prime lesson in an introductory zoology course is that people can learn about man by first studying other animals. From this one develops a truly comparative perspective. When these differences and similarities are clear, we will have come a long way toward fulfillment of the Socratic maxim, "know thyself."

Many courses in zoology begin with the **CHORDATA,** the phylum that includes humans, in order to approach the unfamiliar from the familiar. But, unexpectedly, we often find that we are less familiar with ourselves and with our own phylum than we had supposed. Hence we follow here an evolutionary sequence from structurally simple to more complex animals, so that when the phylum **CHORDATA** is reached, we are beginning to wonder how man fits into the overall picture.

Many of us will be surprised to find that the large and varied group of chordates includes certain wormlike or sac-shaped animals that are not even vertebrates. It may even be disconcerting to learn that among true vertebrates, most are fishes—some 75 percent, in fact. All of this should orient our thinking along strictly biological lines, free from preconceived human-oriented views.

2. Classification

I. General

The phylum **CHORDATA** should first be examined as one of several phyla in the superphylum **ENTEROCOELA**[1] (outpocketing coelom) linked by basic embryological criteria. Just as annelids, mollusks, and arthropods are placed in the superphylum **SCHIZOCOELA,** so echinoderms, chordates, and a few smaller phyla are combined in the **ENTEROCOELA.**

Major phyla in this group are **ECHINODERMATA, HEMICHORDATA** (included as a **CHORDATA** subphylum by some authors), and **CHORDATA. HEMICHORDATA,** as the name "half-chordates" implies, do not possess all of the accepted chordate characteristics. They appear to lie somewhere between echinoderms and chordates. Such chordate features as dorsal hollow nerve cord, paired pharyngeal gill slits, and (possibly) a noto-

---

[1] This assemblage is also called the **DEUTEROSTOMIA** (new mouth), a name derived from another embryological trait of this collection of phyla. Other terms, such as "echinoderm–chordate" line, also refer to the same general group. These terms are to be compared with the **PROTOSTOMIA** for the "annelid–arthropod" line.

chord are found in the hemichordates. At the same time, their larval form (*tornaria*) is so close to some echinoderm larvae (*bipinnaria*) that the two have often been confused, even by experts.

■ **What embryological traits distinguish** ENTEROCOELA **from** SCHIZOCOELA?

■ **Define the phylum CHORDATA. What unites it with** ENTEROCOELA?

■ **Why is it considered a distinct phylum?**

II. Related Phyla

One of the least known of the invertebrate phyla linked with chordates in the ENTEROCOELA is the **POGONOPHORA,** a recently discovered phylum of wormlike creatures with an appearance strikingly similar to the younger stages of hemichordates. The other small but distinct phyla united in this assemblage have also been placed there on embryological grounds. A good example is the phylum **CHAETOGNATHA,** the arrow worms (*Sagitta*)—active, common, torpedo-shaped predators of marine plankton. The wormlike phylum **PHORONIDA** is also thought to be in the entercoel complex, although this position is still conjectural.

III. Phylum CHORDATA

The chordates consist of the following subphyla:

Group **ACRANIATA** (without head)

1. **Subphylum TUNICATA (or UROCHORDATA), tunicates or sea squirts**   These are highly modified, sessile, filter-feeding animals with motile larvae. Most of the chordate characteristics (notochord, dorsal hollow nerve cord, pharyngeal gill slits) are lost during metamorphosis from larva into adults.

■ **Can you account for a possible adaptive aspect to the loss of chordate features in the** *sessile* **adults of this group?**

2. **Subphylum   CEPHALOCHORDATA, lancelets or amphioxus**   These small fishlike mud- or sand-dwelling filter feeders are important animals from a theoretical standpoint. They represent an evolutionary stage in which chordate characteristics (notochord, dorsal hollow nerve cord, gill slits) are well developed, but vertebrate characteristics absent. *Amphioxus* is a chordate of great interest, a prototype of how an ancestral vertebrate might have looked. Except for annelidlike ciliated nephridia, the lancelet's organ systems are similar to those of vertebrates or other simple chordates.

The above subphyla form the group **ACRANIATA.** Together with the hemichordates, they often are called the *protochordates* (first chordates), in contrast to higher chordates in the group **CRANIATA** or **VERTEBRATA.**

Group **CRANIATA** (with head)—higher chordates

These are the true vertebrates, including ourselves. Here is where we can say animals truly are animals in the layman's sense.

CRANIATA are chordates with visceral arches, vertebrae, and a brain. They are divided into two *subphyla*: **AGNATHA**—vertebrates without jaws, and **GNATHOSTOMATA**—vertebrates with jaws.

3. **Subphylum AGNATHA**   These jawless fishes are the most primitive vertebrates and lack limbs and jaws. The first records of vertebrates from early fossil deposits show they were abundant members of the sea-bottom fauna about 400 or more million years ago. Fossil **AGNATHA** (**OSTRACODERMI**) were characterized by heavy bony armor and probably a mud-sucking, filtering mode of feeding. Their highly specialized descendants, the lampreys and hagfishes (**CYCLOSTOMATA**), are still extant. Survival of modern remnants of this ancient class is probably related to their specially adapted rasping and bloodsucking mouthparts, which enable them to feed on other fish. This parasitic way of life frees the lamprey from competition with later evolved forms. Larval Lampreys, *Ammocoetes*, illustrate well the primitive vertebrate form.

4. **Subphylum GNATHOSTOMATA, vertebrates with jaws**   With the development of jaws, new avenues of adaptations and possibilities for evolutionary advancement were open to the vertebrates. The jawless vertebrates were restricted to certain modes of life, while the jawed forms could exploit a wide variety of habitats. The earliest jawed vertebrates appeared around the Silurian period. Jaws were apparently derived from the *gill arches*.

In this subphylum are several classes of

fishes, and the classes of amphibians, reptiles, birds, and mammals. The old term for fishes, PISCES, is now used as a superclass designation to include three distinct classes. The other superclass, TETRAPODA, includes the four-legged vertebrates, although limbs have secondarily been lost in some groups, such as the snakes.

a.   *Superclass* PISCES

        i. *Class* PLACODERMI, *ancient armored fishes, now entirely extinct*. The class is marked by development of a primitive type of jaw suspension that enabled PLACODERMI to replace the more primitive jawless fishes (**AGNATHA**), although the former group died out in the Permian period.

        ii. *Class* CHONDRICHTHYES, *cartilaginous fishes; sharks, rays, skates, and chimaeras*. It is doubtful whether these fishes preceded true bony fishes in evolution, as is often assumed. In fact fossil evidence opposes this view, since earliest remains of modern fishes are actually bony fishes (class OSTEICHTHYES). Sharks and other cartilaginous fishes probably evolved from bony fishes rather than the reverse, or perhaps the two groups evolved in parallel fashion from placoderm ancestors.

        The all-cartilaginous skeleton therefore appears to be a comparatively recent adaptation (though still several hundred million years old). Another important adaptation in the shark and sharklike group is a capacity to retain *urea* in blood and tissues, thus raising internal osmotic pressure to a point nearly equal to that of the salt-water environment.

■   **Why is this such a valuable adaptation?**

Sharks, as typical fish, fearsome predators, or modern derivatives of an ancient lineage, are always interesting. We shall study a common offshore form, the dog shark or spiny dogfish, *Squalus acanthias*.

        iii. *Class* OSTEICHTHYES, *bony or higher fishes*. As noted earlier, the enormously varied and numerous OSTEICHTHYES represent some 75 percent of described vertebrate species, including nearly all freshwater fishes and the preponderance of marine species. Most members of the class in turn belong to the teleost group, an assemblage of some 29 orders of so-called higher bony fishes.

        One subclass of OSTEICHTHYES of special importance to students of evolution is the CHO-ANICHTHYES, fish with notsrils connected to the

mouth cavity and with paired fleshy or limblike fins. These are presumed precursors of the AMPHIBIA, very ancient groups which have changed little since. Familiar CHOANICHTHYES include the lungfish and the remarkable "living fossil," *Latimeria*, discovered recently off the coasts of South Africa and Madagascar. Until its discovery, this fish was presumed to have been extinct for over 75 million years. After careful study of over 50 specimens, *Latimeria* and its group properly are relegated to an important primitive side group in evolution, rather than a true "missing-link" or progenitor status for the higher fishes and tetrapods.

■   **What even more ancient invertebrate "living fossil" can you name? To which phylum does this belong?**

b.   *Superclass* TETRAPODA

        i. *Class* AMPHIBIA, *salamanders, frogs, toads,* and *caecilians* (or APODA, tropical, limbless amphibians). These are the first tetrapoda, or land-dwelling, limbed vertebrates. *Amphi*, meaning "both," implies life both in water and on land.

        Amphibians require abundant moisture, as their soft skin is not protected against water loss. Nearly all amphibians depend on water for egg laying and early development, though some frogs breed in damp moss instead of water. Certain toads are remarkably resistant to desiccation; other amphibians such as the Mexican axolotl and a few frog species, never leave the water. In general, we think of amphibians as transitional creatures, perfectly adapted neither to water nor to land, but somewhat adapted to both—the ancestors of purely terrestrial animals.

        In a later section we will study the anatomy of the familiar leopard or grass frog, *Rana pipiens*, as a typical member of this class. Although salamanders are more truly "typical" amphibians, they are less easily obtained than frogs.

        ii. *Class* REPTILIA, *includes snakes, lizards, turtles, crocodiles,* and *alligators*, as well as numerous extinct forms (many different groups are linked by the common name "dinosaur"). They represent the completed transition from water to land, principally due to adaptations that reduce water loss. These include dry scaly skin; efficient encased (internal) lungs for air respiration (as opposed to the external gills or skin of amphibia); internal fertilization; and eggs protected from desiccation by a relatively impervious, often limy shell and by embryonic membranes that enclose the developing embryo within an aquatic environment. A free larval stage is lacking. The egg, usually placed in a warm, sheltered spot, provides protection and nourishment (yolk) for prolonged embryonic development. The young thus pass through the fragile early stages of life without excessive hazard from violent temperature and humidity changes characteristic of the land environment. They can then emerge essentially as young adults. Internal adaptations that probably accompanied these changes were the bellows method of breathing and improved systems of blood circulation and heart structure. More efficient limbs and superior muscular coordination are also related to survival in an air medium where rapid, controlled movements are highly advantageous.

      iii. *Class* AVES, *birds.* Structurally, birds are feathered, beaked, generally flying, reptile-like animals with physiological adaptations for temperature control. The evolutionary transition from reptiles to birds was gradual, marked by a number of intermediate stages. Evidence for these changes comes from fossil, embryological, and anatomical studies. Structural and physiological adaptations of modern birds, highly specialized for aerial life, make them among the most successful, efficient, and complex living chordates.

      iv. *Class* MAMMALIA, *warm-blooded, fur-bearing tetrapods with mammary glands.* A very early offshoot from reptilian ancestors, mammals stemmed not from specialized but from very primitive reptile stock, at about the time the earliest reptiles were themselves evolving. Temperature-regulating mechanisms gradually evolved, with hair as the chief insulator. Warm-bloodedness (*homeothermy*) probably developed in mammals about the same time it appeared independently in birds. The story of mammal evolution is really one of progressively greater degrees of protection from, or control over, the environment. Homeothermy provides freedom from rapid fluctuations in internal temperature. A greater constancy of the internal environment results. *Viviparity* (young born alive) and *lactation* allow longer incubation and greater protection of embryos and young over longer periods. Unusually complex adaptive behavior patterns and parental *training* are associated with these longer periods of contact between parent and offspring and with the high degree of integration in the central nervous system(CNS) that developed. Beyond this, one can envisage gradual evolution of *social* and *learned patterns of behavior.* Among primates at least, this resulted in an extraordinary flowering of mental development and an extreme degree of environmental control, as demonstrated by man.

      Mammals represent a peak of efficiency in nerve–muscle coordination and responsiveness, ultimately resulting in the high order of intelligence and adaptability that characterize the class. Structural specialization, correlated with different habitats and food habits, is particularly marked in various patterns of tooth specialization. In fact, differences between types of teeth and changes in jaw and middle-ear formation are primary structural criteria for separating mammals from reptiles. Major organ systems also reflect this increased degree of control and coordination with outward structural improvements.

      We will study one of the most abundant and successful members of the class, the Norway rat, *Rattus norvegicus* (Chapter 14). Wily, prolific, and omnivorous, this rat and its close relative *Rattus rattus* are found almost everywhere that man is, and sometimes in far greater numbers. The mantle of successor to man as a dominant mammal may yet fall on these adaptable and widespread members of the largest order of mammals, the RODENTIA.

3. Laboratory Instructions

I. Subphylum CEPHALOCHORDATA, Amphioxus (scientifically named *Branchiostoma*)

1. **Observations** This animal lives in sandy offshore and intertidal areas where it feeds on small organisms filtered through its pharyngeal gill basket. Such a function implies that the most primitive chordate gills were used for feeding rather than for respiration, a supposition apparently confirmed by fossil evidence of earliest vertebrates. Amphioxus possess the chief chordate characteristics in almost diagrammatic clarity.

2. **Anatomy** Stained whole-mount specimens of plastic embedded specimens are excellent for studying both external and internal structures (Fig. 12.1).

      Orient your specimen—the blunt end is anterior, the dorsal surface is more pointed than the flattened ventral surface. Find the *oral hood* enclosing the mouth, and an upper *vestibule* equipped with stiff *cirri* which the

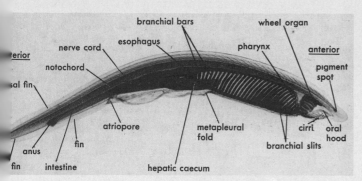

Figure 12.1. Amphioxus, whole mount, lateral view. (Courtesy CCM: General Biological, Inc., Chicago.)

animal uses as strainers for food selection. A single posterior *fin* passes from a medium opening, the *atriopore*, and extends to the tail and around to the dorsal surface. Here it forms the *caudal fin* and then the *dorsal fin*. The *anus* is on the mid-ventral line anterior to the tail.

Observe the V-shaped muscle segments, *myotomes*, with *myosepta* (connective tissue partitions) between them.

■ **Can amphioxus properly be called a *segmented* animal?**

Trace out the digestive tract. Locate the oral hood and cirri, and the wheel organ (fingerlike projections at the posterior inner surface of the hood). The mouth leads directly into the *pharynx*, which is a complex food-sifting chamber of alternating *gill bars* (or *branchial bars*) and *gill slits*. The bars, skeletally supported to form a feeding gill basket, are markedly similar to those in *Balanoglossus*, the most primitive protochordate (phylum **HEMICHORDATA**).

The *atrium* surrounding the gill basket receives water which passes from the pharynx through the gill slits, and then out the atriopore.

The digestive tract posterior to the pharynx continues as a short *esophagus* and wider *stomach* (midgut). The *liver* (or *hepatic cecum*) is an outgrowth from the midgut, which then narrows into the *intestine* (hindgut) and terminates in the anus.

Find the notochord dorsal to the digestive tract.

■ **How far anteriorly does the notochord run?**

Locate the *neural tube* above the notochord posteriorly; it is marked by a row of black spots, which are very simple eyes consisting of one ganglion and one pigment cell. Notice the extremely small anterior vesicle forming the brain.

Draw your specimen and identify as many structures as you can.

3. **Histology** (cross section) Study a stained amphioxus cross section taken through the pharynx (Figs. 12.2 and 12.3) or the intestine (Fig. 12.4). Under low power, you should be able to identify: *epidermis, dorsal fin* with its supporting *fin ray, meta-*

Figure 12.2. Amphioxus, transverse section through pharyngeal region female.

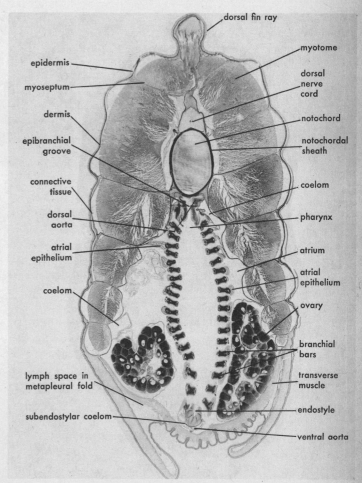

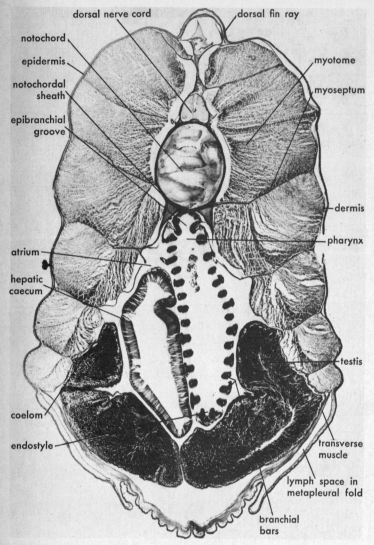

Figure 12.3. Amphioxus, transverse section through pharyngeal region male.

pleural folds (or *ventrolateral* folds), *myotomes* (filling much of the dorsal and lateral spaces), *neural tube, atrium, pharynx, endostyle, hyperbranchial groove, aortae, coelomic spaces, gill bars, gill slits, liver, gonads* (if present, ovary cells have large nuclei, testes appear striated or streaked), and *nephridia* (portions are attached to outer surface of dorsal gill bars).

Draw what you can identify in this section.

■ **Review evidence for relationships between hemichordates, tunicates, and primitive vertebrates. Consider embryological and structural** similarities, and general organization of digestive, nervous, circulatory, and excretory systems.

■ **Does any evidence oppose this conclusion of a phyletic relationship?**

■ **Which view seems stronger? why?**

II. Subphylum AGNATHA, Class CYCLOSTOMATA, Lampreys and Hagfishes

1. **Adult lamprey**
a. *External anatomy*

i. We begin our survey of the vertebrates by examining both the adult and the *ammocoete larva* of these fishes. The simplest and perhaps most primitive living vertebrates, lampreys are contemporary representatives of the **AGNATHA**, the most ancient of fishes.

■ **Define the term AGNATHA.**

■ **Why are these fishes considered primitive?**

Figure 12.4. Amphioxus, transverse section through intestinal region.

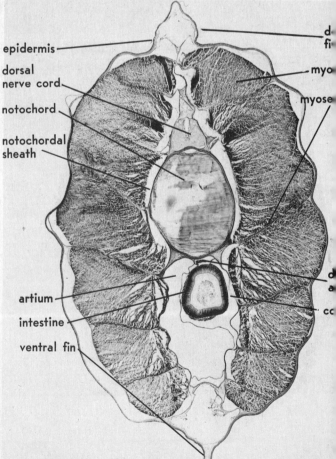

■ **What structures or characteristics enabled the cyclostomes to survive for so long?**

Observe a demonstration specimen of an adult lamprey. Find the *dorsal* fin and *caudal fin* with *fin rays*. Note the absence of paired fins.

ii. Study the head. Note the absence of a lower jaw and the presence of a toothed sucker and toothed tongue (an effective sucking and rasping device). The lamprey attaches by the sucker, then rasps away the host's flesh with the toothed tongue.

Note, too, the single dorsal opening to the olfactory sac instead of paired nasal openings. This is a primitive characteristic typical of living and fossil agnathids.

■ **How many gill slits are there behind each eye?**

iii. The musculature, like that of amphioxus, is divided into myotomes (segments) with myosepta between them. A *cloacal pit* lies on the midventral line in front of the trunk–tail juncture. Inside it are an *anus* and posteriorly a *urogenital papilla* in the male.

b. *Sagittal section* (anterior end). If a plastic-embedded or demonstration section of the preserved animal is available, study it for structural details of feeding and respiratory apparatus.

Find the *buccal funnel, teeth,* and *tongue.* Observe the large muscles that control the rasping tongue. Follow the buccal cavity to the *esophagus* (upper, smaller tube) and *pharynx* (lower, larger tube), which open externally through seven *gill clefts* or slits. These slits open into larger gill pouches, which in turn connect to external slits.

■ **How does the lamprey respire? Trace the path of the water current produced.**
■ **Trace the passage of food.**

Next, find the *notochord.* In lampreys this structure is still an essential supporting element, as the "vertebrae" are only small arches straddling the neural tube. A cartilaginous cranium and complex gill basket form the rest of the skeletal system.

The *nervous system* consists of a dorsal hollow nerve cord over the notochord and an anterior enlargement, the brain (just above the anterior edge of the notochord).

2. **Ammocoetes larva** (Figs. 12.5–12.8)

a. *Introduction.* This peculiar larval form is sufficiently distinct to have once been named *Ammocoetes*, as it was originally thought to be a separate genus of adult fish. It actually is a long-lived cyclostome larva that burrows in stream beds and filters out small food particles or organisms, in a manner similar to that of amphioxus. Of considerably more evolutionary interest than the adult, this larva illustrates many primitive vetebrate characteristics that are later masked by specialized adult cyclostome characteristics.

■ **Why are larval forms often of more phylogenetic interest than adult structures?**
■ **Name three other examples in which larvae, and not their adult forms, are of primary importance for determining phylogeny.**

b. *Anatomy.* Study a *whole mount* preparation (Figure 12.5) and observe the similarities between the ammocoete and amphioxus (Fig. 12.1)

Figure 12.5. Ammocoetes, whole mount, lateral view.

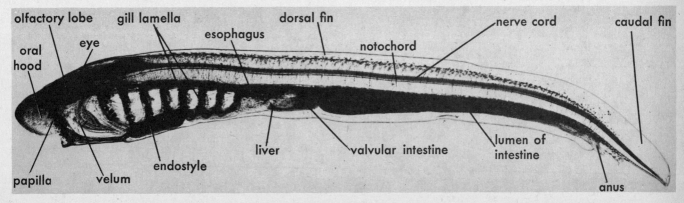

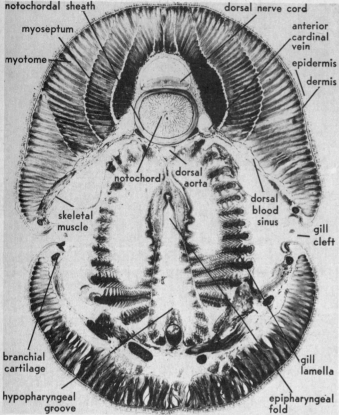

Figure 12.6. Ammocoetes, transverse section through pharyngeal region.

Figure 12.7. Ammocoetes, transverse section through pharyngeal region but posterior to area shown in Fig-

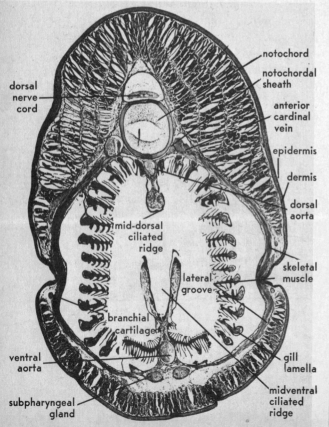

■ **What does this similarity imply?**

Notice the oral hood with papillae that serve as food strainers, then observe the gill region with the brain above it. The brain is divided (front to rear) into *olfactory bulb* and *olfactory lobe* (parts of the *telencephalon*). The *eyes*, the *infundibulum* below the olfactory lobe (which equals the posterior lobe of the *pituitary*), and the *pineal body* above the olfactory lobe are parts of the *diencephalon*. The midbrain (*mesencephalon*) consists largely of the *optic lobes*. The hindbrain, *rhombencephalon*, the largest division, consists mostly of the *medulla oblongata*, the contact center for automatic reactions of the body. From the medulla, the *spinal cord* extends posteriorly above the supporting notochord.

These parts of the CNS are characteristic of all vertebrates, but you will find a remarkable change in the vertebrate classes. Keep the basic divisions in mind, however, and develop a critical, comparative viewpoint to help you relate successive evolutionary stages.

Compare the following structures with those in adults and identify as many as you can. Notice the pharynx with its gill pouches, and the gill lamellae in the pouch walls.

c. *Histology, cross section of Ammocoetes* (Figs. 12.6–12.8).

Figure 12.8. Ammocoetes, transverse section through intestinal region.

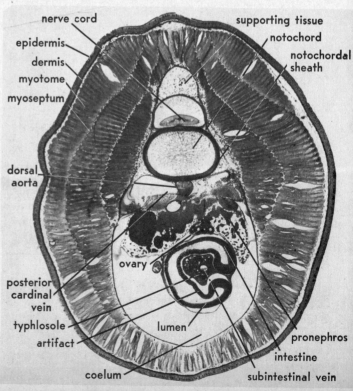

i. *Pharynx* (Figs. 12.6 and 12.7). Locate the *epidermis* or outer surface, then the connective tissue and muscle blocks (*myotomes*), which are especially thick on the dorsal surface.

Find then the *neural canal*, containing the rather flattened *spinal cord* and below it the *notochord* flanked by *cardinal veins*. Below this is the *dorsal aorta*. Most of the section consists of the pharyngeal region with gill lamellae on the pharynx walls. Find the *endostyle*. What is its function? its equivalent in higher chordates?

ii. *Trunk* (Fig. 12.8). This section should show the *dorsal* and *ventral fin fold*, *neural canal* and *spinal cord, notochord, dorsal aorta, cardinal vein*, and *myotomes* enclosing the body cavity (*coelom*). You may also be able to locate the *esophagus* and *pronephric* (excretory) *tubules* and the *heart*; in a more posterior section the *liver* and *intestine*, and possibly the *mesonephric tubules*.

■ **What basic vertebrate characteristics can be seen in the ammocoete larvae?**

■ **What changes occur in these larvae at metamorphosis?**

III. Subphylum GNATHOSTOMATA, Class CHONDRICHTHYES, Cartilaginous Fishes, Sharks, and Rays; *Squalus*

1. **Introduction** The class CHONDRICHTHYES, typified by the spiny dogfish, *Squalus acanthias* or *S. suckleyi*, includes skates, rays, and the relatively uncommon chimaeras or ratfish. These present-day fish are characterized by a cartilaginous skeleton and external gill slits.

The dogfish is often studied as a generalized vertebrate because it shows the essential characteristics of the subphylum with a minimum of specialized structures. A good knowledge of dogfish anatomy will serve as a basis for your introduction to the remaining vertebrate classes.

2. **External anatomy** (Fig. 12.9) Examine both the dorsal and the ventral aspects of the entire animal—*head, trunk*, and *tail*. Note the streamlined shape, the fins, and the rough skin covered by tiny, toothlike *dermal denticles*. These are small *placoid scales*, each of which has a projecting spine whose microscopic dentine structure is the same as that of vertebrate teeth.

Along the sides of the body is a whitish

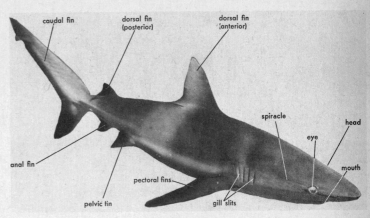

Figure 12.9. Shark, external anatomy. (Photograph courtesy of Mr. Ted Hobson.)

lateral line, which connects anteriorly to an extensive system of cephalic canals. These sensory canals enable the fish to feel water vibrations, equivalent to our sense of hearing. The muscle segments are also readily apparent.

■ **Do you see any resemblance between these and myotomes of amphioxus?**

Find the following head structures: *rostrum* (anterior extremity), *mouth, jaws*, and *teeth*.

■ **How are the teeth arranged?**

There was once a shark—now fortunately extinct— with a 7-ft mouth gape and a sparkling array of 6-in triangular slashing teeth. The structural format, however, was little different from that of your 18-in dogfish.

Next, find the sense organs—*olfactory, visual*, and *auditory*. The paired *nostrils* are small sacs through which water circulates. The *eyes*, unblinking as in all fish, lack a movable eyelid.

■ **Why are eyelids useful chiefly to terrestrial animals?**

■ **Can you find an *external ear opening*?**

■ **Where is the *inner ear* and what is its function?**

Next, find the *spiracle* and behind it five elongated *gill slits* through which respired water exits.

Identify the two types of fins—the *unpaired median fins* and the *paired lateral fins*. Locate the *anterior* and *posterior dorsal*, and *caudal*, *pectoral*, and *pelvic fins*.

■ **Which fins are homologous to our limbs?**

The *heterocercal tail*, asymmetrical, with an extended dorsal portion, is characteristic of all sharks. It is developed to a remarkable degree in the thresher shark, where its enormous, elongate lobe is used to panic and round up schools of fish.

The sex of a shark is shown by its pelvic fin. That of the male is pointed backward and modified into a *clasper* with rolled edges for sperm transfer.

■ **What type of fertilization is found in this group of sharks (in fact, in all CHONDRICHTHYES)?**

Locate the *cloaca*, which is the terminus of the digestive, excretory, and genital systems.

Draw the various systems on an outline of the lateral aspect of the dogfish.

3. **Internal anatomy**

a. *Digestive system* (Figs. 12.10 and 12.11). Pin your specimen out in a dissecting tray, ventral side up. Use long pins through the pectoral fin cartilages to hold it down firmly. Make a midventral incision from the pectoral girdle to and through the pelvic girdle. Use a sharp scalpel or scissors, but be careful not to cut deeply below the skin or into the underlying muscle wall. Expose the viscera by holding back and cutting away the lateral flaps of belly skin. Pin back the pelvic fins to stretch open the cloaca. Identify: *liver, stomach* (*cardiac* and *pyloric* region), *spleen, pancreas, intestine, rectum, cloaca, urogenital papilla, testis* (or *ovary*). Look under the medial lobe of the liver for the *gall bladder* and the *bile duct* (running along the *hepatic portal vein*).

Draw the exposed organs, showing their normal relationships (Fig. 12.10).

Display the viscera (Fig. 12.11) by pinning back the liver lobes and by pulling stomach and spleen to one side. Then make a deep incision up the *intestine* to expose the *spiral valves* that greatly enlarge the internal surface area of this structure.

■ **What is the advantage of increase in intestinal surface?**

Figure 12.10. *Squalus acanthias*, internal anatomy

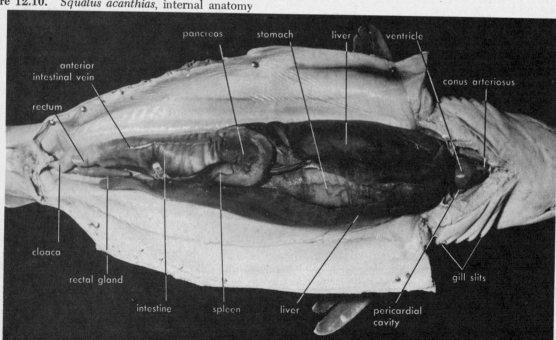

b. *Circulatory system*

i. The chief blood vessels serving the alimentary canal lie dorsal to it. They have to be exposed by careful dissection from the supporting mesentery. This work, which can be undertaken as a special project or by classroom demonstration, should include exposure of the *hepatic portal vein, intraintestinal, gastrointestinal,* and *posterior splenic veins.* Arteries in this area include the *hepatic, coeliac, gastric, anterior mesenteric,* and *lienogastric.*

ii. A more rewarding and equally demanding dissection is that of the *ventral aorta* and *afferent branchial arteries.* These, lying in the region of the floor of the mouth, represent the primitive pattern of anterior arterial circulation that has become greatly modified in terrestrial vertebrates. In gill-breathing aquatic vertebrates, arterial arches or branches from the aorta carry blood from the heart into *afferent* (toward) branches that lead to a capillary bed in each of the five gills where respiratory exchange occurs. The freshly oxygenated blood then passes from the capillaries into *efferent* (away from) branches that pass dorsally into the *dorsal aorta* and into the systemic circulation.

Dissection procedure

i. Cut through the cartilage of the pectoral girdle. Locate the *pericardial cavity.*

ii. Remove the skin from the ventral surface between the pectoral girdle and lower jaw.

iii. Cut through the superficial branchial muscles lying under the skin, pull to either side, and remove. The first *afferent branchial* artery can be seen through the remaining muscle layers.

iv. Loosen the two anteroposterior muscles between the lower jaw and the heart (*coraco-mandibular* and *coraco-hyal muscles*). Cut these muscles near the heart, pull the loose ends forward, and cut them off as close to the jaw as possible.

v. Other branchial arteries are now visible. Carefully cut away the central tissue block from which various muscles originate (*coraco-mandibular, coraco-hyal,* and five pairs of *coraco-branchials*). This will expose the coraco-branchials between the *ventral aorta* in the midline and the five arching *afferent branchial arteries.* Cut these muscles off as short as possible and clean away all interfering tissues to expose the entire length of afferent branchials on one side.

vi. Open a gill pouch (Fig. 12.12) by cutting from the first gill slit ventrally to a point between afferents 1 and 2 (to the V-shaped point of their branching from end of ventral aorta).

vii. Cut off the exposed gill lamellae in order to trace out the first afferent branchial artery. Remove the *gill rakers*, which arise posterior to the

Figure 12.11.   *Squalus acanthias*, internal anatomy, with organs displaced to sides to show underlying organs.

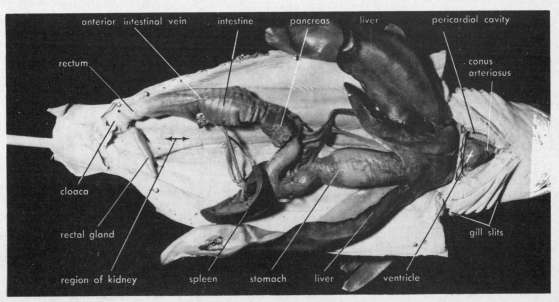

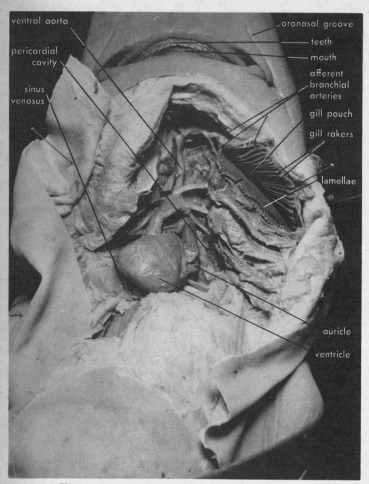

ventral aorta

pericardial cavity

sinus venosus

oronasal groove

teeth

mouth

afferent branchial arteries

gill pouch

gill rakers

lamellae

auricle

ventricle

Figure 12.12. *Squalus acanthias*, heart, afferent branchial arteries, and gill clefts.

branchial artery. Similarly, remove the lamellae and rakers from the other gills along the same side.

viii. Complete your dissection of the entire afferent branchial system on one side and *draw* the exposed vessels. When done with proper care, this dissection, as well as that of the *efferent* arterial system, is an excellent introduction to techniques for studying vertebrate anatomy.

Additional special dissections may be attempted on *afferent branchial arteries, systemic circulation, heart, inner ear,* and detailed *morphology of the eyes.*

c. *Nervous system.* The entire nervous system will not be exposed, but a worthwhile dissection can be made of the *brain* (Fig. 12.13), *brain stem,* and *spinal cord,* which together comprise the *central nervous system* (CNS).

Dissection procedure

i. Cut away the skin and overlying muscles from the area between and anterior to the eyes. Pull these tissues posteriorly toward the base of the head.

ii. Carefully shave away cartilage overlying the *chondrocranium*, which serves as a skull or brain case. Gradually enlarge the exposed area, keeping the scalpel horizontal to avoid damaging the brain.

iii. Expose and identify: *olfactory lobes* and *cerebrum* of the *telencephalon, optic lobes, cerebellum, posterior choroid plexus, medulla oblongata,* and *spinal cord.*

Special dissections

i. Remove an eye and cut away the lateral facial margin of the chondrocranium in order to

Figure 12.13. *Squalus acanthias*, central nervous system, dorsal view.

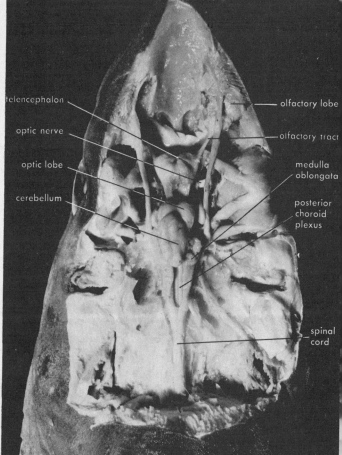

telencephalon

optic nerve

optic lobe

cerebellum

olfactory lobe

olfactory tract

medulla oblongata

posterior choroid plexus

spinal cord

trace out roots of the ten *cranial nerves*. This will require careful technique and foreknowledge of the anatomy to be encountered.

    ii. Cut through the spinal cord and each cranial nerve, and lift out the brain. Identify each nerve root and the major lobes of the brain.

    iii. Finally, make several sagittal cuts through the entire brain and identify the four *ventricles*.

■ **With which parts of the brain are the ventricles associated?**

IV. Class OSTEICHTHYES (ACTINOPTERYGII), Bony Fishes; *Perca flavescens*

1. Introduction This class includes all bony fishes, the most varied and numerous vertebrates. The largest group, the teleosts (superorder TELEOSTEI) includes all so-called modern forms—about 29 orders of higher bony fishes. Ganoid fishes, the bowfin *Amia*, and gars (superorder HOLOSTEI) are combined with teleosts to form one subclass, the NEOPTERYGII. A second subclass, PALAEOPTERYGII, contains the primitive bony fishes, the bichir or *Polypterus*, sturgeons, and spoonbills (superorder CHONDROSTEI). Both of these subclasses, along with the lungfishes and lobe-finned fishes (*Latimeria*) which form the subclass CHOANICHTHYES, are combined into the class OSTEICHTHYES. Features such as bony skeleton; cycloid, ctenoid, or ganoid scales, paired fins supported by fin rays; and first gill slit not reduced to form a spiracle (as in sharks) are the chief characteristics that define OSTEICHTHYES, the bony fishes (see taxonomic outline p.209).

Our example of this class comes from the order PERCOMORPHI, typified by the perch. These fishes all have both spine and soft ray supports in their dorsal and anal fins. They include some 20 families containing many of the best known sport and food fishes: freshwater bass and sunfish (family Centrarchidae); freshwater perch (family Percidae); Cichlidae of South America and Africa; Embiotocidae or surf perches; mackerels (Scombridae); and tuna, bonito, yellow fin, and other of the family Thunnidae.

Our example of OSTEICHTHYES, *Perca flavescens*, the yellow perch, is a suitable and readily available example of a higher bony fish.

2. External anatomy (Fig. 12.14)

a. Observe the laterally compressed but gracefully streamlined shape with no neck and little flexibility along the body (nearly all propulsion is from the tail). Observe the trunk and tail scales and their arrangement. Scrape off a few scales and observe them microscopically. They are *ctenoid* scales—thought to be the most advanced of the various fish scale types—in which bony elements are reduced to thin bony plates under the skin (to be distinguished from *placoid* scales). Ctenoid scales provide a protective but flexible covering, though in some cases heavy bone armor has developed. Observe the grooved concentric ridges (growth rings) with small teeth covering the exposed part of each scale (from which the name *ctenoid* is derived).

Some of the large flattened bones of the skull are thought to be derived from sunken enlarged scales of some ancient ancestor, perhaps an agnathid, and are therefore known as *dermal bones* (a term also related to the mode of embryonic origin of these bones).

Observe the *lateral line*, a row of dash-shaped pits running along the side of the animal.

Authorities differ in their view of the function of the lateral line. Some consider it to be an organ of sensitivity to vibrations, others to water pressure, to body equilibrium, or to smell.

b. Notice next the position and kinds of fins—two unpaired *dorsal fins* with *spiny rays* supporting the anterior fin, and *soft rays* supporting the posterior fin. The caudal fin is called *homeocercal* because of its symmetry (compare with the shark's *heterocercal* tail). Location of

Figure 12.14. The perch, *Perca*, external anatomy.

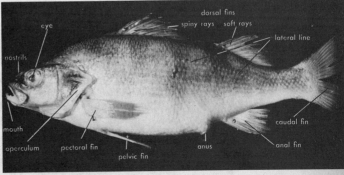

the paired lateral fins is somewhat different from their position in the shark—the pectorals are behind the *operculum* (the shield covering the gills); the rather small *pelvic fins* have migrated forward nearly to the level of, and ventral to, the pectorals. In front of the anal or ventral fin is a large *anus*—not a cloaca as in the shark, since the intestinal and urogenital openings are separate.

c. Cut away the bony *operculum* from one side to expose a *gill chamber* and its four *gills*. Remove an individual gill to study its structure. The hard bony support is the *gill arch*, with posteriorly directed *gill filaments* (containing which blood vessels?). The hard anterior fingerlike projections, the *gill rakers*, prevent coarse material from passing through the gill slits. Describe the function of gill filaments.

3. **Internal anatomy** (Fig. 12.15)

a. The viscera of the perch, as in many teleosts, are compressed (as is the body) and are positioned anteriorly. Expose these organs by making filletlike cuts to remove the wall from one side of the body. Start at the anus, cut dorsally, then toward the gills, and remove this section of body wall. Continue to cut and remove portions of flesh ventral to the gills in order to expose the heart.

b. Within the *coelom* (*abdominal* or *visceral cavity*) a large reddish *liver* is first seen. Move it aside and find a short *esophagus* posterior to the pharynx. The fingerlike continuation of the esophagus is the *stomach*. On its anterior portion are three stubby *pyloric ceca* directed posteriorly. Anterior to the ceca is the *duodenum*, in which digestion occurs. The coiled, posterior *ileum* (*small intestine*) is a region for absorption of digested food particles. Posteriorly, the alimentary canal enlarges to form the *large intestine*, which terminates in the *anus*.

The dorsal portion of the body cavity is filled largely with a thin-walled, extremely tough, whitish *swim bladder*. Dorsal to this, along the wall of the coelom, is the black *kidney*. Posterior portions of the body cavity are often filled with gonadal tissue, usually considerably smaller in the male than in the female.

c. The pericardial cavity, below the gills, contains the two-chambered *heart*. Observe the thin-walled posterior *auricle*. Anterior and ventral to it is the thick-walled *ventricle*, the true muscular pump. The portion of the heart projecting anteriorly is the *bulbus arteriosus*. That entering posteriorly is the *sinus venosus*.

■ **Where is the *ventral aorta* found? the *dorsal aorta*?**

■ **How is the distribution of muscle mass related to the type of swimming action?**

■ **What else can be said about the habits of perch from the appearance of their teeth, eyes, form and length of the gut, body shape, and musculature?**

Figure 12.15. Perch, internal anatomy, lateral view.

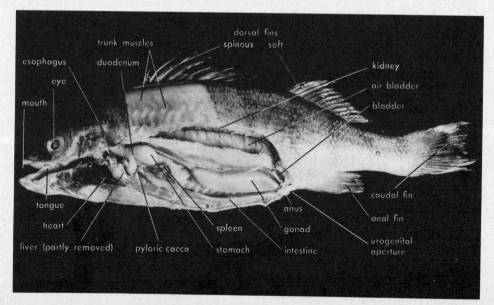

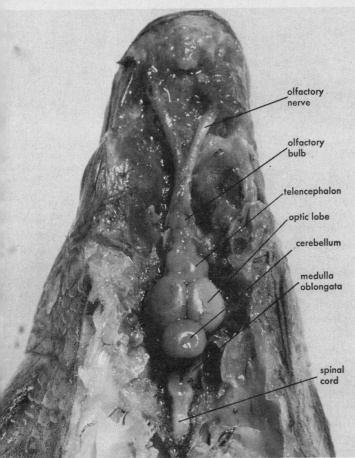

olfactory
nerve

olfactory
bulb

telencephalon

optic lobe

cerebellum

medulla
oblongata

spinal
cord

Figure 12.16. Perch anatomy, central nervous system, dorsal view.

4. **Special projects**

a. *Central nervous system.* Following the procedures used with the dogfish, expose the brain and anterior portion of the spinal cord (Fig. 12.16). Identify as many lobes and cranial nerves as possible by comparison with your previous CNS dissection.

■ **What are the differences between dogfish and perch CNS?**

Draw and label the exposed brain in *lateral view.*

b. *Bony skeleton* (Fig. 12.17). Using your own freshly caught specimen, prepare a demonstration preparation of an intact skeleton using one of the following methods (or another of your own choosing).

i. Bury the specimen or place it within reach of ants or flesh beetles (dermestids).

ii. Tie the specimen to a fixed point on a beach where beach hoppers are abundant.

These biological methods of cleaning will produce beautiful specimens in a remarkably short time (from 6 to 24 hours).

Prepare a finished, neatly dried, and properly labeled skeletal specimen wired to a board.

Figure 12.17. Perch, skeletal system.

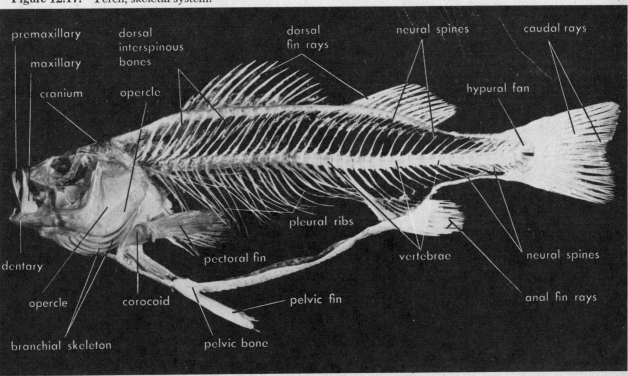

premaxillary

dorsal
interspinous
bones

dorsal
fin rays

neural spines

caudal rays

maxillary

cranium

opercle

hypural fan

dentary

pleural ribs

vertebrae

neural spines

opercle

corocoid

pectoral fin

pelvic fin

anal fin rays

branchial skeleton

pelvic bone

Suggested References

Applegate, V. C., and J. W. Moffett, 1955. "The sea lamprey." *Sci. American*, Apr.

Barrington, E. J. W., 1965. *The Biology of the Hemichordata and Protochordata*. San Francisco, Calif.: W. H. Freeman.

Berrill, N. J., 1950. *The Tunicata*. London: The Ray Society.

——, 1951. "Regeneration and budding in tunicates." *Biol. Rev.*, vol. 26.

——, 1955. *The Origin of the Vertebrates*. New York: Oxford University Press.

Bigelow, H. B., and I. P. Farfante, 1948. "Lancelets," in *Fishes of the Western North Atlantic*. New Haven, Conn.: Yale University, Sears Foundation for Marine Research.

Colbert, E. H., 1955. *Evolution of the Vertebrates*. New York: J. Wiley & Sons.

DeBeer, G. R., 1951. *Vertebrate Zoology*. London: Sidgwick & Jackson.

Eddy, S., 1964. *Atlas of Drawings for Chordata Anatomy*. New York: J. Wiley & Sons.

——, C. P. Oliver, and J. P. Turner, 1964. *Atlas of Drawings for Vertebrate Anatomy*. New York: J. Wiley & Sons.

Gilbert, P. W., 1962. "The behavior of sharks." *Sci. American*, July.

——, (ed.), 1963. *Sharks and Survival*. Boston, Mass.: Heath and Company.

Hyman, L. H., 1942. *Comparative Vertebrate Anatomy*, 2nd ed. Chicago, Ill.: University of Chicago Press.

——, 1959. *The Invertebrates, vol. 5. Smaller Coelomate Groups*. New York: McGraw-Hill.

Jollie, M., 1962. *Chordate Morphology*. New York: Reinhold.

Lennon, R. E., 1954. "The feeding mechanism of the sea lamprey and its effect on host fishes." *U.S. Fish and Wildlife Service Fisheries Bull.* 56.

Millot, J., 1955. "The coelocanth." *Sci. American*, Dec.

Romer, A. S., 1970. *The Vertebrate Body*, 4th ed. Philadelphia, Pa.: W. B. Saunders.

Weichert, C. K., 1967. *Elements of Chordata Anatomy*, 3rd ed. New York: McGraw-Hill.

Young, J. Z., 1962. *The Life of Vertebrates*. New York: Oxford University Press.

CHAPTER 13

CLASS AMPHIBIA

1. Introduction

Rana pipiens, the leopard or grass frog, is "Mr. Vertebrate" himself. Hardly a student anywhere completes a course in biology or zoology without examining one. As happens with many other representative animals, the adjective "typical" is like Mark Twain's explanation of the report of his own death—slightly exaggerated. Perhaps the word should not be used at all: one animal is usually just as typical of the group to which it belongs as any other, and every surviving animal is "specialized" in the sense that it is adapted to its environment. We commonly select representative animals by their availability. The leopard frog, for example, is one of the most common vertebrates in the temperate zones of North America. In addition, it is easy to raise and maintain, and useful in research laboratories.

The frog's ability to jump for escape and its flip-out tongue for insect catching are characteristics around which natural selection and subsequent adaptation have built a rather distinctive animal. Its huge mouth, tongue construction, highly modified backbone, and unique hind legs all manifest this. Actually, an animal more representative of the amphibian evolutionary position, halfway to the dry-land vertebrates, would be one of the elongated, tailed, and less specialized salamanders (waterdogs and newts).

Despite this reservation, the frog is a worthy vertebrate specimen, an evolutionary intermediate between the least and most complex chordates.

■ What anatomical and developmental features prove the frog's dependence on the aquatic environment?
■ What features mark its independence from the aquatic environment?
■ Why are amphibians considered of evolutionary significance?

2. Observations

I. Normal Reactions

Observe a group of living frogs. Notice their normal stance and position of the legs. Touch behind the forelegs with a needle. This stimulus sometimes causes the frog to croak.

■ In breathing, does the chest cavity move, or the throat? How does air reach the lungs?

■ **What other means of respiration is available to the frog?**

Touch the eyes. Note the reaction.

Turn the frog over and observe its righting reaction.

Notice the stop-and-go swimming reaction in an aquarium. Compare the function of fore and hind limbs in swimming. Describe their action. Observe the method of floating and the balancing action of the extended webbing of the hind feet. Disturb the frog and notice the speed of its emergency retreat.

Notice the location of the eyes.

■ **Of what adaptive importance is this location, and that of the nostrils as well?**
■ **What is the importance of the color pattern?**

These amphibians need all their protective devices in order to survive the continual predation by the large variety of animals that feed on them.

■ **What are their predators?**

Compare the rapidly moving frog and sluggish toad in this respect. The toad has highly toxic skin glands that make it highly unpalatable to most predators. Two distinctly different evolutionary paths are illustrated by these animals.

II. Other Reactions

1. Place the frog in a screened cage or terrarium, add some living houseflies or meal worm larvae, and observe feeding reactions.
2. Apply some acetic acid with bits of filter paper to the back or sides of the animal. Note reactions.
3. Study a demonstration of the pattern of nervous responses in a *decerebrate* frog. Various degrees of destruction of the central nervous system can be tested in terms of attitude, swimming control, reaction to tactile stimuli, reaction to acetic acid, and righting reaction.

Decerebrate preparations can be made on lightly etherized frogs by cutting through or dissecting out the cerebrum of the brain. Frogs can also be pithed to destroy the brain and the anterior or the entire spinal column.

These demonstrate clearly the degree to which the frog's actions and responses are under local (spinal) control.

3. External Anatomy (Fig. 13.1)

I. Skin

Notice thickness and feel of the skin. Some of the frog's respiratory gas exchange occurs through its moist skin surface. Contrast this with the skin of *fish*, *reptiles*, *birds*, and *mammals*.

■ **How is the amphibian's skin related to its requirement for external moisture?**

Contrast the pigmentation of the dorsal and ventral surfaces.

■ **Of what adaptive significance is this difference?**

II. Head

On the head, observe the *mouth, nostrils (external nares), eyes*, and *eyelids*, and the *eardrum (tympanic membrane)* behind the eyes.

■ **Are upper and lower eyelids present?**

Figure 13.1. *Rana pipiens*, external anatomy. (Photograph courtesy of Dr. John A. Moore, University of California, Riverside.)

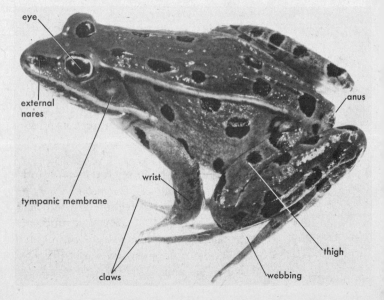

Locate the thin third eyelid, a transparent *nictitating membrane* that protects and cleans the eye without obscuring vision.

■ **Is there a distinct neck?**

The neck gives the head flexibility for feeding or lateral movement.

■ **What feeding mechanism of the frog obviates the need for a flexible or extensible neck?**

III. Appendages

Observe the appendages—*number, position, webbing, toes, claws*. Be sure to notice the lateral position of the legs (as in all amphibians). Mammal limbs, suspended directly under the trunk, provide a direct fore and aft movement, in contrast to the waddling walk of the frog. Relate this movement to the habitats of the two classes.

Notice, too, that the limbs have the same general structure as our own. Find the parts corresponding to our upper and lower arm, wrist, hand, fingers, thigh, lower leg, ankle, foot, and toes.

■ **Can you locate an enlarged thumbpad?**

The thumbpad, absent in females, serves as a male gripping device during *amplexus* (pairing prior to *external* fertilization).

4. Internal Anatomy (Figs. 13.2–13.15)

I. Mouth

Examine the roof and floor of the *buccal cavity*, opening it wide enough to observe the *esophagus*. If necessary, cut the angle of the jaws to permit a better view. The *tongue* lies in the buccal cavity.

■ **Review the tongue's attachment and function.**

Behind the tongue is the *pharynx* with an opening called the *glottis*, which looks like a slit in a slightly raised circular portion of the pharynx floor. Air is carried to the lungs through this opening.

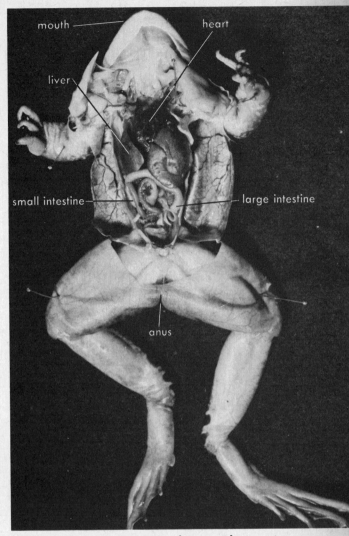

Figure 13.2. *Rana pipiens*, male, ventral view; internal anatomy.

Trace the air passage from the external nares to the buccal cavity through the *internal nares*, to the pharynx, glottis, and lungs.

In the anterior roof of the mouth small *vomerine teeth* (absent in toads) can be felt. Find the teeth by rubbing the upper jaw with your finger. Teeth are named for the jaw or skull bones to which they are attached: *vomer, maxilla, premaxilla,* or *mandible*.

■ **Are there any teeth on the lower jaw?**
■ **What is the function of these teeth?**

II. Body Wall, Coelom, and Viscera (Fig. 13.2)

1. **Dissection procedure** Place your frog ventral side up in a dissecting pan with limbs pinned.

Lift the loose skin of the belly, snip a small hole through it—not through the muscles below—and cut along the midventral line from the posterior end of the trunk to the tip of the jaw. You will notice that the skin cuts easily and that spaces separate it from the skin musculature. In a living frog, this space between skin and muscle is filled with *lymph*, a colorless, intercellular body fluid. Make cuts laterally both behind the forelimbs and in front of the hind limbs, pull the skin flaps down, and pin them. In specimens injected with colored dye or latex shortly after death, *arteries* will be red, *veins* blue. In uninjected frogs, peripheral veins are usually more conspicuous than are the arteries (which closely parallel the veins throughout the circulatory system). This is caused by the passive filling of veins with blood from the tissues, and by the draining of arteries into the tissues without further filling once the heart action has ceased.

Expose the body cavity by cutting through the ventral body wall to one side of the midline. Run the incision from the posterior end of the animal to and through bones of the shoulder girdle (Fig. 13.2). Then make lateral flaps as you did with the skin and pin them.

In cutting through the body wall, you have also cut the *parietal* or *somatic peritoneum*, a thin membrane that forms the outer lining of the *coelom*. Organs lying within the coelom are enclosed by an extension of the inner coelomic lining, the *visceral* or *splanchnic peritoneum*. The somatic peritoneum close to the body wall and the splanchnic peritoneum around the intestine join at the *dorsal mesentery* to form the *coelom* in which the body organs are suspended.

2. Review of organs

a. For general orientation, make a survey of the organs exposed. *Never use scissors* once the body wall has been opened.

The anterior coelom is divided by a transverse partition, part of the pericardium that surrounds the *pericardial cavity* enclosing the heart.

Observe the *heart* and its major vessels. Then look on either side and posterior to the heart to find the *liver*, largest organ of the body. Note the three lobes (two main lobes and a smaller medial one). Lying freely between the median and right lobes is the greenish *gall bladder*, which stores *bile*.

Under the left lobe of the liver is the *stomach*. It may be considerably distended with food or quite shrunken if the animal has not recently eaten. A short *esophagus* connects the mouth and stomach. The *small intestine* leads from the stomach via a *pyloric sphincter* valve into the *duodenum*, which loops forward, runs parallel to the stomach, and then turns back into the remainder of the small intestine (*ileum*). This highly coiled tube, supported by mesenteries, enlarges to form the *large intestine* and the terminal *rectum*, which empties to the outside through the *cloaca*.

b. If your frog was caught during the breeding season, its gonads will be well developed. In fact, much of the female body cavity will be filled with massed black and white eggs. Carefully remove these eggs to expose the organs.

c. Locate the yellow *fat bodies* extending anteriorly from the excretory organs. *Adrenal* (*suprarenal*) glands are yellow bands across the ventral surface of the excretory organs. The term *kidney* properly applies to the typical adult structure in higher vertebrates. The frog's excretory organ is a *mesonephric kidney*, whereas those of mammals, birds, and reptiles are *metanephric kidneys*.

d. A few more organs still to be identified are the *spleen, pancreas, urinary bladder*, and *lungs*. The *spleen* is a small, deep red, spherical organ embedded in the mesentery supporting the small intestine. The pinkish, diffuse, lobulated pancreas lies near the liver within the mesentery connecting stomach and duodenum. The thin-walled, translucent *urinary bladder* is ventral to the large intestine in the most caudal part of the coelom. The *lungs* are covered by the two outer lobes of the liver. A dorsal pocket of the coelom, the *pleuroperitoneal cavity*, encloses the lungs and extends along the dorsal body wall from the lungs to the junction of excretory duct and oviduct near the rectum.

e. During the present and ensuing dissections you may well observe parasitic worms—flukes, nematodes, cestodes—in the lungs, small intestine, large intestine, rectum, and bladder, or even in the mouth and Eustachian tubes. Each worm is highly specific, being adapted to a particular organ as well as host. For example, the fluke *Haematoloechus* and the nematode

Rhabdias are found only in the lungs, the flukes *Gorgodera* and *Polystoma* only in the bladder, and *Proteocephalus*, a tapeworm, in the intestine. Encysted larvae of other helminths are often seen in tissues, indicating that the frog is an intermediate host for these parasites.

■ **What might be their final host?**

3. Digestive system

■ **Where do products of the liver enter the alimentary canal?**
■ **Where do products of the pancreas enter the alimentary canal?**

Note the three *cystic* ducts that carry bile from each lobe of the liver into the gall bladder. Bile reaches its final destination via the *bile duct*. Trace it between stomach and duodenum to its opening halfway down the latter structure.

■ **Can you find the even finer *pancreatic* ducts emptying into the bile duct? (Look in the vicinity of the pancreas where the bile duct passes through it.)**
■ **State the functions of bile and pancreatic juice.**
■ **Do parts of the digestive system other than the pancreas produce digestive enzymes?**
■ **Name some other examples of digestive enzymes (see your text).**

Free the digestive system from its mesenteries; stretch it out but do not disarrange the pancreas in the first loop of the intestine. Arrange the organs so that they do not obscure one another. Draw the extended system.

■ **Describe the sequence of steps involved in digestion from the time a living insect is picked off a leaf by a frog's tongue, to the time a waste parcel passes out the cloaca.**

4. Urogenital system

a. *Excretory system.* Find the *excretory ducts*, light-colored tubes running from the outer and posterior border of each kidney toward the rectum. They can be found along the dorsal surface of the enlarged caudal portion of the oviducts and enter the cloaca through a papilla

and a pore on its dorsal wall. Locate this connection.

■ **How is the bladder filled and emptied?**

b. *Female reproductive system* (Fig. 13.3). Find the *ovaries*, attached to the dorsal body wall by a mesentery, the *mesovarium*. The long white *oviducts* are coiled, posterior to the ovaries. In immature females (or females during the nonbreeding season) the small lobulated *ovaries* near the dorsal midline bear clusters of undeveloped eggs; smaller, less deeply coiled oviducts run the length of the middorsal region alongside the excretory organs. During maturation of the ovaries, the ova fill with yolk and become pigmented on one side; they burst through the thin membranous wall of the ovaries and fill the abdominal cavity. They then

Figure 13.3. *Rana pipiens*, female, ventral view; internal anatomy of urogenital system and digestive system.

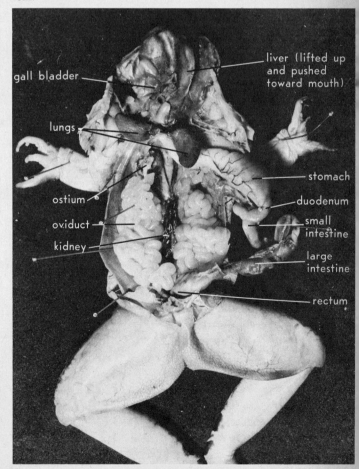

pass into the openings of the oviducts. Trace the coiled oviducts of a mature female anteriorly and find the expanded, funnel-shaped openings dorsal to the lungs at the anterior extremity of the *pleuroperitoneal* cavity. Move the lungs to one side to locate the oviduct openings.

■ **Are ovaries and oviducts directly connected?**

■ **How do eggs pass into the oviduct funnels?**

■ **How are they passed down the tubes?**

At the posterior end of each oviduct, find the expanded thin-walled organ (*ovisac*) for storage of eggs prior to laying and fertilization. It may be necessary to dissect away the dorsal mesenteries to determine the full length and size of the posterior ends of the oviducts. Note the openings of the oviducts in the dorsal wall of the cloaca. Observe that ovisacs, excretory ducts, and bladder have separate openings into the cloaca.

Individual eggs are coated with a gelatinous, albuminous material secreted by glands in the walls of the oviducts. During amplexus the male uses its clasping thumbpads to squeeze the sides of the female. As eggs are extruded into the water, sperm are shed over them, and fertilization follows. A thick, transparent capsule forms from the swelling of the gelatinous layer around each egg as it enters the water. Extruded eggs form a compact mass in which the individual fertilized eggs (*zygotes*) pass the early stages of development.

c. *Male reproductive system* (Fig. 13.4). Male gonads consist of two light-colored, bean-shaped *testes* along the dorsal body wall, which are suspended from the anterior end of the excretory organs by a mesentery, the *mesorchium*. Sperm ducts (*vasa efferentia*) carry seminal fluid through many fine white tubules that cross the mesorchium from the testes and enter the inner border of each kidney. These tubules lead into the *collecting tubules* of the kidneys, which carry the sperm into the *excretory ducts*. Therefore, the *mesonephric duct* of the male frog also serves as a *vas deferens*, conducting sperm and urine to the cloaca. It is thus often termed a urogenital duct. A *seminal vesicle* is attached to the outer side of each urogenital duct near its junction at the dorsal wall of the cloaca.

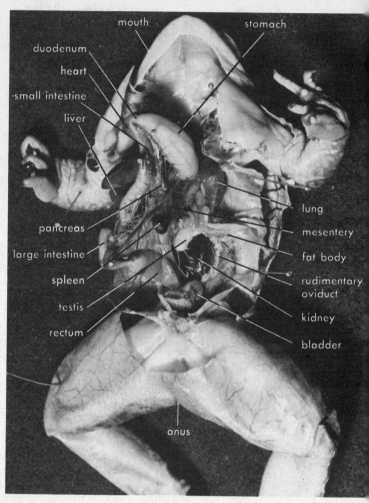

Figure 13.4. *Rana pipiens*, male, ventral view; internal anatomy of urogenital system and digestive system.

Sperm is stored in these vesicles prior to copulation.

■ **Is there a rudimentary female oviduct in your *male* frog? (This organ is present in the leopard frog, but not in the bullfrog.) Though not functional, it corresponds in structure to that of the female.**

5. **Circulatory system** (Figs. 13.5 and 13.6)
a. *Arterial circulation.* Remove the alimentary canal, then the reproductive system, bladder, and fat bodies. Do not damage major blood vessels, nerves, or excretory organs.

Expose the *heart* in its protected ventral position in the pectoral girdle and within two folds of the *pericardium* (Fig. 13.5) that surrounds it. Note the single, heavily muscled pos-

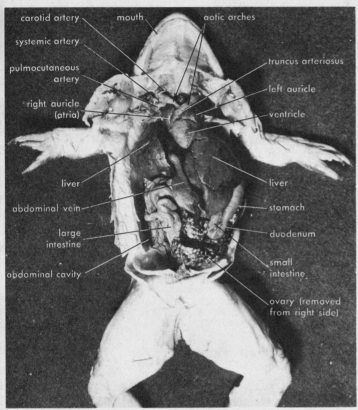

Figure 13.5. *Rana pipiens*, female, ventral view; circulatory, digestive, and reproductive systems.

terior chamber of the heart, the *ventricle*; and two thinner walled, dark-red anterior *atria* which lie on either side of a large bifurcated vessel anterior to the heart, the *conus* (or *truncus*) *arteriorsus*. The latter branches into two heavy walled arterial vessels which in turn are divided into the principal arteries (carrying blood *away* from the heart).

■ **How does the frog heart compare with that of the fish you studied?**

Recall that in the fish *all* blood from the heart has to pass through the gill capillaries to be aerated prior to circulating through the rest of the body. In the frog, however, and in all air-breathing vertebrates, a new structure is involved—the *lung*. Vessels carry deoxygenated blood from heart to lungs, and aerated (oxygenated) blood back to the heart. The heart then pumps this blood through the systemic vessels to all parts of the body. Instead of one

circulation, we have two—*pulmonary* and *systemic*. In the frog an incomplete separation of these two circulations exists: the three-chambered heart does not completely separate blood to be pumped into the lungs from that returning from the lungs and destined to be pumped into the body circulation.

■ **In what way is oxygenated blood kept separated from deoxygenated blood?**

In mammals this evolutionary development is complete, with a *four*-chambered heart that segregates oxygenated blood from deoxygenated and ensures the efficient separation and operation of this double circulatory system.

Trace the conus arteriosus forward and observe the symmetrical division into major ar-

Figure 13.6. *Rana pipiens*, female, ventral view; circulatory and digestive systems.

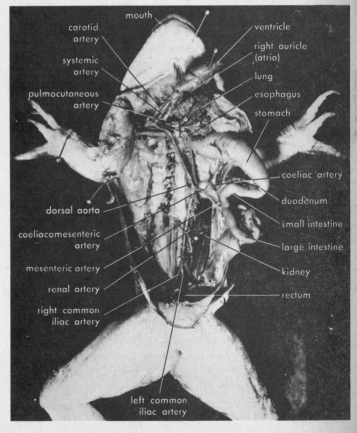

teries. Three branches lead from each of the two primary *aortae* of the conus arteriosus. These are the three principal *arterial arches* on each side of the body: *carotid, systemic*, and *pulmocutaneous*. The most posterior branch, the *pulmocutaneous artery*, supplies blood to the lungs via the *pulmonary artery*; its dorsolateral branch, the *cutaneous artery* to the skin, also serves as part of the respiratory system. The middle branch is the *systemic artery* that passes laterally, arches dorsally, continues caudally, fuses with the corresponding partner systemic artery near the sixth vertebra, and continues posteriorly as the single *dorsal aorta*. The most anterior of the three branches is the *carotid artery* that sends blood to the head. Each carotid branches almost immediately into an *external* and *internal carotid*. Find the systemic artery again and locate its largest branch, the *subclavian artery*, which supplies the foreleg. Continue posteriorly to the united dorsal aorta, which sends a large branch ventrally from the point of junction of the two systemic arteries to the digestive tract. This branch is the *coeliacomesenteric artery* which, as its name implies, passes to the alimentary canal via the dorsal mesenteries (branching into a *coeliac artery* to the stomach, pancreas, and liver, and a *mesenteric* to the small intestine, rectum, cloaca, and spleen). Next, find the many *renal arteries*, which branch from the dorsal aorta near the middorsal line and pass to each kidney, and the small *lumbar arteries* that pass to body wall muscles. Free the margin of the kidney to find the entrance point of these arteries.

■ **How many renal arteries pass into each kidney?**

The most anterior renal artery is the *genital* (*spermatic* or *ovarian*) *artery* that passes to testes or ovaries.

At its terminus, the dorsal aorta divides into two *iliac arteries*, each of which passes into a hind leg as the *sciatic artery* that subdivides into smaller arteries (Fig. 13.6).

Show these major vessels and divisions in a diagrammatic outline.

b. *Venous circulation*. Veins return deoxygenated blood from the capillaries to the heart. These vessels merge into successively larger vessels, from venules to veins, which are thin-walled tubes through which the blood flows in a rela-

tively steady stream (rather than in high-pressure spurts as in the thick-walled arteries). Backflow is prevented by small *valves* located at regular intervals. All venous blood of the body (except blood from the lungs) flows into the *sinus venosus*, the triangular, thin-walled sac on the dorsal side of the heart. Blood from the lungs returns via the *pulmonary veins*, which lead directly into the *left atrium*. Blood collected into the sinus venosus passes into the *right atrium* via the *sinoatrial aperture*, a valvelike opening that prevents backflow into the sinus venosus when the right atrium contracts. The two atria pump blood into the single *ventricle*. Backflow is again prevented by a complex series of valves that also permit the major propulsive effort—contraction of the ventricle —to send blood spurting into the conus arteriosus and into the pulmonary and systemic arterial circulation.

Many major veins will have been damaged or removed in your dissection, but the general pattern should still be discernible. Work out and identify as many of the following veins as you can find, supplementing your search with the illustrations and demonstration dissections available for class use. Three main veins carry the blood from all parts of the body into the sinus venosus—these are the three *vena cavae*. Two *anterior vena cavae* (or *precaval veins*) convey blood from the head, forelegs, and forebody into the anterior connections of the venosus. A single *posterior vena cava* (or *postcaval vein*) carries blood into a single large medial vessel that empties into the posterior apex of the sinus venosus triangle.

Three large veins join in each brachial area, near the arterial arches, to form the anterior vena cavae that drain into the sinus venosus. We shall consider the three branches of the vena cavae individually. The *subclavian vein* drains the forelimb, where it is formed from the *brachial vein* in the basal part of the forelimb and from a large *musculocutaneous vein* that drains the skin and body wall muscles. This vein carries oxygenated blood from the skin and enables the frog to respire through its skin while the animal is immersed and cannot use its lungs. The middle vessel that joins the vena cava is the *innominate vein*, also formed by two tributaries; the *internal jugular* passing dorsally near the angle of the jaw and draining the brain and head, and the *subscapu-*

lar vein, a small vessel from the dorsal surface of arm and shoulder. Most anterior of the three vessels is the *external jugular vein,* which drains the lower jaw and buccal cavity floor (via *mandibular* and *lingual* veins). The single posterior vena cava (in contrast to the paired posterior cardinal sinuses of the dogfish) is formed from a series of paired branches from the kidneys, the *renal veins.* These veins should be compared with the similarly branched renal arteries that supply blood to the kidneys. The posterior vena cava runs anteriorly (ventral to the dorsal aorta and dorsal to stomach and liver), passing to the right of the coeliacomesenteric artery. It is joined by a pair of *hepatic veins* from the two main lobes of the liver, then enters the sinus venosus. Near the anterior end of the kidneys the gonadal veins (*ovarian* or *spermatic*) join the vena cava.

Blood returns from the large hind legs principally through two major vessels, the *femoral veins.* Find them on the outer margin of the legs, near the surface. Trace a femoral vein forward to its division into two branches near the pelvic girdle. The inner branch from each femoral, the *pelvic vein,* passes ventrally. Near the ventral body wall the pelvics join and continue anteriorly along the ventral midbody wall as the single *anterior abdominal vein.* You first saw this vessel in the frog when you made the initial abdominal incision. Small vessels join the anterior abdominal, draining the body wall musculature. Just behind the heart the vessel turns dorsally and branches into the left and right lobes of the liver. These branches become part of the *hepatic portal* system. Other contributors to the hepatic portals are the *gastric vein* from the stomach, *intestinal* from large and small intestines, and *splenic* from the spleen. On an intact specimen you can locate the main hepatic portal in the mesenteries, parallel to the bile duct.

Return now to the two branches of each femoral vein, the pelvic branch having just been traced. The more dorsal branch is the *renal portal vein.* The *sciatic vein* from the inner dorsal surface of each hindleg joins the corresponding renal portal vein. The combined vessel then passes forward to the outer edge of each kidney, sending small veins into the organ. These veins divide into venules and capillaries to produce a capillary bed, the *renal portal system.* The blood flowing through this bed emerges into renal veins that form the posterior vena cava as already described.

Note that all the blood from the femoral veins passes into one of two *portal* systems—directly into the renal portals or into the hepatic portals via pelvic and anterior abdominal veins.

■ What is the definition of a portal system?

■ What is its primary function in kidney and liver, respectively?

■ Can you suggest why portal systems tend to reduce the efficiency of the circulatory system? (Might this account for loss of the renal portals in mammals?)

c. *Capillary circulation.* A fine means to appreciate the fantastic number and branchings of the circulatory system is to observe capillaries in the toe webbing of a living frog. Study such a demonstration closely, noting the regular movements of *blood cells* (*Can you detect the pulse?*), the *arteries, arterioles,* and *capillaries.* Many of the latter are fine enough to allow only a single individual blood cell to squeeze by at a time. This represents a considerable difference from the rate of movement in the dorsal aorta. The frog's toe webbing is a thin membrane in which circulation can be seen, and all tissues and *every living cell* is sustained by such intimate and constant association with a capillary bed. Be sure to notice that all of these vessels still form a single *closed system* (*as opposed to what other type and in what group?*).

■ How can you distinguish veins from arteries in the frog webbing?

■ Can you locate branched *lymph vessels* between the capillaries in the frog membrane?

d. *Lymphatic system. Lymph,* though part of the circulatory system, is very different in appearance and function from the blood in arteries and veins. It is colorless, contains few cells, no red blood cells, and is found outside the blood vessels, bathing individual body cells. It is therefore an *intercellular fluid,* the actual vehicle conveying oxygen from the capillaries into the cells and CO_2 from cells to capillaries. Dissolved food products and cellular wastes—in fact all materials exchanged between cells and blood—are carried by lymph. Lymph flowing into the veins becomes part of the supporting fluid or *plasma* of the blood. It also serves an essential function in maintaining ionic (osmotic) balance and as a transport medium for antibodies and other elements of the body's defense mechanism.

The lymphatic fluid has its own circulatory channels and vessels and joins the bloodstream at junction points with certain major veins. Spaces between cells enable lymph to flow through regular channels. In the frog, these are well developed and are seen as *subcutaneous lymph spaces* between skin and body wall. These connected spaces actually carry the lymph through fine *lymph capillaries* and *lymph vessels* propelled by two pairs of *lymph hearts* that pump the fluid into the venous system. An anterior pair is located near the third vertebra, pumping into the *vertebral vein* (a tributary of the internal jugular). The posterior lymph hearts lie near the end of the vertebral column and pump into the *iliac veins. Semilunar valves* prevent backflow into the lymph channels. These hearts beat independently of one another and of the true heart. Details of this system cannot be studied here, but its functional importance should be understood in order to follow the general conception and unity of the complete circulatory system.

6. Respiratory system Remove the heart of your frog, being careful to avoid damage to the *lungs*. Trace the lungs anteriorly to their entrance from the *larynx*, dissecting away any obstructing tissue.

■ **What is the function of the** *Eustachian tubes*, **which pass into the** *middle ear* **from the roof of the mouth (near the juncture of the jaws)?**

Make a complete dissection of the respiratory system. Raise the stump of the bisected esophagus and cut the corners of the mouth to remove intact: esophagus, lungs, lower jaw, and floor of the mouth. Cut through the glottis to expose the *larynx*, a small, boxlike cavity supported by two pairs of ringlike cartilages. Stretched longitudinally across the larynx are a pair of elastic bands, the *vocal cords* (Fig. 13.7). Air passing through a slit between the cords sets them into vibration, producing the characteristic resonant croaking of the frog (aided in the male by an expanded *throat sac* of skin that serves as a resonating chamber for croaking during the spring breeding season).

Cut through the larynx, the paired *bronchi*, and into the attached lung. Notice the lung's pleated inner surface, which greatly increases its total surface area.

■ **Of what advantage is this increase?**

Review how air passes into and out of the lungs, including the related role of the circulatory system.

■ **How is the** *skin* **also involved as an essential respiratory organ in amphibians?**
■ **Did you find parasites in the lungs?**
■ **Was there evidence of damage to the lungs?**
■ **What does such damage (or lack of it) imply about a physiological host–parasite balance between these parasites and their anuran host?**

7. Nervous System
a. *Sense organs*
     i. *Eye*. Remove both eyes carefully from their sockets. Cut around the eyelids, sever the muscles that hold the eye in position, and cut the optic nerve behind the eyeball.

Identify the *upper* and *lower eyelids* and the inner or *nictitating membrane*; the smooth *conjunctiva*, underlying the eyelids; the eyelids; and the *cornea*, light-focusing transparent surface of the eyeball. Behind the *cornea* lies the pigmented *iris diaphragm* with its central oval opening, the *pupil*. Continuous with the cornea is the *sclera*, a tough white membrane that covers the back of the eyeball. Locate the white *optic nerve* passing posteriorly through the eyeball and socket.

■ **To what part of the brain does this nerve connect?**
■ **What is the function of the structures named above?**

Cut a vertical section through each eye (use a new razor blade); make one cut along the anteroposterior axis, the other perpendicular to the first.

In front of the *lens* is the *anterior cavity* (with fluid *aqueous humor*). The lens, behind the iris, is transparent in a living specimen and forms a flattened sphere. Behind the lens is the *posterior cavity*, with gelatinous *vitreous humor*.

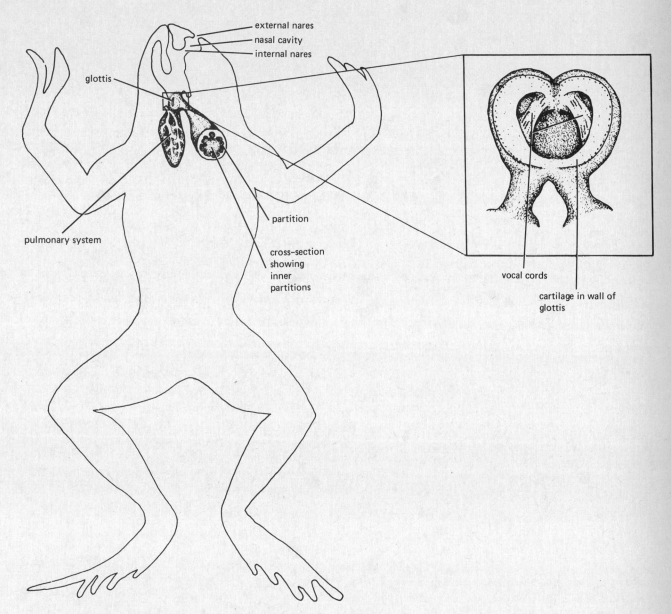

Figure 13.7. Frog respiratory organs. Note relatively small lungs.

Inside the sclera lining the back of the eyeball is the *retina*, the true photosensitive organ. Between the two coats is the blackened, vascular *choroid coat*.

When stimulated, the light-sensitive retinal cells transmit impulses along the optic nerves. These impulses are interpreted in the higher brain centers as *vision*. It is believed that each retinal nerve fiber stimulates a discrete cell in the visual center so that the retinal pattern of stimulation is copied in the visual center of the brain. Photographs from the retina of the frog have actually been made, in which the retinal picture was protected by killing

the frog in the dark, and the retina subsequently "developed" to show an image reproducing the same pattern that had been used experimentally to stimulate the retina. The ingeniously simple method employed a dark-adapted frog that was briefly exposed to a light through a window and then quickly sacrificed in the dark—quite as a photograph would be exposed and then protected from additional light. The frog's retina was removed in the dark and the cells that had been stimulated formed a pattern of a window with cross markings and a window shade exactly as it actually appeared.

ii. *Ear.* Expose the ear by carefully remov-

234   CLASS AMPHIBIA

ing the skin covering one of the two *tympanic membranes* of the eardrum. Open the cavity into the middle ear by cutting around the membrane. Find the *columella*, a slender bone that is connected from the center of the tympanic membrane to the opening of the inner ear (the *fenestra ovalis*). This bone conducts sound vibrations from the tympanic membrane to highly sensitive cell endings (*organ of Corti* cells) that transmit stimuli to delicate nerve endings of the *auditory nerve* in the inner ear. The auditory nerves in turn transmit impulses according to the pattern of stimuli and the result is interpreted in the brain as sound.

The *Eustachian tube* connects the mouth cavity with the middle ear and serves to equalize the air pressure on the two sides of each tympanic membrane, thus ensuring a constant tautness of the membrane, essential for proper auditory reception. Locate the tube by careful probing. Review in the laboratory models, charts, or your text the functions of the *middle* and *inner* ear in shark, frog, and man.

■ **What are the relationships between the sense of *balance* or *equilibrium* and *sound perception*?**

b. *Spinal nerves.* On the dorsal wall of the pleuroperitoneal cavity are the shiny white or yellowish cords you probably have already seen. Ten pairs of these spinal nerves pass out from the vertebrae, the largest being the second pair paralleling the subclavian arteries. This relatively small number is a reflection of the frog's reduced number of vertebrae and loss of the tail. The first three nerves have interconnecting branches that form a *brachial plexus*; the next three are small and innervate the back muscles. The seventh, eighth, and ninth have many interconnections, forming a *sciatic plexus* from which a single large *sciatic nerve* passes into each hind leg, where it is seen on the inner dorsal side embedded among the muscles and near the sciatic vein. *Trace* one of these out into a leg along with the sciatic artery and sciatic vein that accompany it.

Ventral to the spinal nerves and parallel to the spinal cord are the *sympathetic trunks*, a pair of chains of ganglia, connected by lengthwise fibers, one on each side of the cord. This is part of the *autonomic nervous system* by which various organs, smooth muscles, and involuntary bodily reactions are controlled. A *ramus communicans* connects each ganglion to a nearby spinal nerve. Ganglia in turn innervate various visceral organs, glands, and blood vessels via fibers that unit in *plexuses,* then connect to the target organs. Search for some of these trunks and nerves. (Keep your specimen under water, and use a hand lens or dissecting microscope.) Observe that the two sympathetic trunks are in close association with the systemic arteries, approach each other where the arteries merge, and then run parallel to the dorsal aorta. Usually the chains are covered by a large layer of black pigment cells.

c. *Central nervous system—the brain* (Fig. 13.8) *and spinal cord*

i. *Dissection.* Expose the brain of your specimen by making careful cuts anteriorly from the base of the skull (point of attachment of first vertebra). Remove bits of the dorsal skull plate, using only the points of your scissors and fine forceps. Expose the entire brain with its covering of protective *membranes.*

Similarly, expose the spinal cord by cutting posteriorly through the dorsal parts of a few vertebrae. Pull the exposed vertebrae apart to view the cord.

ii. *Anatomy of the brain.* Remove the covering membranes and blood vessels. Note the abundant blood supply to the brain. Identify the following major parts of the brain (Fig. 13.8).

Figure 13.8. *Rana pipiens*, dorsal view; central nervous system.

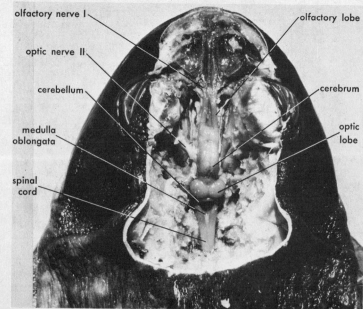

a. *Cerebrum* (two *cerebral hemispheres*), or forebrain, with the *olfactory lobes* extending anteriorly. Find the *olfactory* or *first cranial nerves* extending from the olfactory lobes to the nasal cavities.

b. *Diencephalon*, a small medial structure of the midbrain covered by a thin vascular roof called the *anterior choroid plexus*. On its dorsal surface arises a small *pineal stalk*, ending in the *epiphysis* (*pineal body*).

c. *Optic lobes*, swollen lateral lobes, characterize the dorsal surface of the *midbrain*.

d. *Cerebellum*, highly crenulated thin ridge of tissue running dorsally across the anterior border of the medulla.

e. *Medulla oblongata*, the hindbrain, roofed over by another thin vascular coat, the *posterior choroid plexus*. The medulla is ventral and continuous with the spinal cord. It marks the base of the brain.

The brain is actually a hollow structure, consisting of *ventricles* (cavities) within the portions noted above and all connected with one another. The *lateral* or *first* and *second ventricles* lie in the cerebral hemispheres; the diencephalon contains the *third ventricle*, followed by optic ventricles of the optic lobes, and then the medulla oblongata, the fourth *ventricle*. The *foramina of Monroe* connects the first two ventricles to the third, and the latter leads to the fourth by the *iter* or *aqueduct of Sylvius*. The same passage leads to the optic ventricles.

iii. *Cranial nerves*. The ten structurally specialized *cranial nerves* each innervate specific structures, as discerned by tracing their pattern of embryological development. An important evolutionary assumption therefore follows. If specific muscles in two different vertebrates are innervated by the same cranial nerve, these structures are probably *homologous*, in spite of extreme functional differences of the muscles in the adult stage of the animals being compared. Particularly useful in this study of vertebrate relationships are the *olfactory* (I), *optic* (II), *trigeminal* (V), *auditory* (VIII), and *vagus* (X) nerves.

Draw the intact CNS from the dissection you have completed. Then remove the brain and cord intact by lifting them out gingerly. Cut away nerve roots or membranes that block the removal.

Place the dissected CNS in a syracuse dish and study it from all views to check your drawing and earlier observations.

Keen observers may locate the *optic chiasma*

(ventral crossing of optic nerves under the diencephalon); the *infundibular lobe*, posterior to the chiasma; and the *pituitary body* or *hypophysis* connected to the infundibulum, but often left adhering to the floor of the cranium when the brain is raised.

Make thin dorsal slices, then a careful sagittal section to help you visualize the position and interconnections of the ventricles.

8. **Muscular system** (Figs. 13.9–13.12) Our review of the muscular system of the hind leg (perhaps the only remaining intact part of your frog) is intended as a sample of the organization and function of the system. Use the leg not dissected for the sciatic nerve branches.

Figure 13.9. *Rana pipiens*, dorsal view muscular system.

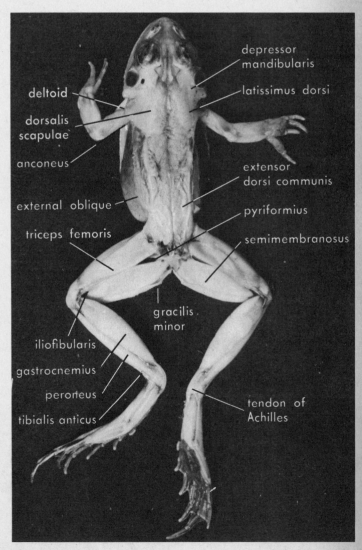

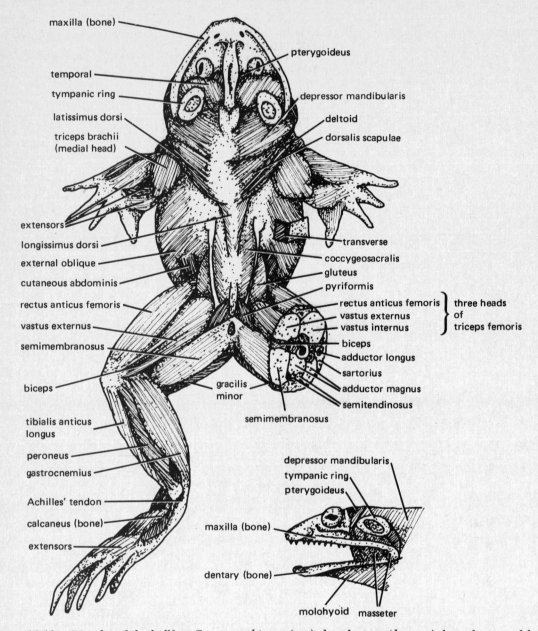

maxilla (bone)
pterygoideus
temporal
tympanic ring
depressor mandibularis
latissimus dorsi
deltoid
triceps brachii
(medial head)
dorsalis scapulae
extensors
longissimus dorsi
transverse
external oblique
coccygeosacralis
cutaneous abdominis
gluteus
rectus anticus femoris
pyriformis
vastus externus
rectus anticus femoris
semimembranosus
vastus externus
vastus internus
biceps
biceps
adductor longus
sartorius
adductor magnus
tibialis anticus
longus
semitendinosus
gracilis
minor
peroneus
semimembranosus
gastrocnemius
depressor mandibularis
tympanic ring
pterygoideus
Achilles' tendon
calcaneus (bone)
extensors
maxilla (bone)
dentary (bone)
molohyoid   masseter

three heads
of
triceps femoris

Figure 13.10. Muscles of the bullfrog *Rana catesbiana*: (*top*) dorsal view (*bottom*) lateral view of head.

Muscles are in opposing units, mainly *flexors* and *extensors* (*flex* your biceps; *extend* your arms); *adductors* and *abductors* (adduct suggests *ad*jacent, hence to bring alongside or flex, as opposed to *ab*duct, to move *away* from the body axis). The movements are under delicate nerve control and give balanced or *opposed movements*.

Skin one of the legs and separate the individual muscles with a blunt probe.

Find the *gastrocnemius*, the large calf muscle. Identify the *fascia*, the tough membrane of connective tissue which encloses the muscle, and the *belly*, the fleshy part of the muscle.

At each end of the muscle identify the tough cord or band, the *tendon*, the means by which the muscle is attached firmly to a bone or other fixed structure.

■ **Which is the tendon of Achilles?**

Determine its *origin*, or *point of attachment to the more stable base*. The gastro-

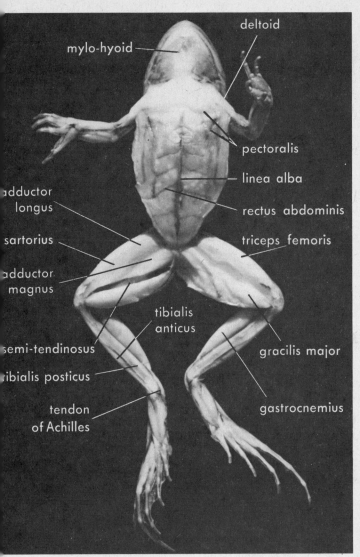

Figure 13.11. *Rana pipiens*, ventral view; muscular system.

mylo-hyoid

deltoid

pectoralis

linea alba

rectus abdominis

adductor longus

triceps femoris

sartorius

adductor magnus

tibialis anticus

semi-tendinosus

gracilis major

tibialis posticus

tendon of Achilles

gastrocnemius

cnemius has two points of attachment that you should trace.

Determine the *insertion,* or *point of attachment to the part being moved* by the muscle. Through its insertion, the gastrocnemius can bend the entire leg below the knee (*flexion*). The *action* of a muscle is its function—the effect of contraction on the structure on which it is inserted.

Find muscles that oppose the movement of the gastrocnemius.

■ **Are they *flexors* or *extensors*? of what structures?**

Along with the gastrocnemius and tendon of Achilles, the lower leg or *shank* has in dorsal view the *peroneus muscle,* which originates on the femur and inserts on the tibiofibula, extends the leg. The *tibialis anticus,* which also originates on the femur and inserts on the tibiofibula, extends the leg and flexes the foot.

■ **What is the action of the gastrocnemius? of the tendon of Achilles?**

The upper leg or *thigh* has several large muscles as seen in dorsal view: the *vastus externus, pyriformis,* and *semimembranosus.* In ventral view one can add the *sartorius, triceps, femoris, adductor magnus,* and *gracilis major.*

With the help of Figs. 13.9–13.12 identify as many of these muscles as possible. Then prepare a *chart* showing each muscle you have identified, its *origin, insertion,* and *action.* Remove the muscles as needed to establish these facts.

■ **Which of these muscles can properly be termed paired or opposed?**

Identify the muscles involved in the swimming action of the hind leg, that is, in extension and withdrawal of the leg.

■ **How does the frog jump?**

9. **Skeletal system** (Figs. 13.13–13.15) Prepare a skeleton by removing all tissue possible from a freshly killed frog. Dip what remains in hot water; brush away the flesh until the bones are clean. Use the hot water sparingly, or the bones will fall apart. Or examine a prepared skeletal mount of a bullfrog to see relationships of the different bones. Study the photograph (Fig. 13.13) and drawings available (Figs. 13.14 and 13.15) to learn the principal bones. Keep in mind that the skeleton is divided into an *axial* part, which includes the skull, vertebral column, and the visceral skeleton; and an *appendicular* part which includes bones of the appendages, girdles, and the sternum.

■ **How can you identify points of muscle attachments?**

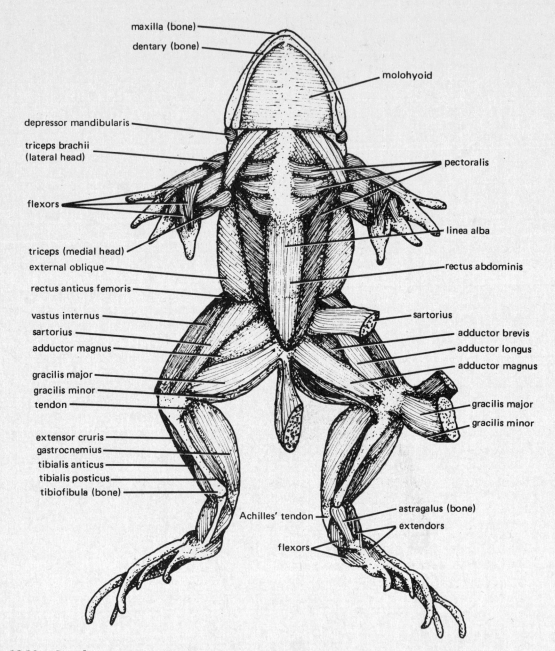

Figure 13.12. Semidiagrammatic ventral view of muscles of the bullfrog *Rana catesbiana*.

Identify from a mounted specimen as many of the following bones as you can: *skull, cranium, upper* and *lower jaw* (individual bones of the skull need not be learned except as a special assignment) (Fig. 13.14), *pectoral girdle* (Fig. 13.15), *scapula, suprascapula, sternum, vertebra, urostyle, humerus, radioulna, carpals* (six), *metacarpals, phalanges* (fingers), *pelvic girdle* (Fig. 13.15),

ilium, ischium, pubis, femur, tibiofibula, tarsals (*calcaneum, astragalus*, plus several smaller bones), *metatarsals, phalanges* (toes), *clavicle*, and *coracoid*.

Make a chart listing the corresponding bones from the *shark, frog, cat,* and *man*.

■ **What are the essential differences among the four skeletal systems?**

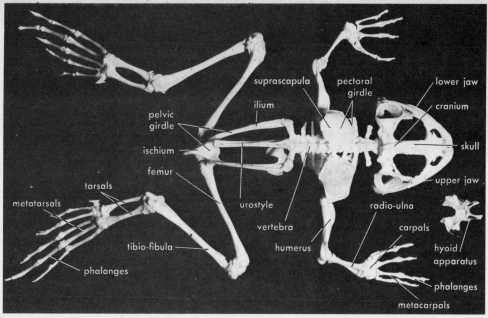

Figure 13.13. *Rana catesbiana*, dorsal view; skeletal system.

Figure 13.14. Semidiagrammatic lateral view of skull and ventral view of hyoid of frog.

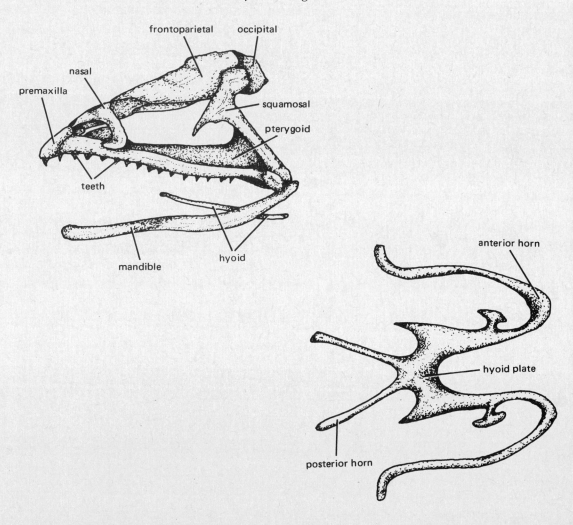

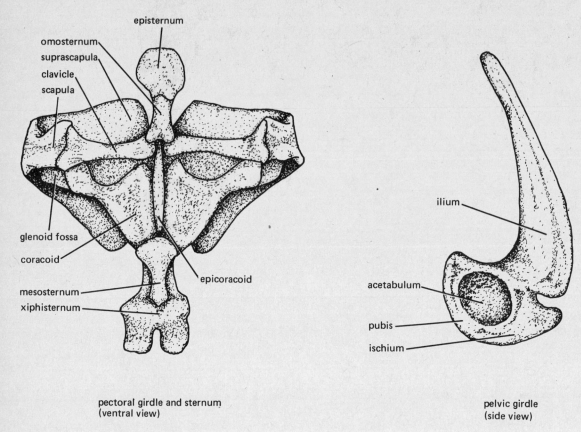

Figure 13.15. Semidiagrammatic view of girdles of the frog.

■ Are these differences fundamental structural ones or adaptive modifications from a common structural archetype? Defend your position with specific examples.

■ What unusual skeletal specializations are found in the frog?

Suggested References

Conant, R., 1958. *A Field Guide to Reptiles and Amphibians of Eastern North America*. Boston, Mass.: Houghton Mifflin.

Frieden, L., 1963. "The chemistry of amphibian metamorphoses." *Sci. American*, Nov.

Lee, A. E., E. R. Savage, and W. L. Evans, 1959. *Development and Structure of the Frog*. New York: Holt, Rinehart and Winston.

Moore, J. A. (ed.), 1964. *Physiology of the Amphibia*. New York: Academic Press.

Oliver, J. A., 1955. *The Natural History of North American Amphibians and Reptiles*. Princeton, N.J.: Van Nostrand Reinhold.

Stebbins, R. C., 1954. *Amphibians and Reptiles of Western North America*. New York: McGraw-Hill.

Wright, A. A., and A. H. Wright, 1949. *Handbook of Frogs and Toads of the United States and Canada*, 3rd ed. Ithaca, N.Y.: Comstock Publishing Associates.

Young, J. Z., 1962. *The Life of Vertebrates*, 2nd ed. London: Oxford University Press.

CHAPTER 14

CLASS MAMMALIA

1. Introduction

Why are mammals said to be "highest" among living things? What characteristics permit us to place mammals at an evolutionary peak? Is it because mammals are the most specialized, numerous, large, or widely distributed animals? Or do none of these factors adequately explain their success?

Authorities agree that we must look elsewhere for a key to understanding the successful evolution of mammals. One must examine physiological, neurological, and reproductive factors especially carefully. Specialization, or improvement of these systems in mammals, has made possible a remarkable degree of plasticity and adaptability to a wide range of external conditions, retaining at the same time a highly efficient constancy of the internal environment. Structural and functional stabilizing adjustments have been developed, which are summarized in the word *homeostasis*; a dynamically balanced, self-regulating system of internal controls —the original feedback control, a principle now applied to a remarkable degree in self-regulating computing machines.

Chief among the evolutionary developments leading to homeostasis is *internal temperature regulation* (warm-bloodedness or *homeothermy*), aided by body insulation of hair, fat, and elaborate peripheral vasomotor controls. In birds the insulating material is feathers, but the result is much the same. Temperature regulation is associated with many physiological and anatomical controls to maintain an internal temperature of 37°C, the most efficient temperature for most physiological processes. Among the evolutionary developments that ensure this are rapid and efficient circulation of blood; sweating, governed by precise circulatory and metabolic controls; the level of biochemical activity of individual cells; and hormonal and neurological controls.

The cold-blooded (*poikilothermic*) vertebrates—fishes, amphibians, reptiles—have developed behavioral reaction patterns to control their body temperatures to some extent. Among reptiles particularly, basking in the sun, hiding underground, and similar behavior patterns permit a considerable degree of temperature control. But these are external measures, far different from the exquisitely balanced internal controls typical of homeothermic animals.

A second important evolutionary step is the

development of means to protect the young and to permit long periods of both early development and later training. *Intrauterine development* of the embryo is followed by varying periods of nourishment of the newborn young by maternal *mammary glands* and by parental *protection* and *training*. This extended development facilitates learning and permits further growth and specialization of the central nervous system, chiefly the brain—the most important coordinating center of all. Highly organized, the brain most clearly separates mammals from reptiles and amphibians, as well as from birds. Although birds possess a similarly highly developed homeothermic system, the physical demands of flight may have prevented the development of a brain as heavy or complex as that of the higher mammals—though birds are more intelligent than the phrase "flying feathered reptiles" implies. Intelligence permits *adaptive behavioral response* to environmental demands and, most important, the ability to *vary* these responses, to adjust, and to change as a result of experience and learning. The ultimate expression of this is the capacity to modify the environment, best seen in man. Man has even escaped to a considerable degree the rigors of natural selection. Whether or not he has truly escaped natural selection and substituted his own is a question that inescapably faces us today and that will grow increasingly critical in the future.

Whether through physiological ability to withstand climatic and other environmental extremes, or through patterns of behavior, intelligence, or learning to avoid or to alter the environment, mammals as a group are less dependent upon temperature than are reptiles and amphibians. As a result, they are less restricted by geography. *Individual* mammalian species or groups may, of course, be highly restricted in distribution, having become adapted to a set of specialized conditions (for example, monkeys to the tropics, koala bears to certain Australian *Eucalyptus* trees, the okapi to the deepest Congo rainforests). Yet, as a *class*, mammals retain an extraordinary capacity to withstand a great variety of conditions, as witnessed by their vast distribution and numbers. Some demonstrate as *species* this capacity to flourish in many climates and environments, notably man, certain rodents, many domestic animals, the mountain lion, deer, coyote, and others.

Along with their capacity to respond and to cope with the environment, mammals have become highly specialized morphologically, thus becoming adapted to specific ways of life and

restricted habitats. To appreciate mammalian types of adaptation, it is convenient to review mammalian *orders*, since each order represents a fairly distinct pattern. Some 26 orders are recognized among living groups of higher (*placental*) mammals. The most primitive mammals comprise two orders usually placed in a distinct subclass: PROTOTHERIA, the monotremes or egg-laying mammals—echidna and duckbilled platypus. A second subclass is the extinct ALLOTHERIA or multituberculates. Higher mammals are placed in the subclass THERIA, with two divisions of living animals. Of these the division METATHERIA or marsupials are more primitive (kangaroo, opposum, and other pouched animals). More familiar are the placental mammals in the division EUTHERIA, which includes the following ten major orders: INSECTIVORA (shrews, moles), CHIROPTERA (bats), PRIMATES (lemurs, tarsiers, monkeys, apes, man), RODENTIA (rodents), LAGOMORPHA (rabbits, hares), CETACEA (whales, porpoises), CARNIVORA (dogs, cats, weasels, and the like), PROBOSCIDEA (elephants), PERISSODACTYLA (odd-toed, hoofed mammals such as horses, zebras, rhinoceroses), and ARTIODACTYLA (even-toed mammals such as pigs, hippos, camels, deer, giraffes, goats, cattle, buffaloes).

The subclasses and divisions are distinguished by the degree of evolution of internally controlled development by *placentation*; hence the distinction between *egg-laying, marsupial,* and true *placental* mammals.

■ **What distinctive adaptations enable each of these ten orders to survive in a particular habitat?**

Probably no terrestrial vertebrates are more abundant or more wide-ranging, rugged, adaptable, and prolific (all distinct criteria of biological success) than *Rattus norvegicus* and *Rattus rattus*, the common rats. Aside from highly specialized teeth (*What mammals do not have specialized dentition?*) the rat can be thought of as a fairly generalized representative mammal. In several of the following sections the rat will be used for study; in the section on the skeletal system, however, the cat will be substituted. A final example—the fetal pig —will be used to illustrate the embryological stages in a representative mammal. The fetal pig is a useful tool in studying this class as it has most of the adult mammalian characteristics, while its still-soft bones and tissues make it easy to dissect.

2. External Anatomy of the Rat (*Rattus*)

Carefully examine your specimen to observe the general body shape, elongated tapering head, thick neck, heavy body, and long scaly tail. Then observe its external features more closely. Rodents are characterized by long, gnawing, continually growing *incisor teeth* that must be worn down as rapidly as they grow. This wearing process removes the *dentine* surface on the inner face more than it does the harder outer *enamel*.

■ What results from this differential wearing?

■ What types of teeth are found in the rat?

■ How many are there?

■ How are types of teeth and food habits correlated?

The *vibrissae* (whiskers) project laterally from the snout. These tactile hairs are considerably wider than the animal. In a fraction of a second the rat can tell whether the hole he is diving into is large enough to accommodate his body. Almost any information discernible by touch reaches the rat through his vibrissae.

Notice the *mouth*, its anteroventral position, cleft upper lip, and the elongated lower lip. Examine the external *nares* (*to what do they connect*?), the *eyes*, and *eyelids*.

In the anterior corner of the eye is a vestigal skin fold, the *plica semilunaris*, apparently homologous to the extra eyelid or nictitating membrane of birds and reptiles.

The *pinna*, or cartilaginous *concha* of the outer ear, directs sound waves into the ear opening (*external auditory meatus*) and assists in the determination of the direction or location of the sound source.

Find the *anus* below the base of the tail. In the male, the *scrotum*, which contains the *testes*, will be seen near the anus. During the season of reproductive activity, the testes descend from the abdominal cavity into the scrotum, which becomes quite conspicuous. At other times, or in immature males, the scrotum is reduced and the testes are withdrawn into the body cavity. This external location, when testes are spermatogenically active, is related to the poor tolerance of spermatozoa for normal body temperature. Male gametes are not formed at $37°C$ or higher, and the scrotal sac is a mammalian adaptation providing a lower temperature for them. Just anterior to the scrotum is the opening of the male urogenital system terminating in the *penis*, which is usually withdrawn into a skin covering, the *prepuce*. In the female, there are three openings: (1) *anus*, (2) *vaginal opening* in front of the anus, and (3) *urinary opening* on the tip of a papillate structure in front of the vagina.

Check the feet of your specimen. All digits except the thumb, or *pollex*, of the forefoot end in claws. Pads, or *plantar tubercules*, protect the bottom of the feet and provide a gripping surface for walking or climbing, a useful combination with claws. Notice the occasional bristles between the epidermal scales of the tail.

■ What is the function of the tail?

3. Internal Anatomy

I. Skeletal System of the Cat (*Felis domesticus*)

The vertebrate skeleton conforms to a uniform pattern, although still showing highly varied adaptations in different groups. A 2-in shrew has as many neck vertebrae as a 17-ft giraffe or a 100-ft whale—but what a difference!

We can learn the vertebrate skeleton pattern from any example, although the cat (Fig. 14.1), a representative carnivore, is especially useful because of its availability and the absence of extreme skeletal specialization such as is found in the frog (see Fig. 13.13). However, the frog skeleton still can be used—especially that of the very large bullfrog (*Rana catesbeiana*). The marked differences between frog and cat should be noted (ribs, sternum, forearm, tarsals, pelvis, vertebrae) with reference to specialization for jumping in the frog.

Locate and work out the movements and attachments of the cat bones (Fig. 14.1) and processes listed below:

a. Skull (*Which bones form the jaw? the cranium? the face?*)
1. premaxilla
2. maxilla
3. nasal
4. lacrimal
5. frontal
6. jugal
   i. zygomatic process

7. parietal
8. squamosal
    i. zygomatic process of squamosal
9. pterygoid
10. palatine
11. occipital
12. sphenoid complex (*How many bones?*)
13. nasal aperture
14. internal choanae
15. foramen magnum
16. tympanic bulla
17. external auditory meatus
18. mandible
19. teeth (Compare the *type, number*, and *form* of teeth of the cat, rat, and man. *What does this tell us of their food habits and probable body structure?*)
    i. incisors (*Number?*)
    ii. canines (*Number?*)
    iii. premolars (*Number?*)
    iv. molars (*Number?*)
b. Thoracic cage
1. ribs (*How do they articulate dorsally? ventrally?*)
    i. true ribs
    ii. floating ribs
2. sternum (*Is it all bone?*)

c. Pectoral girdle (1 and 2 below) and anterior appendages (3–8 below)
1. scapula
    i. spine
    ii. acromion process
    iii. coracoid process
    iv. glenoid fossa (*Articulating with what?*)
2. clavicle
3. humerus
4. ulna
    i. olecranon process (*Another name for this structure?*)
5. radius
6. carpals (*How many?*)
7. metacarpals (*How many?*)
8. phalanges
d. Pelvic girdle (*Articulating with which vertebrae?*) and posterior appendages (2–8 below)
1. innominate—consisting of three fused elements (i–iii)
    i. ilium
    ii. ischium
    iii. pubis
    iv. obturator foramen
    v. acetabulum (*Formed from which bones? Function?*)

Figure 14.1.   Cat skeleton.

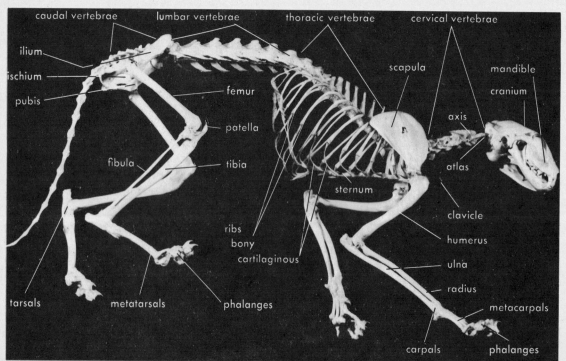

2. femur
   i. head
   ii. shaft
   iii. greater trochanter
   iv. lesser trochanter
3. patella (*Another name?*)
4. tibia ⎫ (*Corresponding to which arm*
5. fibula ⎭   *bones?*)
6. tarsals
7. metatarsals (*Common name?*)
8. phalanges (*Common name?*)

e. Vertebrae (*How many in the cat? elephant? whale?*)
1. cervical
   i. axis
   ii. atlas
2. thoracic (Identify the following parts:)
   i. neural spine
   ii. neural arch
   iii. neural canal
   iv. transverse processes
   v. pre- and postzygapophyses
3. lumbar ⎫
4. sacral   ⎬ (*How are they distinguished?*)
5. caudal ⎭

II. Muscular System

Our review of muscles will have to be a selective sampling to demonstrate attachments, insertions, and actions in a few characteristic groups. In vertebrates, *skeletal* or *striated* (voluntary) muscles are attached to other muscles or to bones, either directly or by *tendons* (nonelastic cords of connective tissue). A good example is the *Achilles tendon* between heel bone and gastrocnemius.

■ **What happens when this tendon is cut (ham-stringing)?**

Controlled movement of a limb must therefore be handled by two or more pairs of *opposed* muscles, each of which performs a specific contracting function. The muscle that straightens or extends a limb or part is an *extensor*, the opposing muscle, which bends it, is a *flexor*. The degree of tension between the two—*tonus*—is a state of partial contraction of the individual muscle. The balance of muscle tensions maintains the body's position and helps permit controlled movement. Muscles can also be classed into *abductors* (moving a part *away* from the median line) and *adductors* (moving it *toward* the midline).

The complexity of muscle attachments and the use of tendons to economize the serious space problem—there is simply not enough room on our bones for attachment of all our muscles—emphasizes the advantage of arthropods in this respect. Having an exoskeleton, arthropods live inside their skeleton, lack tendons, and utilize the entire outer body wall for muscle attachment. This represents about ten times the relative surface area that vertebrates have.

Each muscle has its own separate point of attachment, the *origin* (point or area of attachment to the less movable or more stable base) and the *insertion* (point of attachment to the part being moved).

We shall examine a group of muscles of the upper arm and shoulder of the rat. The skin should be pulled away from the arm and shoulder, the fascia removed, and the muscles separated by probing with the blunt edge of your scalpel. Separate the muscles, but do not cut them. Leave their origin and insertion intact for review.

Remember that this is a *sample*, the value of which is to give practice in gross dissection and to teach you the appearance and actions of a group of muscles—a lesson in organization. It is not intended as a memory session. For each muscle listed in Table 14.1, work out the *origin*, *insertion*, and *action*. Try to imagine the opposing muscles at work, then answer these questions:

■ **What muscles does the rat use to lift its front leg?**

■ **What arm and shoulder muscles contribute the primary motions in throwing a baseball?**

III. Digestive System of the Rat

1. **Mouth and esophagus** Examine the mouth and the muscular tongue attached to the floor of the mouth cavity. Anterior to this attachment is a more loosely connected portion of the tongue, held by a vertical *frenular fold*. The roof of the mouth is formed by the anterior *hard palate* and posterior *soft palate*. Above the latter is the *nasopharynx*. Into this large cavity the *posterior nares* pass, as do *auditory tubes* that enter the nasopharynx through longitudinal openings near the junction of the dorsal and lateral walls of the *pharynx* and the posterior border of the soft palate.

Table 14.1

	ORIGIN	INSERTION	ACTION
1. *pectoralis major* (triangular shape)	sternum (anterior half)	humerus	?
2. *pectoralis minor*	sternum (posterior half)	?	?
3. *sternomastoids* (straplike muscles on ventral surface of neck)	sternum	mastoid of skull	?
4. *acromiodeltoid* (large flat muscle across back and shoulders)	clavicle and acromium process (lumbar vertebrae and fascia)	humerus	?
5. *trapezius* (three thin muscle sheets covering neck and anterior back)			
a. *spinotrapezius*	dorsal spines of thoracic vertebrae	spine of scapula	?
b. *acromiotrapezius*	cervical and thoracic vertebrae	spine and acromion process of scapula	
6. *triceps brachii*			
a. *long* head of t.b.	scapula	olecranon process	extensors
b. *lateral* head of t.b.	?	olecranon process	extensors
c. *medial* head of t.b.	?	olecranon process	extensors
7. *biceps brachii brachialis* (lies lateral to the biceps and acts with it)	glenoid fossa	forearm	?

NOTE: *The triceps, which are particularly adapted to quadrupedal gait, are very poorly developed in humans. The longest muscles of the quadruped forelimb lie on the posterior side of the upper arm. The three heads a–c have the same action but are so distinct that they are practically separate muscles.*

The pharynx in turn connects with the *esophagus* posterior to the *glottis*, the valve-like action of which keeps food from entering the *trachea* (windpipe). This latter structure, ventrally located in the neck (feel it from the outside), contains the *larynx* (voice box). The glottis is covered by the *epiglottis*, which is raised when air passes into or out of the larynx but is bent backward over the glottis and trachea while food is being swallowed. Anyone who has choked on food knows the function of the epiglottis.

2. **Dissection procedure** (Figs. 14.2–14.6) Lay the animal on its back and make a medioventral abdominal incision, from pelvis to diaphragm. Cut laterally through the body wall near pelvis and diaphragm to form flaps that can be pinned back to expose the viscera (Fig. 14.2). Remove abdominal fat deposits but do not disturb the alimentary canal.

3. **Structures exposed** Locate first the curved *stomach* that consists of two portions —the thin-walled, large *cardiac sac*, anterior; and the opaque, thicker walled *pyloric sac*, posterior. The esophagus empties into the cardiac sac. The food passes into the pyloric sac and then into the anterior part of the intestine (*duodenum*) via a muscular fibrous ring, the *pyloric sphincter*. Find this valve as an externally visible groove.

Insert a flexible probe into the mouth. Locate the route of the esophagus through the neck, thoracic cavity, and diaphragm (before it terminates at the cardiac sac of the stomach).

Trace the coiled *small intestine*, sup-

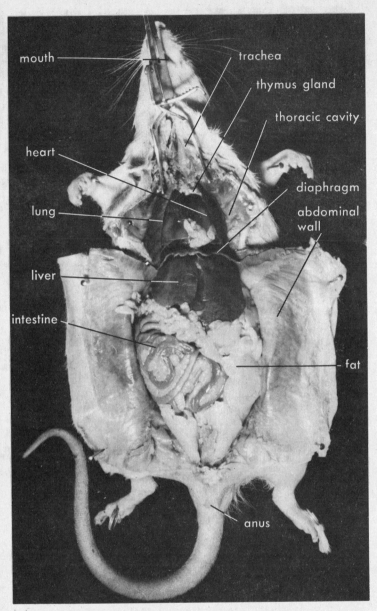

mouth

heart

lung

liver

intestine

trachea

thymus gland

thoracic cavity

diaphragm

abdominal
wall

fat

anus

Figure 14.2.   Rat, female, internal anatomy.

of cellulose and other plant materials is particularly difficult, and herbivores characteristically possess a long intestine and large cecum—carnivores do not.

Trace out the large intestine, which starts at the *ileocecal junction,* to the *ascending colon.* Bulbous objects inside the terminal portion or *descending colon* are fecal pellets. Waste passes through the colon, to a short *rectum* (near and within the pelvis), and finally outside via the *anus,* guarded by an *anal sphincter muscle.*

Observe the large, multilobed brown *liver.* This organ (largest of all the organs) produces, among numerous other substances, *bile,* which enters the duodenum through the *bile duct* found in the *hepatoduodenal mesentery.* The rat lacks a bile storage organ or *gall bladder,* which is present in the mouse and many other animals including man. Find the point near the pyloric sphincter where the bile duct joins the duodenum.

The *pancreas,* lighter in color than the liver, is a rather diffuse organ in the first (duodenal) loop of the small intestine. Its large dorsal lobe can be seen near the pyloric sac. Pancreatic juice with a number of digestive enzymes flows through many small pancreatic ducts that enter the bile duct, and passes into the gut through the latter tube. (Look near the dorsal surface of the pancreas for the pancreatic ducts.)

Trace a food bolus from mouth to anus, reviewing all major structures passed along the way.

Find the dark, red, spherical *spleen,* posterior to the greater (outside) curve of the stomach. This organ is functionally a part of the circulatory system and a major phagocytic and white-blood-cell-forming center.

IV. Respiratory System

1. **Trachea and bronchi**   If possible, study a mammal skull cut in median saggital section. Observe the complex passages in the *nares* leading to the *pharynx.*

Dissect away the neck muscles of the rat to show the ringed *trachea,* the *glottis, epiglottis, larynx,* and *pharynx.* Review the relationships of these parts and the passage of air through each.

ported by mesenteries containing numerous blood vessels. Do not damage these structures; simply follow them. The *duodenum, jejunum,* and *ileum,* successive divisions of the small intestine, are inseparable externally. The ileum empties into a saclike *cecum,* which in man is reduced to the vestigial *vermiform appendix.* The cecum may function as a storage organ for the bulk of ingested food, for bacterial fermentation, or as a special absorptive structure. Digestion

■ How is air prevented from being sucked into the esophagus when food is swallowed, or food prevented from passing into the trachea?

■ What happens when one has the hiccups?

On the outer ventral surface of the trachea, near the attachment of the larynx, is the H-shaped *thyroid gland*, an essential endocrine organ.

Dissect out the larynx and epiglottis; separate them from the pharynx. Slit the larynx ventrally and study the *vocal cords*, a pair of dorsoventral folds in the lateral walls.

Trace the trachea into the thoracic cavity. Note its division into two *primary bronchi*. These bronchial tubes enter the lungs and divide repeatedly into *secondary* and *tertiary bronchi* and thence into still smaller *bronchioles*. They ultimately branch into the smallest units of the lung—innumerable, microscopic, bulb-shaped termini of the smallest bronchioles, the *alveoli*. Each of these is supplied by an equally small knot of capillaries connected to pulmonary venules and arterioles. Gas exchange occurs by diffusion through capillary walls into and from the extremely thin-walled aveoli—some three fourths of a billion of them in man, with a surface area of over 1000 ft^2.

2. **Thoracic cage and lungs** In the thoracic cage the lungs lie in two distinct *pleural cavities* lined by a fine membrane forming a *pulmonary pleura* over the lungs and a *parietal pleura* over the inner surface of the thoracic wall.

Review the mechanism of breathing—the role of the diaphragm, thoracic cage and intercostal muscles, the pleural cavities, and relative pressures inside and outside the pulmonary cavity and the highly elastic lungs.

V. Circulatory System (Fig. 14.3)

1. **Introduction** By far the largest number of vessels with the greatest total blood-holding capacity will not even be seen in this exercise. For the complete system, we must include not only the heart and its major vessels, but miles of fine *capillaries* and *lymph channels* in which the real work of circula-

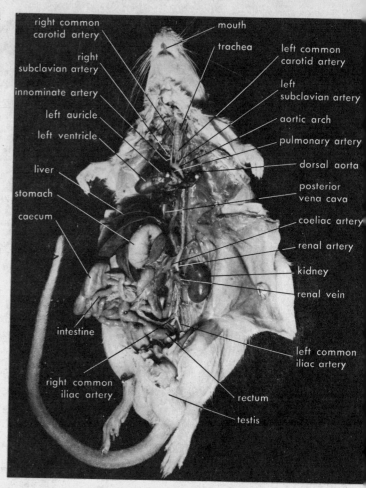

Figure 14.3. Rat, male, internal anatomy: Digestive and circulatory system.

tion and the exchange of gases, nutrients, and waste materials are performed with each individual cell. The heart is the vital pump, as is well known. *Arteries* and *veins* are the principal conduits, functioning primarily to get the blood to and from the tissues. Allowing blood to flow to individual cells is the job of capillaries. These minute ducts, only one cell thick, form a vast interconnecting network between arteries and veins and their respective branches. Blood from arteries passes through successively smaller branches, the *arterioles*, before entering the capillary bed. Gases, nutrients, and wastes diffuse through capillary walls and pass via intercellular *lymph* to and from each cell.

Red blood cells serve as vehicles for carrying oxygen to the intercellular fluids and carbon dioxide from them. Only under accidental conditions (shock or hemorrhage) do red blood cells pass through capillary walls. Capillaries converge into larger ducts, *venules*, which carry the deoxygenated blood into veins and eventually back to the heart.

2.   **Dissection and orientation**   Injected animals showing red dye or latex in the arteries and blue material in the veins should be used for this exercise. Continue the abdominal incision through the skin to the throat. Pull back the cut skin with your fingers to expose the thoracic musculature. Carefully insert your scissors and cut through the thoracic cage and body wall, continuing to cut toward the throat. Use short strokes with the point of your scissors to avoid damaging the heart and lungs. Continue the cut through the center of the clavicle until you see the trachea and esophagus as it enters the thorax. Bend the forelimbs back, snapping them at joints if necessary. Carefully use your fingers and a blunt probe to free the membranes from the rib cage and remove bits of fascia and clots of latex that obstruct your view. Look for injected blood vessels as they enter the thorax and expose them carefully. Pin back the walls of the rib cage and orient yourself with respect to exposed structures.

Locate: *rib cage* and *sternum; diaphragm* (wall of muscular tissue separating thoracic and abdominal divisions of the coelom); *pericardium* (membrane surrounding heart); and on either side of the *pericardial cavity* surrounding the heart, a *pleural cavity*, each surrounding a lung.

■   **Which lung is the larger?**
■   **Is this uniform in other students' dissections?**

The *mediastinal septum*, a membrane between the pericardium and the floor of the thorax, runs from the diaphragm to the anterior wall of the thoracic cavity. See if any remnants of this membrane have escaped destruction. The *mediastinum* is actually the space between the pleural cavities in which the heart, the major vessels, the trachea, and the esophagus are situated.

■   **Describe the function of the *diaphragm muscle*.**

Now you can explore the heart and great vessels of the cardiopulmonary circulation. Remove the double layer of pericardium surrounding the heart. The pericardial cavity is the space between the two layers of pericardium. First note the triangular shape of the heart with its pointed posterior apex. The small, thin-walled pair of anterior chambers, *atria* or *auricles*, appear as flaplike structures on the heart, being collapsed because they lack sufficient musculature to be held rigid. Actually, the *left ventricle* comprises most of the heart tissue, the *right ventricle* being much smaller. The four chambers, separated by *interauricular* and *interventricular septa,* permit a distinct *double circulation* (pulmonary and systemic) with complete separation of oxygenated and deoxygenated blood.

■   **How does this compare with the circulatory pattern in the frog? in the perch?**
■   **Why is the mammalian pattern said to be more efficient?**
■   **Is this efficiency related to homeothermy and to the high metabolic rate of mammals?**

Clear away any fat or tissue interfering with your view of the major blood vessels. Notice the *thymus gland*, a mass of fatty tissue anterior to the heart. This structure is best developed in young mammals and regresses after maturity. The thymus performs a key role in activating *lymphocytes*, which are formed in the *hemopoietic* — blood-forming—centers, then passed through the thymus, after which their function is so distinctive that they are called thymus-activated lymphocytes or T cells. Populations of these T-cells become essential elements in *cellular immunity (cell-mediated immunity)*, which is an essential part of the body's immunological capacity.

3.   **Pulmonary circulation**   Find the *pulmonary artery*, which arises from the right ventricle and then divides to form *right* and *left* pulmonary arteries to the lungs. Newly oxygenated blood returns from each lung via *right* and *left pulmonary veins* into a single *pulmonary vein* that empties into the *left* auricle. (Notice that distinction between *vein* and *artery* is based upon *direction of blood flow* to or from the heart.) From the left auricle, blood is pumped into the left

ventricle, at the same time that the right auricle is filling the right ventricle for another spurt of blood into the pulmonary artery. Simultaneous contraction of the two ventricles is perfectly timed with their state of filling and sends blood out into the pulmonary circulation from the right ventricle and into the systemic circulation from the powerful left ventricle. Backflow on the right side of the heart (into the right auricle when the right ventricle contracts) is prevented by the three-flapped *tricuspid valve*. On the left side, the same function is performed by the *bicuspid* (or *mitral*) *valve*. *Semilunar valves*, three between the right ventricle and pulmonary artery and three between the left ventricle and aorta, prevent back flow into the ventricles after their contraction. Ventricular contraction is called *systole*. (*Diastole* is the relaxation phase of the heartbeat.)

4. **Major arteries**   The *aorta*, a large, thick-walled, whitish vessel, passes forward diagonally to the right, then arches dorsally to the rat's left side after it leaves the left ventricle. This is the unpaired *aortic arch* from which five important branches arise. (Note that there is no corresponding *right* arch, as in the frog.) First of these branches is the *innominate artery*, massive but very short, actually the stem of the aorta that gives rise to the second and third major branches— the large *right subclavian* to the right foreleg, and the somewhat smaller *right common carotid artery* that runs along the right side of the trachea. The latter vessel later divides into the *right external* and *right internal carotid* arteries (Fig. 14.3).

The fourth principal branch from the arch is the *left common carotid artery*, which also runs along the trachea, the left side, and forms the *left external* and *left internal carotids*. The *left subclavian artery* is the last trunk; it leads anteriorly, passing into the left foreleg (*brachial artery*) after giving off one branch to the neck and skull (*vertebral artery*) and another to the ventral wall of the thorax (*internal mammary artery*). Similar branching of the *right subclavian* occurs.

At the base of the aorta, *coronary arteries* branch. These are the vital arteries supplying the muscles of the heart itself.

The dorsal arch then curves dorsally toward the vertebral line where it penetrates the diaphragm and forms the main systemic trunk, the *dorsal aorta*. In birds it is the *right* aortic arch that persists, rather than the *left* as in mammals.

Move the viscera to one side and follow the dorsal aorta caudad as it branches into its principal arteries (Fig. 14.3). First find the *coeliac artery*, whose three branches supply blood to the stomach, liver, and spleen. Next is the *anterior mesenteric artery*, slightly posterior to the coeliac. Its various branches supply the pancreas, the small intestine (remember the many red blood vessels in the intestinal mesenteries?), the cecum, and the large intestine. Locate the large *renal arteries*, branching at different levels, one to each kidney, followed by branches to the gonads that are called either *ovarian* or *spermatic arteries*. Near the genital arteries are small *iliolumbar arteries* which send blood to the dorsal body wall musculature.

Near the caudal end of the aorta is the unpaired *posterior mesenteric artery*, which supplies the large intestine and rectum. This artery often forms confluent branches (anastomoses) with branches of the anterior mesenteric artery.

Finally, the dorsal aorta ends in a bifurcation to form the *right* and *left common iliac arteries*. These in turn send branches dorsally to the back, and then divide into *external iliac arteries* to the thighs, *internal iliac arteries* to the pelvic areas, and a small single *caudal artery* to the tail.

5. **Major veins**   Notice that most arteries are paralleled by veins that return deoxygenated blood to the heart. Starting with the major veins in the neck, find the *right* (or *left*) *external jugular*, the principal vessel at the side of the neck. It is formed from the *anterior* and *posterior facial veins* near the corner of each jaw. Look along the margin of the trachea for the small *internal jugular*, which joins the external jugular and the *subclavian* from the foreleg, to form the *anterior vena cava*. The *left subclavian* receives the principal vessel of lymph circulation, the *thoracic lymph duct*, near the junction of the left subclavian and left internal jugular. *Anterior intercostals* and *internal mammary veins* join the anterior vena cavae (right and left) before the latter enter the right auricle.

The *azygos vein* joins the right anterior vena cava near its entrance into the right auricle. It is an asymmetric vessel, bringing blood from posterior intercostal tissues, passing to the right of the middorsal line. It is thought to be a remaining portion of the *posterior cardinal* system as studied in the dogfish, just as the anterior vena cavae (or precavae) evolved from *anterior cardinal* veins.

All returning blood from the posterior end of the body passes into the right atrium via the single *posterior vena cava* (derived from the right posterior cardinal vein). Try to determine the successive blood vessels feeding into the posterior vena cava by dissecting back and working out the source of the major veins supplying blood to it. Find the *hepatic veins* (*How many?*) entering the posterior vena cava when the latter becomes embedded in the liver near the diaphragm, where it receives the *phrenic veins*.

The short but large paired *renal veins* enter the vena cava at different levels, corresponding to the position of each kidney.

■ **Where do the *left ovarian* (or *left spermatic*) *vein* and the *right ovarian* (or *right spermatic*) *vein* join the blood flow returning to the heart?**

Observe that in the posterior trunk the vena cava is *dorsal* to the dorsal aorta, opposite to the condition in the thoracic region.

Similar to the branching of the arterial system, a pair of *iliolumbar veins* drains the dorsal body wall and joins the posterior vena cava caudad to the gonadal veins. The vena cava in turn begins at the junction of the *common iliac veins*, each of which is formed from an *internal iliac* (from the dorsal side of the hind limb) and from an *external iliac vein* (from the *femoral* and other vessels on the inner margin of the legs).

6. **Hepatic portals** (dyed yellow in triply injected animals) This important *capillary system within a vein* distributes food-laden blood from the digestive tract into the liver, where its load of dissolved food materials can be exchanged and physiologically processed, passing from innumerable capillaries to the liver cells. The *hepatic portal* is formed from four vessels: the *lienogastric* (from stomach and spleen), *duodenal* (from duo-

denum), *anterior mesenteric* (from small intestine, colon, and cecum), and *posterior mesenteric* (from rectum). The hepatic portal vein can be seen in the supporting mesenteries parallel to the bile duct. The portal capillaries recombine into venules and veins, and then enter the posterior vena cava from hepatic veins, as you have already observed. Note that *no renal portal system is found in mammals*.

■ **How would you explain the advantage of such a loss?**

VI. Urogenital System

The U-G system, as it is called, is considered a combined unit, since its component parts are closely related embryologically and evolutionarily. In the mammalian male reproductive system, tubules and ducts (*Wolffian ducts*) of the *mesonephric kidney* are utilized, whereas ducts corresponding to female or *Müllerian ducts* disappear or remain vestigial (Fig. 14.4).

In females, on the other hand, the Müllerian ducts give rise to *oviducts* and *uterus*, and the mesonephric or Wolffian ducts and tubules disappear (Fig. 14.5).

1. **Kidney (external)** Kidneys are the "filters" of the circulatory system. They are bean-shaped organs covered with peritoneum ventrally and attached dorsally to the wall of the peritoneal cavity. It is best to think of them not only as waste-removing organs but also as organs that control the essential balance between body salts and water. They remove from the blood nitrogenous wastes— urea, nonvolatile foreign substances, excess salt, and excess water.

Three types of kidneys occur in vertebrates, the *pronephros* (seen in adults only in certain primitive fishes); the *mesonephros* (in adult cyclostomes, most fishes, and amphibians); and, finally, the *metanephric kidney* in all adult reptiles, birds, and mammals. The embryological sequence of these three kidney types, especially development of the mesonephric kidney, shows a close relationship between the gonads and the adult kidneys, despite their anatomical separation.

Examine the kidneys of your rat. Note that the right one is not in line with the left

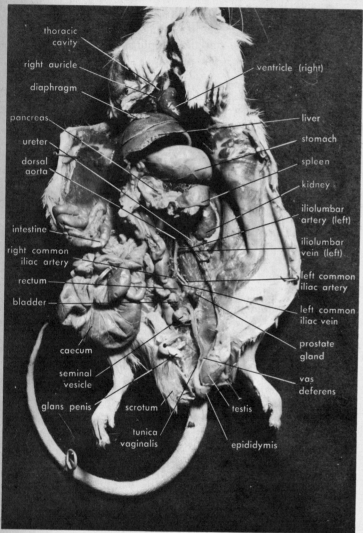

thoracic cavity
right auricle
diaphragm
pancreas
ureter
dorsal aorta
intestine
right common iliac artery
rectum
bladder
caecum
seminal vesicle
glans penis
tunica vaginalis
scrotum
ventricle (right)
liver
stomach
spleen
kidney
iliolumbar artery (left)
iliolumbar vein (left)
left common iliac artery
left common iliac vein
prostate gland
vas deferens
testis
epididymis

Figure 14.4. Rat, male, internal anatomy: digestive, urogenital, and circulatory systems.

but extends forward into the right lobe of the liver.

■ **Can you offer an explanation for this asymmetry?**

Along with the ureter, the renal artery enters and the renal vein leaves at the medial margin (*hilus*) of each kidney. Filtration of blood from specialized capillaries or *glomeruli*, excretion of excess salts or water, and removal of other materials occur in the many *tubules* in the kidney. Each tubule is enmeshed in capillaries from the renal arteries, which then pass their blood into the renal veins. Waste fluid, or *urine*, passes from the tubules to a larger duct, the *ureter*, which carries this fluid from kidney to *bladder*. The bladder, a distensible, thin-walled sac, is attached by a strong ligament to the ventral body wall. Two ureters join the bladder, which drains its contents to the outside via the single *urethra*. In the male the urethra is a *urogenital duct*, as in the male frog, since it carries both urine and seminal fluids.

Locate the *adrenals*, small oval structures at the anterior end of each kidney (Fig. 14.3). These important endocrine glands produce adrenalin (or epinephrine) and numerous other hormones.

2. **Kidney (internal)** Remove one kidney and cut it lengthwise to expose the hilus and to view the major divisions. Observe the funnel-shaped ureter that opens in the hilus to form the internal *pelvis* of the kidney. Next, observe the pelvis closely and see how it is divided into sections, each of which is a *calyx* (plural: *calyces*).

The kidney tissue, enclosed by a connective tissue *capsule*, is divided into outer *cortex* and inner *medulla*. Observe the numerous fine lines in the medulla, converging toward the hilus and forming triangular *renal pyramids*, the outer layers of which show radial *medullary rays*. These striations are actually *collecting tubules*, each an important portion of the *nephron*, the basic functional unit of the kidney. Examine these units in tissue slices or prepared slides under a dissecting microscope. Darker radial striations and scattered dots in the cortical region consist largely of *renal corpuscles*; the other striations are the *tubules*.

Prepared slides should be consulted for histological details of this complex and remarkable organ. The renal corpuscle consists of a *glomerulus* or knot-like mass of capillaries, largely enclosed by a *Bowman's capsule*, like a deeply indented balloon, which forms the origin of each tubule, the length of which is surrounded by a fine capillary net.

■ **How do these corpuscles function in filtering blood?**

Find the *proximal* and *distal convoluted tubules* in your preparation. (In cross section, the lumen of the distal tubules has a *testes* and *sperm ducts*. The ventral groove between the testes marks the *septum scroti*, and coincides with an internal partition between them. Slit the scrotum ventrally, and observe the overlying sac, the *tunica vaginalis*, formed of connective tissue in which the testes appear to be suspended. Remove the tunica carefully. The next layer or sac exposed is the *tunica albuginea*, and under it a highly twisted mass of *seminiferous tubules*. Remove a few tubules and study microscopically a teased out preparation.

■ **Can you find bundles of spermatozoa within the tubules?**

The *epididymis* is a tubular storage system in which spermatozoa are retained after formation in the seminiferous tubules. This compact structure, closely adherent to the wall of the testes, is divided into an anterior or head portion (*caput epididymis*); a slender midportion or body along the dorsal surface of the testis (*corpus epididymis*); and a tail portion (*cauda epididymis*) spreading over the posterior (caudal) end of each testis. Find the large *vas deferens*, which collects the spermatozoa from the epididymis and passes anteriorly into the body cavity. This duct, along with the associated blood vessels and a nerve, comprises the *spermatic cord*. Trace this cord into the penis.

■ **Where does the cord join the urethra?**

In man and most mammals, testes remain in the scrotum, permanently descended from the body cavity. In rodents, insectivores, bats, and marsupials, this descent occurs periodically before each breeding season, after which the testes are withdrawn into the body cavity by the *cremaster muscle*.

■ **What anatomical evidence do we have for the descent of the testes?**

Next, examine the male intromittent organ, the *penis*, through which spermatozoa and accompanying seminal fluids are transferred to the vagina during copulation (coitus). Insert the point of your scissors into the external opening of the penis and cut the skin along the length of the structure.

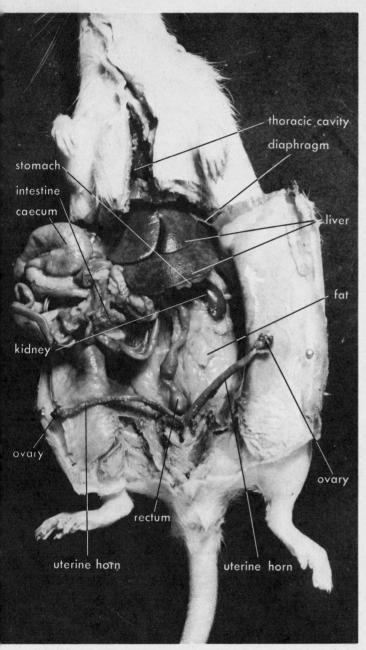

Figure 14.5. Rat, female, digestive and urogenital systems.

characteristic star shape, of the proximal tubules a circular shape.) Locate the *loops of Henle*, which function chiefly for water resorption. With the aid of laboratory models and charts, review the organization and function of the entire nephron: the glomerulus, Bowman's capsule, capillaries, and tubule.
3. **Male genital system** (Fig. 14.4) Observe again the large *scrotum* containing the

Clear away the *prepuce* (foreskin) that covers it. The penis is directed posteriorly in rodents, whereas in most mammals it projects cephalad. Remove the muscle and connective tissue, and locate the spongy, erectile tissue that fills with blood to give the penis rigidity during copulation. Along the ventral surface are two *corpora cavernosa penis* that lie dorsally when blood fills the organ, causing it to be directed anteriorly. A single *cavernosum urethra* lies in a groove between the two corpora cavernosa. Notice that the cavernosum urethra ends in an enlarged head, or *glans penis*, within the prepuce. The urethra passes from the bladder through the cavernosum urethra and to the outside via the urethral opening in the glans.

The remainder of the male genitalia, semen-producing and sperm-storing glands, lie in the pelvic area of the body cavity. Two irregularly shaped or curved and rather large *seminal vesicles* and two pairs of smaller *prostrate glands* lie nearby, one pair dorsal and the other ventral to the spermatic cord.

4. **Female genital system** (Fig. 14.5) The vagina, the external orifice of the female genitalia, opens anterior to the anus. Anterior to the vagina, the urethra opens separately through a papillate structure at the base of which is the *clitoris*, an erectile structure homologous to, though less well developed than, the male penis, and lacking an opening.

The rest of the female U-G system is internal, near the kidneys. First find two round, pea-sized *ovaries*, each invested in a capsule, the *bursa ovarica*, merging with the special ovarian mesentery, the *mesovarium*.

■ **Where is this mesentery attached?**

The *oviduct* forms a tight coil at one end of the mesovarium. The proximal opening within the bursa ovarica receives an egg freed by rupture of an *ovarian follicle* and passes it by ciliary action into the *funnel* or proximal end of the oviduct. In other mammals the funnel lies free in the peritoneum and sweeps in the eggs. In the rat the connection is much closer, as the funnel of the oviduct lies within the bursa of the ovary.

The egg passes down the tightly convoluted tubule of the oviduct and enters one *horn* of the uterus.

■ **What happens to the egg from this point?**

Notice that the uterus consists largely of the horns. In the rat the horns open separately into the *vagina*. This development of horns rather than body of the uterus is an adaptation for multiple births.

■ **How does this compare with the human uterus? with that of the dog? the horse?**

VII. Central Nervous System (Fig. 14.6)

Our review of the nervous system will be restricted to an examination of the *brain* and its associated *cranial nerves*. This is the most specialized and distinct part of the nervous system, and, in fact, the part that appears to be of the greatest evolutionary significance throughout the class MAM-

Figure 14.6.   Rat, central nervous system, dorsal view.

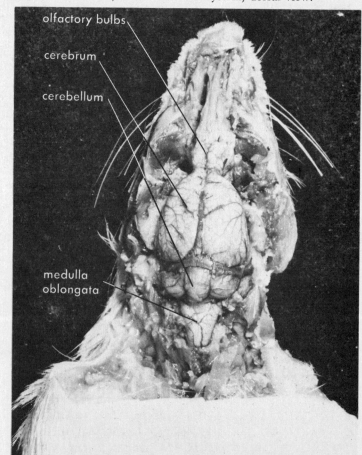

MALIA. The skull should be soaked first in 10 percent nitric acid to soften the cranium, then washed thoroughly to remove acid before beginning the exercise.

Dissection of the skull requires care, skill, patience, and complete control of your tools to avoid damaging the delicate structures within.

A good point of entry into the braincase is the middorsal *sagittal suture*. With a careful puncture as a start and judicious use of the scalpel (firmly held flat so as to shave thin portions off the skull roof), you can enlarge the hole, chipping away small pieces of bone to expose the dorsal surface of the brain. Learn to control the pressure of your instrument so that you can direct considerable force and leverage with complete control and reasonable freedom from a fatal slip into the brain.

First you will observe the two membranes, *dura mater* and *pia mater*. Expose the *medulla oblongata, cerebellum, cerebrum,* and *olfactory bulbs* in posterioanterior succession. Break through the occipital portion of the skull and expose the vertebral column. The *atlas,* lying against the base of the skull, serves as a convenient marker for separation of the brain from the spinal or vertebral column with its *spinal cord.*

Individual vertebrae can be separated from one another and pushed posteriorly to expose the anterior portion of the cord; or parallel dorsal cuts can be made through the vertebrae to expose the cord below.

Next, sever the spinal cord posterior to the medulla and very carefully remove the intact brain from the skull. Work up from the cord, using the stump of the cord as a lever. Concern yourself primarily with the *cranial nerves,* to prevent pulling them from the brain. To facilitate later identification of these nerves, each cut should be made as far from the brain as possible. Be particularly cautious with the *olfactory lobes* the most difficult part to remove intact. Place the isolated brain in water and identify each of the 12 cranial nerves, the *optic chiasma,* the *pineal* and *pituitary bodies,* and the major divisions and chief subdivisions of the brain. The latter include *frontal* and *temporal lobes* of the cerebral hemispheres, *Sylvian fissure, corpus callosum* (transverse sheet between the hemispheres), *hippocampus* (floor of lateral ventricles), and *choroid plexus* (vascular cover of the third ventricle). In the midbrain find the *corpora quadrigemina* (two divisions of each optic lobe, seen posterior to the pineal body on the dorsomedial surface), and *crura cerebri* (swellings on ventral surface of midbrain). In the hindbrain find the *medulla oblongata* which tapers into the spinal cord. Then locate the *cerebellum,* dorsal outgrowth of the medulla, with its dorsomedial lobe, the *vermis,* and two lateral, irregularly convoluted lobes, the *flocculi.*

Study illustrations or sections of the brain to locate the major *commissures* (groups of fibers connecting one side of the brain to the other), the ventricles (including the third, fourth, and lateral ventricles), and their connecting openings.

■ How does the rat brain differ from that of man in relative size and in the development of its major divisions?

4. External Anatomy of the Fetal Pig (*Sus scrofa*)
(Fig. 14.7)

The fetal pig—an unborn pig removed from its mother's uterus—displays both adult and embryonic characteristics of the class MAMMALIA. Observe the general shape of the body. Note the *head, neck, trunk,* and *tail* regions.

■ What is the relation of head size to total body size in the fetus?

Locate the following structures in the head region: the *tongue* and *mouth* (you may need to force the jaw open in order to reveal the tongue), the *snout* with paired *nostrils,* the *eye,* the *nictitating membrane* (*What other animals have this?*), the *eyelids,* the *external ear* (*pinna*).

■ Does your pig have eyelashes?
■ If there is hair on your specimen, is it restricted to certain parts of the body?

Note the strong muscles on the dorsal side of the short neck. In what activity are these muscles important?

The trunk can further be divided into (1) the *thorax* (chest) including the ribs, (2) the *abdomen,* in which the gastrointestinal tract, urinary and genital systems are found, and (3) the *iliosacral* region. (*Are there observable differences in the paired rows of nipples on the abdomen of the male and female?*)

Find the *anus* in your specimen. In the

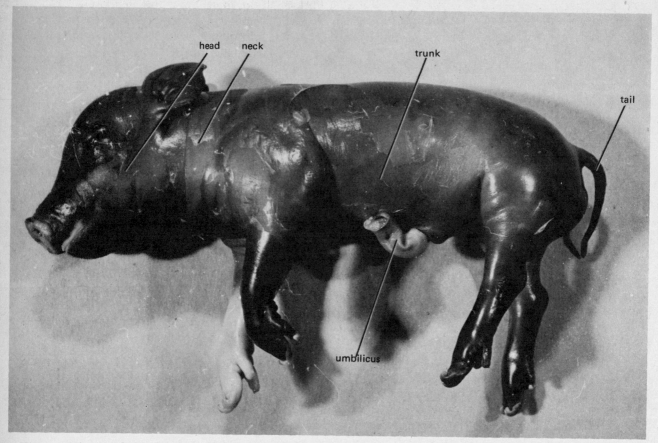

head   neck   trunk   tail

umbilicus

Figure 14.7.   Fetal pig, *Sus scrofa*.

female, the *vulva*, urogenital opening below the anus, can be seen by locating the protruding *clitoris*.

■ **As we saw in the rat, the clitoris is homologous to what male organ?**

In the male, the penis is just posterior to the *umbilical cord* on the abdomen. If your fetus is sufficiently mature, the *scrotal sacs*, which would normally receive the paired testes as the animal developed, are found in the region below the anus.

Note the appendages, the two fore- and two hindlimbs, and the most conspicuous embryonic structure, the umbilical cord. The umbilical cord is the fetus' connection—life line—to the placenta, from which the unborn animal receives food and oxygen, and into which it excretes its wastes. *The blood of fetus and mother never mix.* Rather, the exchange of gases and small molecules occurs at the capillary interfaces in the thickened wall of the placenta.

By cutting off a few pieces of the umbilical

cord (large cross sections) you will be able to observe its internal structure (Fig. 14.8). The cord has one large vein and two arteries, its three major blood vessels.

■ **How does food- and oxygen-rich blood reach the fetus from the maternal circulation if the two systems do not mix directly?**

■ **Would you consider the fetus–mother relationship to be similar to a parasite–host relationship?**

5. Internal Anatomy of the Fetal Pig

I. Digestive System (Figs. 14.9–14.11)

1. Position your fetal pig in a dissecting tray with ventral side up; spread the legs carefully so as not to tear the skin of your

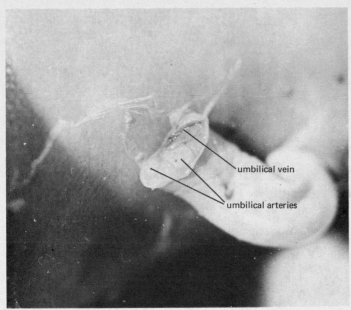

Figure 14.8. Internal structure of umbilicus.

specimen. If possible, pin the limbs down to the wax in the tray. (This will depend on the stiffness of the tissues.) If this fails, tie a string around each forelimb and tie the loose ends tightly beneath the tray. Do the same for the hindlimbs.

With scissors, start at the breast bone and make a medial cut through the body wall, stopping at a point just above the umbilicus. Cut *only* the body wall in order not to damage the underlying organs. Cut around on both sides of the umbilicus, and continue just anterior to the pelvic girdle. At this point make transverse incisions to both sides. Then make a transverse cut through the ventral body wall, posterior to the forelimbs. Pin back these flaps of tissue and place your specimen under running water to flush out preservative fluids in the abdominal cavity.

Beneath the outer layer of skin and the underlying sheets of muscle locate the *peritoneum*, the membrane lining the peritoneal cavity.

2. **Structures exposed** First locate the *liver* under the *diaphragm* in the anterior portion of the abdominal cavity.

■ **How many lobes are there in the liver?**

Figure 14.9. Abdominal cavity of fetal pig.

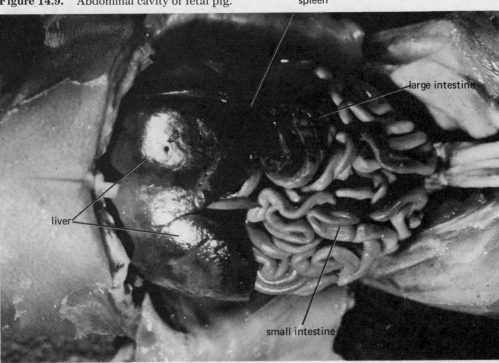

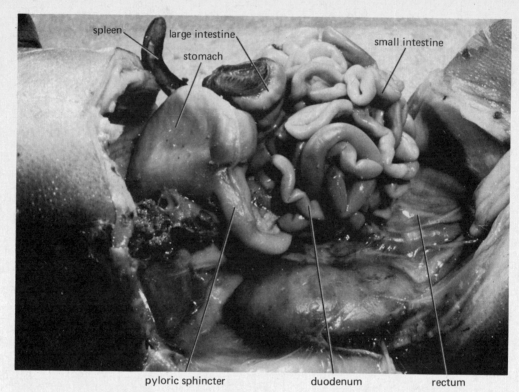

Figure 14.10. Digestive system of fetal pig with liver removed.

Figure 14.11. Digestive system of fetal pig with part of liver removed.

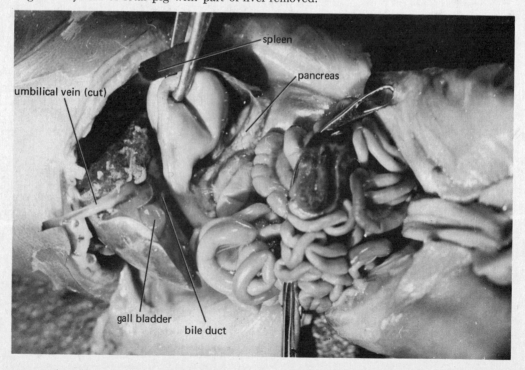

Underlying the liver is the large saclike *stomach*. As in the rat, the stomach of the fetal pig is composed of two parts—the larger anterior *cardiac* region and the posterior *pyloric* region. The contents of the *esophagus* (when functional) will empty into the cardiac portion; the food then will pass into the pyloric region. then to the anteriormost portion of the *small intestine*, the *duodenum*, via a muscular valve, the *pyloric sphincter*. The three divisions of the small intestine, *duodenum, jejunum*, and *ileum*, are often difficult to distinguish. The duodenum, starting at the *pyloric sphincter*, is about one inch long, and the following portions, jejunum and ileum, are approximately of equal length. Connected posteriorly to the small intestine is the flatly coiled *large intestine* (*colon*). It has a blind pouch, the *cecum*, which in some mammals is reduced to a vermiform appendix. (*Example?*)

■ **To what structure is the cecum connected in the rat?**

Wastes pass from colon to rectum (located against the dorsal wall), and the formed fecal pellets then leave the body by the *anus*.

Near the right central lobe of the liver, find the long greenish *gall bladder*. The rat, you may recall, lacks this organ. Carefully remove small pieces of the liver, leaving intact the *dorsal* portion of the organ through which large blood vessels run, which will be examined in our study of the circulatory system. Now locate the *bile duct*.

■ **To what portion of the small intestine is this duct connected?**

Find the *pancreas*, located in the mesentery between stomach and duodenum. Note its irregular shape, pale color, and granular appearance. The *pancreatic duct* leads to the duodenum (though this may be difficult to discern in your specimen) and enters the small intestine posterior to the bile duct entrance. (In the frog, the bile duct and pancreatic duct unite *before* reaching the duodenum).

The *spleen*, reddish in color, lies close to the stomach and should be easily observable.

■ **To what body system does this organ belong?**

■ **What is its function?**

Draw an outline of your fetal pig preparation. Include the organs you identified that comprise the digestive system. Refer to the photographs to aid in organ identification, but use your own specimen from which to sketch out the digestive tract.

■ **What organ is most conspicuous?**

■ **How does the digestive system of the fetal pig differ from that of the frog? from the rat?**

II. Urogenital System

Studying the pig's urinary and reproductive systems together is convenient and appropriate, as these systems are so closely related (as in the rat). Be sure to observe the reproductive systems of both male and female specimens.

1. **Kidney** (Fig. 14.12)   As in all mammals, the pig has a *metanephric kidney*.

■ **What other types of kidneys occur in vertebrates?**

■ **What analogous structures occur in invertebrates?**

Locate the two kidneys along the mid-dorsal line of your animal. Free one of the kidneys from the peritoneum, and cut away the fat surrounding the organ. Note the *hilus*, the concave side of the kidney from which the *ureter* (urinary duct) arises. This duct leads posteriorly to the *urinary bladder*, a thin-walled sac attached to the ventral body wall. Once you have found the urinary bladder, locate the *urethra*, the organ through which the bladder drains its contents to the outside. Care should be taken not to damage reproductive organs close by. Ventral to the anterior end of each kidney is the long, yellowish-brown *adrenal gland*. (Fig. 14.12)

■ **To what body system does this gland belong? What other glands have you seen that are part of this system?**

diaphragm   kidney   ureter   urinary bladder

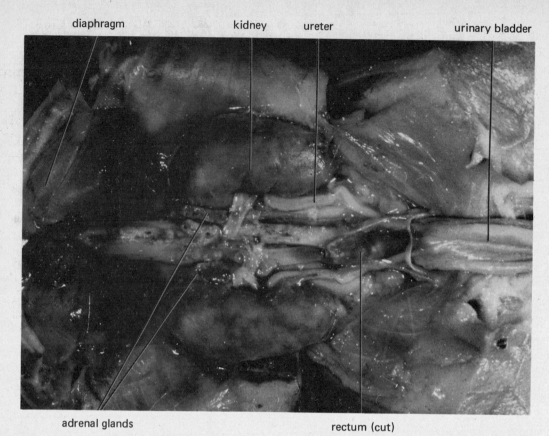

adrenal glands   rectum (cut)

Figure 14.12.   Excretory system of fetal pig.

2. **Male reproductive system** Spread the hind legs well apart, and with a sharp scalpel cut slightly to one side of the midventral line of the pelvic girdle, being careful not to dissect the delicate penis or scrotal sacs. Continue your incision through the skin and heavy muscles until you reach the *pubic symphysis* of the pelvic girdle, a cartilaginous structure that you will wish to cut through in order to completely expose the urethra.

First locate the *scrotal sacs* (Fig. 14.13). You will want to cut one open in order to see if the testes have descended into the scrotal sacs. If the testes have not yet descended in your specimen, you will be able to locate them either in the *inguinal canal* or on the dorsal wall of the abdominal cavity, below the kidneys. Along the inner edge of the testes there is a coiled mass of tubules—the *epididymis*. The epididymis merges with the *vas deferens*, the long sperm duct that eventually leads through the inguinal canal into the abdominal cavity. Locate the epididymis and the vas deferens from each testis, and trace each to its junction with the *urethra*.

■ **How does the role of the urethra as a** *urogenital* **duct differ between the male and female fetal pig?**

Near the bladder you will find the *seminal vesicles*. Just anterior to the seminal vesicles, but extremely difficult to see because of its small size, is the *prostate gland*. Posterior to the vesicles and relatively large is the pair of *Cowper's glands*.

Note that the urethra becomes the *penis*, the structure which carries both urinary and reproductive fluids to the outside.

Draw the male pig dissection, including the organs of the urogenital system that you were able to identify.

Construct an outline sketch that shows the pathways of urinary and reproductive products through the body.

3. **Female reproductive system** (Figs. 14.14 and 14.15) Using a scalpel, cut along the midventral line of the pelvic girdle. Spreading the hindlimbs farther apart, continue your incision through the cartilaginous *pubic symphysis* as you did in the male. Now

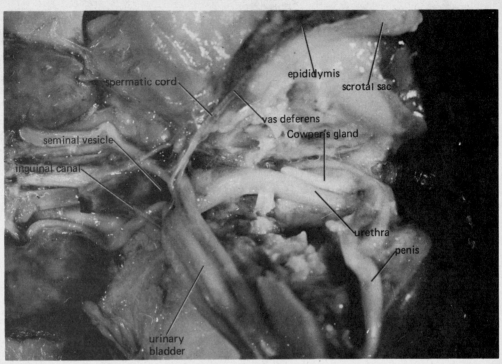

Figure 14.13.   Reproductive system of male fetal pig.

Figure 14.14.   Reproductive system of female fetal pig.

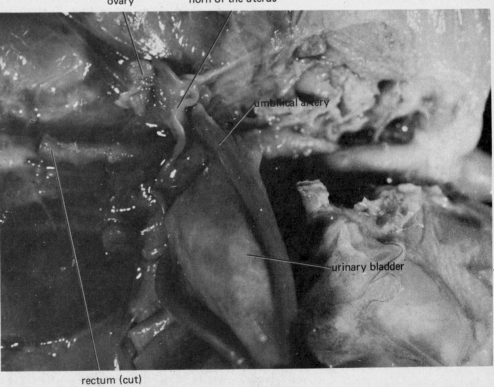

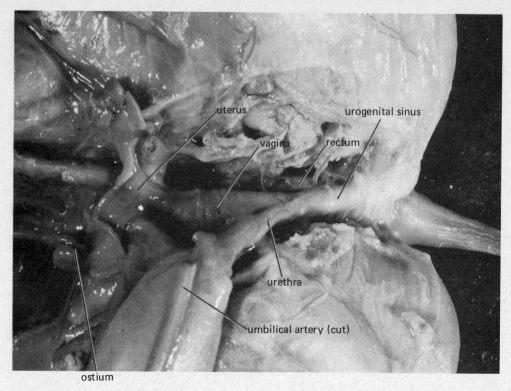

Figure 14.15.   Reproductive system of female fetal pig with umbilical vein cut.

carefully tease away the tissues from the rectum and reproductive tract using a blunt needle. Located posteriorly to the kidneys you will find the *ovaries*. Surrounding each ovary is the convoluted *oviduct* (*Fallopian tube*) which opens to the abdominal cavity. The open end of the oviduct, the *ostium*, is held close to the ovary, supported by *mesenteries*. You will be able to locate the ostium by probing with your needle around the inner margin of the ovary. Each oviduct is connected posteriorly to one of the two *horns of the uterus*. The two horns eventually unite to form the *common uterus*. The constriction at the posterior end of the common uterus is the *cervix*. The cervix separates the uterus from the *vagina*. The thin-walled vagina opens into the *vestibule*, a cavity which lies between the lips of the external genital organs. Cut the vagina and uterus along the ventral line by inserting a point of your scissors into the vestibule and cutting anteriorly. Find the *urethra* which, along with the vagina, opens into the vestibule (further support for calling this a urogenital system). Finally locate the *clitoris*, arising in the vestibule.

III. Anterior Digestive Organs and Respiratory System

1.   **Mouth** (Fig. 14.16)   With scissors, starting at the corners of the mouth, cut through the cheeks toward the neck. Using bone shears, cut through the jaw bones near their junction with the skull. In order to enlarge the mouth cavity, cut through the muscles at the corner of the mouth. Then, holding the snout and pushing down on the tongue, expose the mouth cavity fully.

Locate *lips*, *tongue*, and *teeth*. Unlike most land mammals, the pig has some of its permanent teeth (those of the second set) at birth. Identify *soft palate* and *hard palate* on the upper roof of the mouth.

■   **Can you find a line separating the two palates?**

2.   **Pharynx** (Fig. 14.16)   The cavity that lies between the soft palate and the base of the tongue in the back of the mouth is called the *pharynx*. It has two posterior openings. First is the *glottis*, the ventral opening that leads to the *larynx* (voice box). The *epi-*

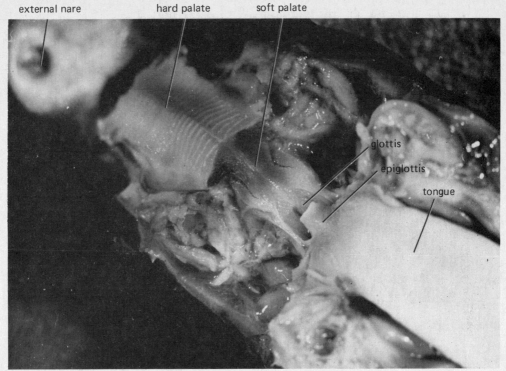

external nare   hard palate   soft palate

glottis
epiglottis
tongue

Figure 14.16.   Mouth and pharyngeal area of fetal pig.

glottis, which you should be able to locate at the base of the tongue, is a cartilaginous structure that prevents food in the mouth cavity from entering the larynx. The second pharyngeal opening (the dorsal one) leads into the *esophagus*, entry for ingested food into the digestive tract.

By cutting away the soft palate, you will be able to locate the *tonsils* (palatine lymphoid tissue) on the walls of the pharynx.
3. **Throat** (Fig. 14.17) Beginning below the chin, cut along the midventral line to a point approximately 3 cm caudal to the forelimbs. Separate the skin from underlying muscles using a blunt probe. Make a transverse cut on each side of the neck and pull the loose flaps of tissue aside, exposing the throat region. Cut through the muscles and connective tissues to reveal the *thymus gland*.

■ **What is the function of this organ?**

With forceps or scalpel, continue to cut away connective tissue overlying the throat region. Locate the *larynx* and the respiratory structure to which it leads, the *trachea* (windpipe).

■ **What structures support the trachea and hold it open?**

The trachea branches into two *bronchi*, each supplying air to one of the two *lungs*. By moving the trachea to the side you should be able to locate the *esophagus*. Ventral to the trachea is the *thyroid gland*, an endocrine organ whose primary product is the hormone *thyroxin*.
4. **Thoracic cavity** (Fig. 14.18) Starting at the breast bone, cut anteriorly through the tissues along the midventral line until you reach the dissected throat region. By a few transverse cuts, pull the loose flaps of skin to the sides, exposing the chest. Spread the forelimbs further apart and cut through the pectoral muscles to free the limbs; then cut through the rib cartilage 1 cm on each side of the breast bone. Free the structures dorsal to the breast bone and remove the bone.

Observe the separation between thoracic and abdominal cavity by the *diaphragm*.

■ **What are the principal organs or systems contained in each cavity?**

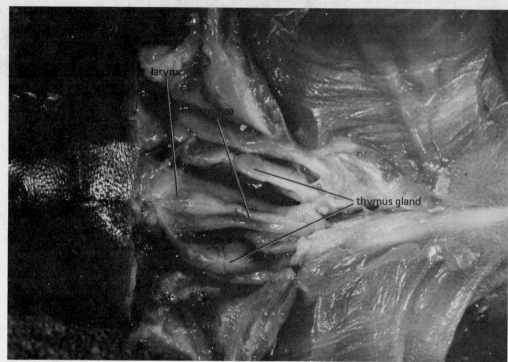

Figure 14.17.   Throat area of fetal pig (ventral view).

Figure 14.18.   Thoracic cavity of fetal pig.

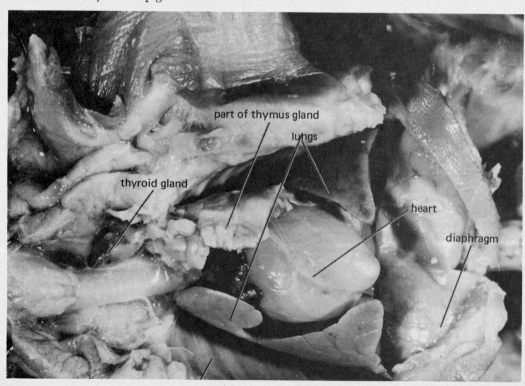

■ **Describe the *shape* of the diaphragm and its functional role.**

The chest wall consists successively of *skin, fat, muscles, ribs,* and *peritoneal lining* of the thoracic cavities, filled by the lungs lined with *pleura (Splanchnic peritoneum).* Examine the pleural cavities and the lungs more carefully.

■ **Are the lungs *lobed*? Explain what effect this would have on respiration.**
■ **Where are the lungs attached? How is this related to respiration?**

Remove a lung by cutting through the mesentery connecting it to the body.

■ **Describe the arrangement of the *bronchi* and its subdivisions.**

IV. Circulatory System

1. **Introduction** Refer to Fig. 14.19 as you study the circulatory system in your pig, but recognize that your specimen may differ substantially from the photographed example. After your review of the heart and network of arteries and veins, *draw* these structures to show the circulatory pattern and their relationships. You may find it helpful to sketch the circulatory vessels as you examine them.
2. **General view** In front of the diaphragm, the *heart* is found in the pericardial cavity. Cut the pericardium to reveal it more clearly. Then dissect out the thymus and the thyroid gland, which you located earlier. Locate the *right* and *left ventricles*, and *right* and *left atria (auricles).*

■ **Which of these parts of the heart is largest? Why?**

The two large veins opening into the right atrium, the *precava* and the *postcava,* carry the CO_2-rich (deoxygenated) blood from tissues of the body back to the heart.

■ **Where does the blood go next? (Keep in mind this is the *embryonic* phase.)**

Turn your attention to the two major arteries. The *dorsal aorta,* the largest artery, arises from the anterior portion of the left ventricle, curves to the *left*, and extends dorsally through the thoracic and abdominal cavities. The curved portion is referred to as the *aortic arch.*

■ **How would you suggest that pressure changes in the thoracic cavity (in the *postnatal,* normally breathing pig) would facilitate blood movement?**

The second largest artery—the *pulmonary artery*—arises anteriorly from the right ventricle, bends to the left, and divides into two branches, each branch passing to one lung. Separate the dorsal aorta from the pulmonary artery in the area of the aortic arch. Find the *ductus arteriosis,* the vessel that connects the two arteries during fetal life, shunting past the pulmonary circulation.

■ **How does the connection between these major arteries affect the function of the heart in the fetal pig? How does it explain fetal circulation?**
■ **How does this circulatory path compare with that through a two-chambered heart?**

3. **Systemic arteries** Trace the dorsal aorta from the heart. Find the *brachiocephalic trunk,* a large aortic branch that then divides into the *right subclavian artery* to the right forelimb, and two *common carotids* to the left and right sides of the neck and head. Past the brachiocephalic trunk, the dorsal aorta divides to form the *left subclavian artery,* which supplies the left forelimb. The *coeliac axis,* the next branch, found posterior to the diaphragm, supplies liver, stomach, and spleen. The *anterior mesenteric artery,* the next branch, supplies small intestine and pancreas. The *renal arteries,* smaller branches of the aorta, send blood to the kidneys and adrenals. Locate the *spermatic* or *ovarian arteries* in this area. Following down the aorta, find the *posterior mesenteric artery* connecting to colon and rectum. In the pelvic region, locate the *external iliacs* arising from the aorta and passing to the hindlimbs where they become the *femoral arteries.* Posterior to the external iliacs are the two *umbilical arteries. Recall these arteries as they occurred in your cross section of the umbilical cord.* The extension of the aorta into the tail is the *caudal artery.*

4. **Major veins** Locate the *jugular vein* on each side of the trachea. Follow them posteriorly. The larger of the two is the *external jugular*; the smaller, the *internal jugular*. These are joined by the *subclavian veins* from the forelimbs. These four veins unite to form the *anterior vena cava* (*precava*). The two *sternal veins* coming from the ventral portion of the thoracic cavity and the *costo-cervical-vertebral*, a vein from the neck and back area, join the anterior vena cava before it enters the right atrium. Locate this series of veins in your specimen. Now find the *azygos-unpaired vein*, which also enters the right atrium.

In the posteriormost portion of your specimen, locate the *femorals*, the *external* and *internal iliacs*, and the *caudal veins*. These unite (see illustration) to form the *posterior vena cava* (*postcava*). Find the *renal veins* that pass from the kidneys to the postcava. Trace the course of the *umbilical vein* from the umbilical cord into the liver where it joins the postcava and the *hepatic portal vein*. The hepatic portal vein passes through the liver near the gall bladder. This large vein unites many smaller branches from the various organs of the gastrointestinal tract. In the fetal pig, much of the blood flows from the liver through this vein, directly to the right atrium.

■ **What happens to the hepatic portal vein in the pig after birth? (Consult your text references at end of chapter.)**

Finally, locate the *phrenic veins*, the veins that join the postcava anterior to the liver.

5. **Heart** (Fig. 14.19) In order to study the heart, remove it from the body by severing the postcava, precava, pulmonary veins, and aorta. Locate the *coronary arteries* and *veins* on the surface of the ventricles. The veins unite in the *coronary sinus*, opening into the right atrium. Trace the origin of the coronary arteries by opening the dorsal aorta.

■ **Outline the coronary circulation, starting with the blood that enters the heart from the coronary sinus.**

Dissect the dorsal aorta further and the pulmonary artery and examine their internal structure.

Figure 14.19.   Heart of fetal pig.

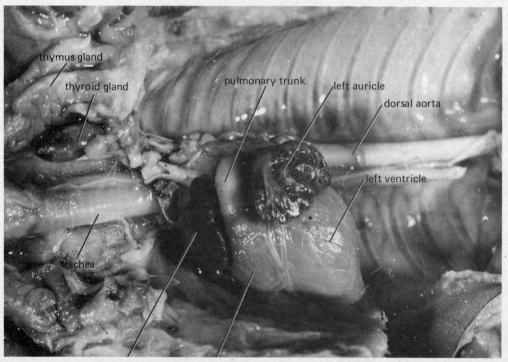

Make a midventral cut to open the heart. Locate the *tricuspid valve* between the right atrium and the ventricle. Now find the *bicuspid valve* on the left side of the heart, located in the corresponding position. Using a blunt needle, find the opening between the two atria—the *foramen ovale*.

V. Nervous System (Figs. 14.20 and 14.21)

We will confine ourselves in this section to the gross anatomy of the cranial and anterior spinal portions of the central nervous system (CNS).

First, clear the skull and anterior neck region of all integument, muscle, and connective tissues. Using strong scissors and a hard scalpel, carefully slice and chip away the portion of the *cranium* covering the top and sides of the *brain*. Try to avoid breaking the *meninges* (membranes covering the brain and spinal cord). When the upper portion of the brain has been exposed, use scissors and tweezers to remove the meninges, noting the *dura mater* (tough, outer covering), the *arachnoid* (a very thin membrane just under the dura mater and possibly indistinguishable from it), and the *pia mater* (the membrane directly covering

the brain and spinal cord). Peel off these membranes. Remove the *atlas* (uppermost vertebra), and remove the meninges covering this portion of the spinal cord. Looking down on the brain (Fig. 14.20), you should now be able to see four distinct structures. Most anteriorly are the *olfactory lobes,* responsible for the reception of smell. Posterior to them and occupying most of the cranium (in dorsal unsectioned view) is the *cerebral cortex,* composed of the two cerebral hemispheres. This area is concerned mostly with "conscious" integration and control of lower functions. More posteriorly is the *cerebellum,* a trilobed structure concerned with fine motor coordination. The central lobe is the *vermis,* and the lateral lobes are the *flocculi.* Posterior to the cerebellum, and partially uncovered when you removed the atlas, is the *spinal cord.*

Now remove the eye and ear capsules of the skull to expose the sides of the brain. You may be able to see some of the cranial nerves, which may appear as fine white threads extending out from the ventral surface (some extend more ventrolaterally). Section the spinal cord as far posteriorly as you can (to the anterior margin of the *axis,* the second vertebra). Remove the brain from the braincase as carefully as possible in order to

Figure 14.20. Dorsal view of fetal pig brain.

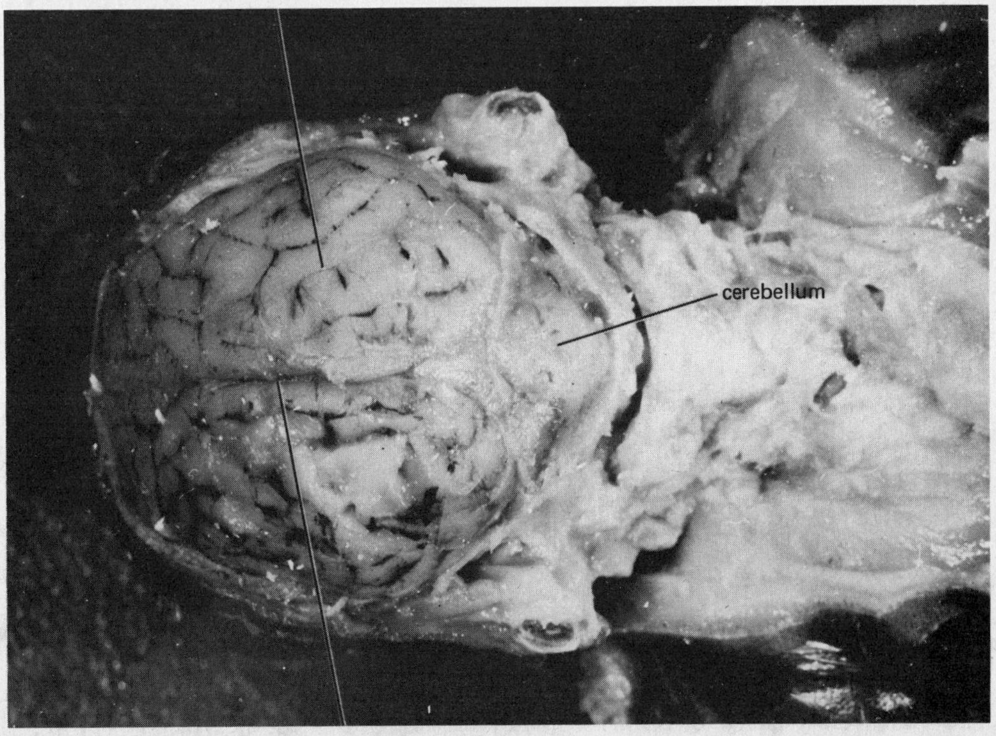

cerebrum

cerebellum

longitudinal fissure

leave intact the *cranial nerves*. The brain is connected to the *optic chiasm* in the anterior area, so try to cut the optic tracts as ventrally as possible.

Looking at the intact brain in side view (Fig. 14.21), note the two "bulges" appearing ventral to the cerebellum. The first bulge (just anterior to the spinal cord and posterior to the cerebellum) is the *medulla oblongata*, which has important regulatory functions. Anterior to this is a second, more prominent bulge, the *pons* or *midbrain*. The pons is an important relay center between the spinal cord and "higher" CNS structures.

In ventral view, note the points of entrance of the optic tracts. If you were unsuccessful in saving the chiasm itself (a difficult feat—congratulations if you succeed!), examine the braincase and note where the optic tracts enter the skull, converge at the chiasm, cross over, and diverge into left and right optic tracts. The twelve sets of cranial nerves should now be visible. Cranial nerves I through V exit anterior to the pons. The lobe hanging midventrally from the brain is the *pituitary gland* or *hypophysis*. If the pituitary did not come out with the brain, then you will only be able to see the stalk, the *infundibulum*.

Now make a careful midsagittal section all the way through the brain, starting by separating the *cerebral hemispheres* (they need not be cut apart) and working downward. Looking at the sagittal section, note first the whitish treelike structure in the vermis, the *arbor vitae*. Ventral to the cerebellum is a space, the *fourth ventricle*, which continues anteriorly into the *third ventricle*. The *lateral ventricles* are the hollows within the cerebral hemispheres. The whitish structure in the center of the brain, extending anteroposteriorly in midsagittal section, is the *corpus callosum*, which communicates between the hemispheres. Try to locate the *pineal body* or *pineal gland*. (Descartes thought the pineal was not connected to the rest of the brain, and that therefore it was the organ of communication between body and "soul," the "seat of the soul.") The small *commissures* (that appear to be circular as this is a midsagittal section) are bundles of communicating neurons that pass in a lateral axis.

Return to the fetal pig and prepare a cross section of the spinal cord. Can you differentiate *gray* from *white* matter? Carefully removing vertebrae, see if you can find *dorsal* (*sensory*) and *ventral* (*motor*) roots of the spinal cord. Note that the

Figure 14.21.   Lateral view of fetal pig brain.

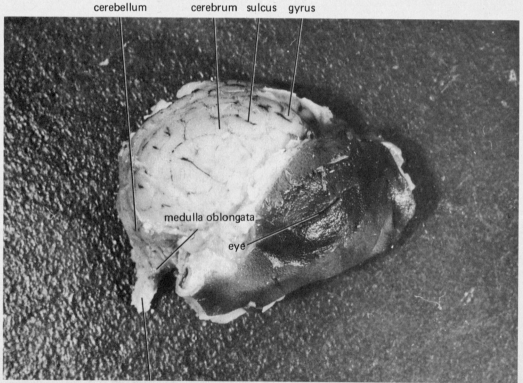

lumen of the cord is *continuous with the ventricles* of the brain itself. In life it is filled with the protective bathing *cerebrospinal fluid*, the exact pressure and constitution of which is a vital clue in humans to their state of health or possible abnormal states, such as *meningitis* (inflammation or infection of the meninges), *encephalitis* (inflammation or infection of the brain), or presence of a pathogenic organism such as the African trypanosomes, cause of trypanosomiasis or African sleeping sickness.

Suggested References

Booth, E. S., 1961. *How to Know the Mammals*. Dubuque, Iowa: William C. Brown.

Bourliere, F., 1954. *The Natural History of Mammals*. New York: Alfred A. Knopf.

Bradley, O. C., and T. Grahame, 1943. *Topographical Anatomy of the Dog*, 4th ed. New York: Macmillan.

Burt, W. H., and R. P. Grossenheider, 1952. *A Field Guide to the Mammals*. Boston, Mass.: Houghton Mifflin.

David, D. E., and F. B. Golley, 1963. *Principles in Mammalogy*. New York: Reinhold.

Greene, E. C., 1935. "Anatomy of the rat." *Tri. Amer, Philos. Soc.* (N.S.), 27:1.

Hall, E. R., and K. R. Kelson, 1959. *The Mammals of North America*, 2 vols. New York: Ronald Press.

Howell, A. B., 1926. *Anatomy of the Wood Rat*. Baltimore, Md.: Williams & Wilkins.

Marrable, A. W., 1971. *The Embryonic Pig. A Chronological Account*. Baltimore, Md.: University Park Press.

Reighard, J. E., and H. S. Jennings, 1935. *Anatomy of the Cat*, 3rd ed. New York: Holt, Rinehart and Winston.

Romer, A. S., 1959. *The Vertebrate Story*. Chicago, Ill.: University of Chicago Press.

Sanderson, I. T., 1955. *Living Mammals of the World*. New York: Doubleday.

Simpson, G. G., 1945. *The Principles of Classification and a Classification of Mammals. Bull. Amer. Mus. Nat. His.*, 95:1.

Walker, E. P. (ed.), 1964. *Mammals of the World*. Baltimore, Md.: Johns Hopkins Press.

Walker, W. F., Jr., 1972. *Dissection Guides* (a series of separate loose-leaf exercises): *Fetal Pig; Frog; Rat*. San Francisco, Calif.: W. H. Freeman.

Wischnitzer, S., 1972. *Dissection Guides* (a series of separate loose-leaf exercises): *Amphioxus; Cat; Dogfish Shark; Lamprey; Mud-Puppy Necturus*. San Francisco, Calif.: W. H. Freeman.

Young, J. Z., 1957. *The Life of Mammals*, 2nd ed. Oxford: Clarendon Press.

——, 1962. *The Life of Vertebrates*, 2nd ed. New York: Oxford University Press.

CHAPTER 15

CYTOLOGY AND HISTOLOGY

1. Introduction

In previous chapters we have been for the most part involved in understanding a macroscopic view of the zoological world. We now turn our focus to the microscopic counterpart of that world, comprised of *cells* and *tissues*. An impressive amount of research has been directed toward uncovering the mechanisms and wonders of life by observing its functional/structural units, cells, and their component parts, the organelles. Further understanding has been gained by examining cell activities on a larger scale, the study of how *groups* of cells, or tissues, operate.

In this exercise we will examine several cell and tissue organizations, the distinguishing characteristics of which will be our main concern. As you become familiar with important cell types and tissue groups, keep in mind that you are approaching the study of our basic objective, the *intact living organism* through successive levels of structural organization, from *cell* to *tissue* to *organ*. Along the way, conflicting definitions, new interpretations, and new concepts will surface. The occurrences of such inconsistencies lend impetus to different views and interpretations of the biological evidence, and ultimately to new syntheses and greater understanding.

Before commencing a detailed comparison of specialized vertebrate cells and their function, it would be useful for you to review your text on the typical cell and its components; then consider the definitions and review the outline of some common cell types that are described in the ensuing sections.

The following definitions are basic and generally accepted; however, do not allow them to rule out alternate definitions or to avoid marginal exceptions. A questioning attitude should make it easier for you to approach new ideas and to abandon cherished interpretations gleaned from this or other texts.

2. Definitions

1. **Cytology**[1] Study of cells, their structures and functions.
2. **Histology**[1] Study of tissues, their structures and functions.
3. **Cell** The smallest living integral unit,

---

[1] Protein synthesis, gene action, developmental processes, cell specialization and control mechanisms, cellular immunity, comparative physiology, aging processes, cell and tissue pathology, and recovery mechanisms are other areas of active contemporary research in cytology and histology.

characterized by growth at some stage and by metabolism throughout its life.

4. **Tissue** A group of coordinated cells with one or several common structural and functional specializations.

5. **Organ** A complex of tissues forming a functional unit and usually having a stable structural organization.

6. **Organism** A highly organized unit that is living and capable of growth, differentiation, assimilation, metabolism, reproductive constancy, and an integrated response to its environment.

The vast majority of cells are capable to some degree to perform all essential cellular functions, including *reproduction, metabolism, conductivity*, and, to a lesser degree, *contractability*. Cells specialized for one or more of these functions can form different tissues or organs, each of which performs a specific activity in the intact organism (for example, a germ cell, liver cell, nerve cell, or muscle cell).

Examine your slides with these purposes in mind: (1) to be able to identify a few cell types, both generalized and specialized; (2) to recognize tissues and organs composed of these cells; (3) to integrate this information into an appreciation of the *organism* as a harmoniously functioning balance, whether in one or many cells; (4) to provide background information for appreciation of current research probings into genetic control mechanisms, differentiation, embryogenesis, and overall control processes in the intact organism; and (5), an added aim, to gain an appreciation for the painstaking steps needed to prepare the slides you are studying, the skill required to reduce an organ or a tissue to minute, meticulously stained, selected cross sections permanently affixed to a slide. Review with your instructor the steps involved (the process is known as *microtechnique*) in killing, fixing, embedding, sectioning, staining, clearing, and mounting tissues, each with associated periods of hydration, dehydration, and other chemical treatments.

3. Outline of Basic Tissue Types—Examples from Vertebrate Tissues

I. Epithelium (covering or lining tissue)

Cell arrangement of the three types listed below may be *simple* (one layer), *stratified* (lay-

ered, or *pseudostratified* (cells of different sizes, but all of basically one layer). Epithelia may have specialized outgrowths, such as undulipodia (cilia or flagella). Functional classification may also be employed, rather than a descriptive one as employed here.

1. *Squamous epithelium* Forms lining of blood vessels (Fig. 15.1), peritoneum (Fig. 15.2); and epidermis (Fig. 15.3).

2. *Columnar epithelium* Found in lining of digestive tract, mucous membranes, excretory ducts, and trachea (Fig. 15.4).

3. *Cuboidal epithelium* Glandular tissue, as in the lining of the kidney tubules, in the thyroid (Fig. 15.5), or in other glands.

II. Blood and Lymph (vascular and intercellular tissues; a liquid (lymph) matrix with supported cellular elements)

1. **Blood** Circulating vascular fluid with various formed elements of blood cells (Fig. 15.6).

a. *Red blood cells* (RBCs or *erythrocytes*). Circular, biconcave disks, nonnucleated in mammals, contain hemoglobin that carries oxygen and carbon dioxide to and from cells.

b. *White blood cells* (WBCs or *leucocytes*). Nucleated, ameboid corpuscles of blood, lymph, pus, or tissue (lymphoid wandering cells); in-

Figure 15.1. Squamous epithelial cells of the blood vessel. (*Courtesy* CCM: General Biological, Inc., Chicago.)

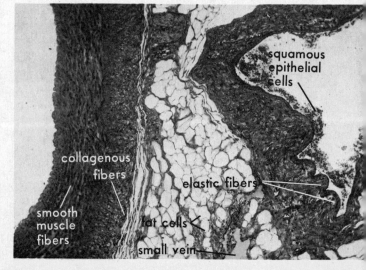

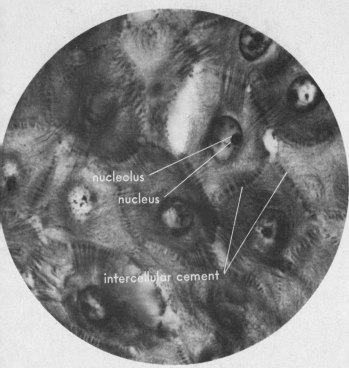

Figure 15.2. Squamous epithelial cells of the peritoneum. (Courtesy CCM: General Biological, Inc., Chicago.)

Figure 15.3. Squamous epithelial cells of the epidermis. (Courtesy CCM: General Biological, Inc., Chicago.)

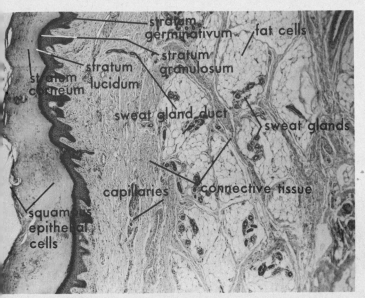

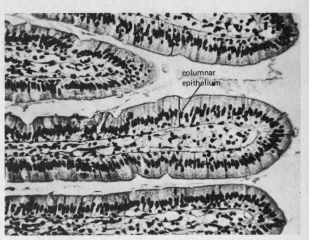

Figure 15.4. Columnar epithelium from a human intestine.

cluded are *lymphocytes, monocytes, plasma cells, mast cells, eosinophils, basophils,* and *neutophils* (the last three sometimes combined under the term *granulocytes*); an important function of most of these cells is to recognize, ingest, destroy (or encapsulate), or isolate harmful elements, foreign substances, senescent, infected or damaged cells, and to clear away cellular debris within the organism. A critical requirement of these cells is to recognize what is normal to the organism (self) from what is not normal, i.e., sick or foreign (not self).

c. *Blood platelets.* Small, colorless disks that play a role in blood clotting, occur only in mammals.

d. *Plasma.* Blood fluid *including* clotting elements but not cells.

Figure 15.5. Cuboidal epithelium from human thyroid gland. The epithelium lines the spaces in this photograph.

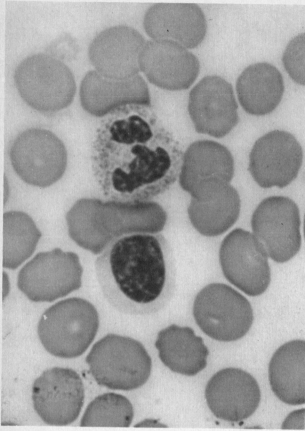

Figure 15.6.   Human blood cells.

e. *Serum.* Blood fluid *without* either clotting elements or cells.

2. **Lymph**   Transparent intercellular or lymphatic vessel fluid, includes white blood cells, chiefly lymphocytes, and important proteins, such as the immunoglobulins, from which protective *antibodies* are derived.

III. Connective Tissue (supportive tissue)

These are various cells embedded in a matrix of *fibrous* or *mucoidal* substances produced by the cells themselves, which bind together and support body structures. Types of connective tissues are the following:

1. **Loose connective tissue**   Differentiated by the type of intercellular fibers present: *collagenous* (white) or *elastic* (yellow). Loose connective tissue may contain such cells as *fibroblasts, macrophages, lymphoid wandering cells, mast cells, eosinophils, plasma cells, pigment cells, fat cells,* and *undifferentiated cells* (Fig. 15.7).

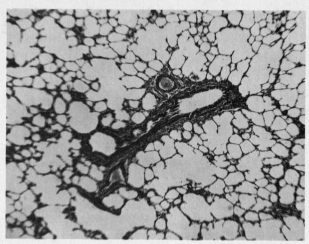

Figure 15.7.   Loose connective tissue (adipose tissue).

2. **Dense connective tissue**   Chiefly in the skin and in the submucous layer of intestine. Fibers are thick, compactly woven collagenous bundles. The cells are difficult to identify but probably the same as those in loose connective tissues (Fig. 15.8).

Examples of connective tissues are the following:

1. **Regular connective tissue**   Collagen bundles arranged in a definite manner.

a. *Tendon.* Connects muscle to bone or muscle.
b. *Ligament.* Less regular form than tendon, a tough fibrous band between bones or supporting viscera (Fig. 15.9).

Figure 15.8.   Dense connective tissue from human tendon.

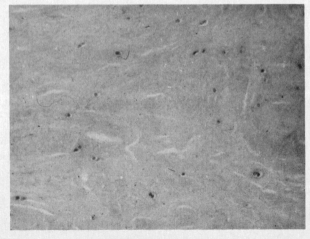

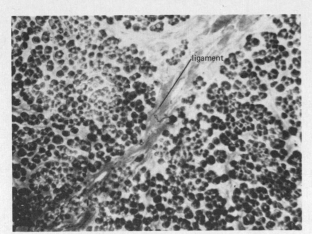

Figure 15.9. Ligament.

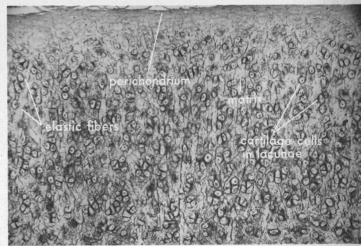

Figure 15.11. Hyaline elastic cartilage. (Courtesy CCM: General Biological, Inc., Chicago.)

2. Special connective tissue

a. *Mucous connective tissue*. Found in embryos.

b. *Elastic tissue*. Found in vocal cords, arteries (Fig. 15.1).

c. *Reticular tissue*. In spleen, liver.

3. Cartilage
Specialized fibrous connective tissue in skeleton of embryo and joints of adults (Fig. 15.10).

a. *Hyaline cartilage*. In embryo and adult joints (Fig. 15.11).

b. *Fibrocartilage*. In intervertebral disks.

4. Bone
Calcified fibrous connective tissue with cells connected by thin cytoplasmic strands (Fig. 15.12).

IV. Muscle Tissue (contractile tissue)

1. Striated muscle
Voluntary, forms skeletal muscle; made up of long, cylindrical, multinucleate fiber cells containing fibrils,

Figure 15.12. Bone, a calcified fibrous connective tissue. (Courtesy CCM: General Biological, Inc., Chicago.)

Figure 15.10. Cartilage, a specialized connective tissue. (Courtesy CCM: General Biological, Inc., Chicago.)

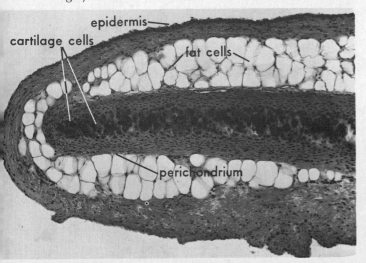

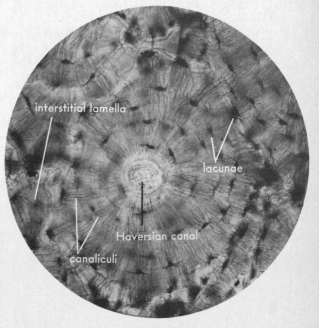

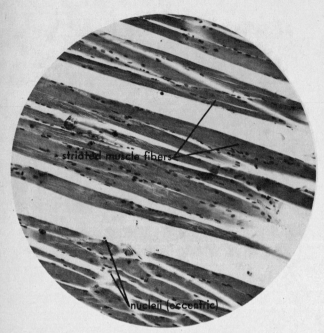

Figure 15.13. Striated muscle showing the multinucleate fibers with characteristic cross striations. (Courtesy CCM: General Biological, Inc., Chicago.)

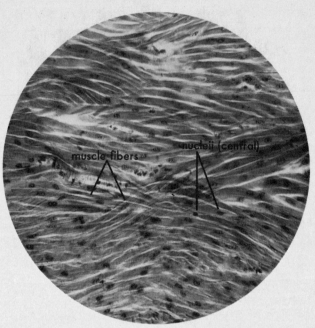

Figure 15.14. Smooth muscle: Dark, oval elements represent the nuclei; the longer elements are the muscle fibers. (Courtesy CCM: General Biological, Inc., Chicago.)

characteristic cross striations; nuclei lie outside fibrils (Fig. 15.13).

2. **Smooth muscle** Involuntary, chiefly in internal organs, principally in the digestive tract, hence also called *visceral* muscle; long, spindle-shaped, mononucleated fibers (Fig. 15.14).

3. **Cardiac muscle** Involuntary heart muscle made of striated fibers that branch and anastomose to form a single interconnected network; nuclei interior, fibers with *intercalated* disks (function uncertain) (Fig. 15.15).

V. Nerve Tissue

This is composed of *neurons*, cells that are highly specialized for irritability and conduction (Fig. 15.16), and of *neuroglia* (cells in close association with neurons, serving probably for nutrition and support).

Nervous tissue should be thought of as a closely coordinated single system with specialized units.

1. **Central nervous system** Brain and spinal cord, including the *cranial* and *spinal* nerves and their *association interneurons*.

2. **Peripheral nervous system** Nerve tissue outside CNS; the system of nerves that connects the peripheral *receptors* and *effectors* with the CNS.

3. **Autonomic nervous system** Formed

Figure 15.15. Cardiac striated muscle showing interior nuclei, cross striations, anastomosing network of fibers, and intercalated disks. (Courtesy CCM: General Biological, Inc., Chicago.)

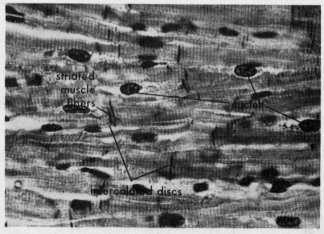

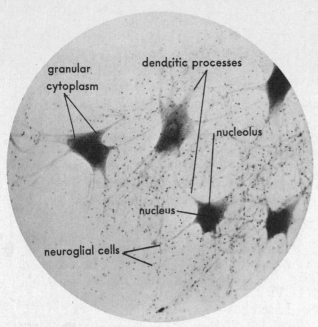

Figure 15.16. Nervous tissue of the spinal cord showing five nerve cells and numerous neuroglia cells (small dark dots). (Courtesy CCM: General Biological, Inc., Chicago.)

from two chains of ganglia found along spinal cord and scattered among body tissues; concerned chiefly with regulation of visceral activity; divided into two units, named for location of their ganglia, with balanced (opposed) function. Actually a portion of the peripheral nervous system, but listed separately because of its distinctive form.

a. *Sympathetic nervous system* (or *thoracolumbar* n.s.).

b. *Parasympathetic nervous system* (or *craniosacral* n.s.).

■ **How do the two major divisions of the autonomic nervous system function as a "balancing of opposites"?**

VI. Adipose Tissue (fat tissue)

This is tissue specialized for body insulation and for energy storage. Fats yield twice the energy per gram as do carbohydrates, hence are the most economical form of energy reserve storage. Adipose tissue may be either white or brown; the former is more common and found in numerous areas and tissues of the body, while the latter is found mostly in hibernating organisms and thought

to provide the slowly released energy during the hibernation period.

Adipose tissue also serves as an insulator, and in some cases, as in marine mammals, provides moderate support as well (Fig. 15.7).

VII. Other Specialized Cells

There are other specialized cells that are of interest but cannot be considered as distinct tissues. These are the *reproductive* or *germ* (*sex*) *cells*—*ova* and *spermatozoa*.

■ **Why are these not properly classed as tissues?**

Other specialized cells may form a well-defined tissue, such as the *pigment* cells that form part of the choroid and sclera at the eye.

4. Laboratory Instructions

I. Stained Slides

Study as many examples of stained tissue preparations as are available. Certain slides will be studied at your desk, others at demonstration microscopes.[2] Refer to your text and to laboratory charts for views of comparative material and detailed discussions of the more specialized cell and tissue types, checking them against the outline in the preceding section.

In your slide preparations, learn to judge differences due to the angle of cut of the microtome knife, the intensity of stain, or to factors such as shrinkage or pressure resulting from the technique employed. *Be particularly careful to locate portions of tissues that demonstrate representative cells.* Consult your instructor to be certain you have judged correctly. From available slides *select, draw,* and *label* (1) a "typical" cell, and (2) a few cells characteristic of as many of the basic tissue types described in Section 3 as are available.

Select good examples and draw grouped cells showing sufficient detail to illustrate their

_____

[2] Proper use of the microscope is of utmost importance. Review the discussion of control of illumination in Chapter 1. Use low-level illumination with low-power objective, especially for living material. Glare masks the image, whereas reduced illumination (controlled by closing down iris diaphragm or lowering the condenser) enhances outlines and brings out tissue pockets or folds that are entirely lost in the scattered rays of brighter light.

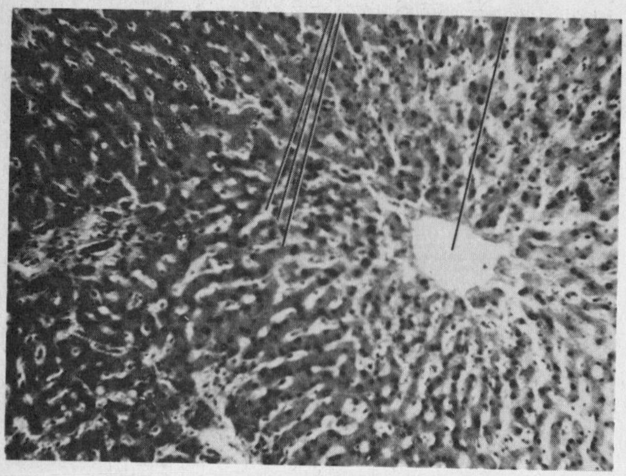

central
vein
nuclei   area

Figure 15.17.   Cells of the liver.

specialized characteristics. State the magnification,[3] the portion of organ or tissue drawn, and the type of tissue illustrated.

The following are examples of prepared slides that are commonly available.

1. **Liver cross section**   This slide is useful for examining the structure of a typical cell. Select a cell showing *cell membrane, cytoplasmic granules, mitochondria, vacuoles* or other inclusions, *nucleus, nuclear membrane, chromatin* (Fig. 15.17).

2. **Cartilage**   Note the *hyaline matrix* with embedded *chondrocytes* (cells).

■ **Can you find collagen fibers?**
■ **Distinguish between the *white* fibers of hyaline cartilage and the *wavy* fibers of elastic cartilage (Figs. 15.2 and 15.10).**

3. **Bone** (Fig. 15.12)   Study a transverse section from a long bone. Living cells are absent, as the slide was made from specially prepared dried bone. This method of preparation fills all cell spaces and canals with fine powder causing them to appear as black areas. The outermost cover is the *periosteum,* a fibroelastic sheath that surrounds the outer

bone surface. Below this are concentric bone layers, *circumferential lamellae,* parallel to the outer surface. *Concentric lamellae* surround each central pore (*Haversian canal*), once a cavity for blood vessels. *Interstitial lamellae* fill the space between Haversian canals. Together they form a *Haversian system.* Bone cells (*osteocytes*) originally occupied the spaces between adjacent lamellae, but only lens-shaped spaces (*lacunae*) remain in your prepared specimen. *Bone cell processes* once served as intercellular bridges, passed through *canaliculi,* and connected different portions of living bone. All that remains are microscopic radiating lines of canaliculi. The large central *marrow cavity* connects with bone tissue by a *subperiosteal lymph space.* Note also that *Haversian canaliculi* radiate from each central canal. Sections of decalcified bone, stained to show the cells, may be available and should be examined in conjunction with the ground dried bone preparations.

4. **Skin (integument)** (Fig. 15.3) Rat, frog, or human skin is generally used. Study the successive layers with their associated cell types:

a. *Epidermis*

i. *Stratum corneum.* The outermost or "horny" squamous layer of *keratin* (main constituent of bone, hair, nails).

ii. *Stratum lucidum.* A thin, translucent band, best seen in thickened skin of palm or sole.

iii. *Stratum granulosum.* Flattened cells.

■ **How many layers comprise this stratum?**

The dark stain is probably from keratohyalin, the same substance that forms keratin in the stratum corneum.

iv. *Stratum germinativum.* Basal growth layer.

b. *Dermis (corium).* Connective tissue with blood vessels, nerves, dermal glands; matrix of wavy collagenous fibers. Projections or *papillae* extend from the dermis into the superficial epidermis.

Embedded in the skin are *hair follicles,* consisting of epithelial cells (*inner root sheath* and *outer root sheath*) and a connective tissue

---

[3] The magnification (power of the ocular times power of the objective used, expressed in *diameters*) is for the microscopic image. Your *drawing* will probably represent a still greater enlargement unless you use a *camera lucida* or similar projection apparatus. How would you determine your actual *drawn* magnification from that of the microscopic image?

cover, or *theca*. The *hair shaft* arises from the center of the follicle.

Note the associated blood vessels and nerves in the hair papillae. With each hair is a delicate, smooth muscle, the *arrectores pilorum* (which in man produces "goose pimples") and a *sebaceous gland* usually connected with a follicle. The gland secretes *sebum*, an oily substance that maintains the pliable structure of hair.

c. *Sweat glands.* Coiled tubular glands connected to the surface by a *sweat pore*. (These glands are absent in the rat and frog, but present in man.)

■ **Where are the sweat glands found in the dog?**

5. **Blood**   Study stained slides of frog or rat blood. Compare with fresh preparations. Notice the presence and large size of a *nucleus* in the frog erythrocytes.

■ **How many types of leucocytes can you identify?**

6. **Muscle** (Figs. 15.13–15.15)   Study examples of striated, smooth, and cardiac muscle fibers.

7. **Nerve** (Figs. 15.16, 15.18, and 15.19)

a. *Brain.* Cellular structures are made visible by a special *Golgi stain*, a silver impregnation process for nervous tissue that stains nerve cells black. Find different types of supporting or *glial* cells.

Figure 15.18.   Cross section of spinal cord.

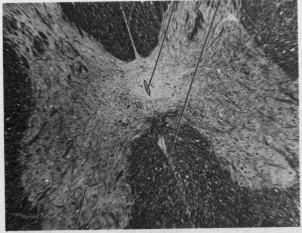

dorsal
medium   central   ventral
medium   central   median
sulcus   canal   fissure

grey matter   white matter

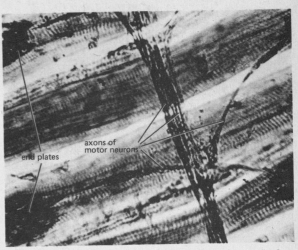

Figure 15.19.   Motor neurons terminating in end plates on skeletal muscle.

end plates

axons of motor neurons

■ **What nerve cell types can you locate?**

b. *Spinal cord.* Locate the central, dense, cellular *gray matter*, and lighter outer *white matter*, containing few cells and many fibers (Fig. 15.18). Find the *motor cells*, very large cells at each corner of the ventral area of the "butterfly"—a trapezoid formed by gray matter, named for its appearance in cross section. Study the *cell body* of a motor cell under high power (Fig. 15.19). Locate its large *nucleus*, the *nucleolus, granular cytoplasm, Nissl bodies*, and nerve cell *processes* (*dendrites* and *axons*).

Define: *nerve, nerve cell body, dendrite, axon, internuncial* or *association neuron*. Find an example of each in your slide.

II. Fresh Material

Use very small bits of tissue in order to get well-flattened preparations under the coverslip. Place the sample on your slide in a drop or two of 0.7 percent saline; if necessary, macerate it with needles to separate the cells before applying the cover. Thick or bulky samples are useless, as you are looking for *cell types*, not organ anatomy. Fresh blood samples can be smeared directly across a *clean* slide by using the end of another slide to drag a spread-out blood drop along the slide surface. The smears should be stained in Wright's or Giemsa's stain. Diluted blood can also be examined in a drop of saline to see the blood cells—and possibly some blood parasites in the frog blood, such as trypanosomes with their waving undulating membrane and anterior flagellum.

From available material, select representatives of the basic cell types described in Section 3. As you did with the stained preparations, select and draw an example of each.

Types of living material readily available (mostly from fresh frog tissue) are the following:

1. **Red blood cells** Use high power to examine blood from human (finger prick), frog (heart puncture or cut toe), or rat (tip of tail).

■ **How do the red cells of these animals compare with respect to cell size and shape, and presence of nucleus?**

■ **Can you find any white cells?**

■ **What happens to blood cells in saline that is too concentrated (producing "star cells")? too dilute (producing "ghost cells")?**

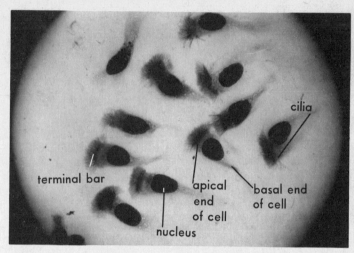

Figure 15.20. Ciliated epithelium isolated from frog mouth. (Courtesy CCM: General Biological, Inc., Chicago.)

2. **Epithelium** Use a toothpick to scrape the lining of your cheek. Mix contents in a small drop of saline and examine.

■ **What did you expect to find?**
■ **What cell types do you find?**

3. **Cartilage** (Figs. 15.10 and 15.11) Find matrix with embedded chondrocytes.
4. **Skin** (Figs. 15.3)

■ **What cell types can you identify?**

5. **Ciliated epithelium** (Fig. 15.20) Scrapings from the roof of the mouth of a freshly killed frog will show ciliary movement under high power.
6. **Muscle** Preserved or fresh tissues from frog or rat are generally used.

a. *Striated (skeletal) muscle* (Fig. 15.13). Macerate a small bit of tissue from leg muscle, teasing it lengthwise to separate the fibers. Find the *sarcolemma* (muscle membrane), *nuclei*, and *fibrils* with their *transverse striations*.
b. *Heart muscle* (Fig. 15.15). Find the branches and anastomes, and intercalated disks (dark bands).
c. *Smooth (visceral) muscle* (Fig. 15.14). Small sections of intestinal wall soaked in chloral hy-

drate are easiest to study. Find the tapering, elongate, smooth muscle fibers, best seen at the edge of your preparation. Vital stains can be used to sharpen details and to accentuate the nuclei.

7. **Neuron** (Fig. 15.19) Carefully examine a portion of nerve from a freshly killed frog. Place it in saline on your slide; separate the minute longitudinal fibers with your needles. Each of these fibers is actually a *nerve cell process*, either an *axon* or a *dendrite*.

■ **What is the functional difference between these types of nerve cell processes? Are they morphologically distinguishable?**

Under high power, study an axon and find its central, transparent *axis cylinder*, often surrounded by the *myelin sheath* and then the *neurilemma*, thin outermost membrane of the fiber.

■ **Which nerve fibers *lack* the myelin sheath?**

8. **Reproductive cells**

a. *Spermatozoa.* These can be obtained from freshly excised testes of the frog, rat, echinoderm, or any other available animal in its breeding cycle. Among invertebrates, the seminal vehicle of an earthworm, the testis of *Ascaris*, or various echinoderms are all excellent sources. Study a diluted sample in saline under high power and find the numerous actively moving spermatozoa.

■ **How do they move? Find the *head*** (shape?) and *tail* (relative length?).
■ **Can you find developing *spermatids*?**
■ **Are any other cells visible? Compare** morphology of frog and *Ascaris* sperm.

b. *Eggs.* Sea star or sea urchin ovaries are convenient, as the eggs are small; the frog egg is good for low-power study. Note the size and form of the membrane, the granular nature of the cytoplasm, and the large nucleus.

■ **Can you locate a *nucleolus*? a *polar*** *body*? **Define each.**

III. Study of Organs—Complexes of Integrated Tissues

A more intensive study of vertebrate physiology than is possible here is needed to relate the activity of specific tissues to the functions of each organ. Yet a review of the tissue constituents of organs will help to illustrate the sequence of organizational levels from cell to tissue, to organ, to organ system, and to the intact living organism. Among organs usually available in stained sections are the following: *liver, pancreas, kidney, esophagus, stomach, intestine, testis, ovary, lung, trachea, spinal cord,* and *brain*. Some of these have already been studied for details of cell or tissue types. Review examples given to you for tissue study under low power to develop your general concept of the construction of these organs.

Select preparations of three different organs; *draw* an outline of the entire structure, label the basic regions, and show tissue and cellular detail in a small portion.

Suggested References

Bloom, W., and D. W. Fawcett, 1968. *Histology*, 9th ed. Philadelphia, Pa.: W. B. Saunders.

DeRobertis, E. D. P., W. W. Nowinski, and F. A. Saez, 1970. *Cell Biology*, 5th ed. Philadelphia, Pa.: W. B. Saunders.

Fawcett, D. W., 1966. *The Cell; Its Organelles and Inclusions*. Philadelphia, Pa.: W. B. Saunders.

Kennedy, D., 1965. *The Living Cell* (*Readings from Scientific American*). San Francisco, Calif.: W. H. Freeman.

Loewy, A. G., and P. Siekevitz, 1969. *Cell Structure and Function*, 2nd ed. New York: Holt, Rinehart and Winston.

CHAPTER 16

CELL DIVISION

1. Introduction

The life cycle of all sexually reproducing organisms consists of two basic steps, (1) *gametogenesis* and *fertilization*, and (2) *differentiation*, followed by *growth* and *maturation*.

Step 1 involves *separation* and *recombination* of units of hereditary material (namely, genes and chromosomes).

Step 2 is the expression of the genetic information contributed by the sperm and ovum in a newly fertilized egg (*zygote*) to produce an individual with characteristics representing *both* parents in varying degrees. Selective expression (including repression) of the inherited message continues throughout the differentiation of cells, tissues, and organs, during later growth, in the fully mature organism, and in the final stages leading to cell senescence and organismic death.

Mitosis, cell division of body (*somatic*) cells (as opposed to *sex* cells) is the chief means of orderly growth. This remarkable mechanism almost always precisely apportions the nuclear substance of one cell to two daughter cells. The result is an exact, programmed duplication of the essential nuclear information. As many as 200 trillion cells are said to comprise the human body. The nuclear constitution of each somatic cell is identical to that of the original cell, the zygote. Exceptions are modified cells that have no nucleus (as in red blood cells that start out with nuclei but lose them in the course of their maturation), or other types that have multiple or polyploid nuclei.

■ **How many mitotic generations would you estimate have occurred in *your* development from the zygote stage?**

One purpose of this survey is to recognize the mitotic mechanism for what it is: fantastically efficient and accurate to convey the precise genetic information imparted by sperm and egg into a zygote, and thence to every living somatic cell in the body, no matter how many or how specialized. As long as cells of an organism retain their nuclear apparatus intact, they remain genetically identical.

A specialized type of cell division occurs in the gonads when sexual cells (*gametes*) are formed. This modification is an essential one, because gametes possess only *half* the normal chromosome complement of their parent somatic cells. They are therefore said to be *haploid* (one set).

■ **Can you tell why this is necessary?**

Somatic cells, in contrast, have *two* complete chromosome sets in their nuclei, and therefore are called *diploid. Meiosis* is the process by which haploid cells are produced from diploid ones in the gonads. It also is an example of precise cellular mechanics. Exclusively part of *gametogenesis* (*gamete production*), meiosis consists of two linked nuclear divisions. The first of these separates each chromosome *pair* (derived originally from the parent gametes) into two separate nuclei, each with *one complete set* of chromosomes. This is the *reductional* division. Each of these separate. This is the *equational* division, resulting in four future gametes. Details of nuclear rearrangement show the extraordinary procedure of chromosomal reduction from diploidy to haploidy and tell us much about how the built-in mechanisms of chromosome (and gene) assortment, and recombination within this precision method ensures *separation* of *complete chromosome sets* and still accounts for *genetic recombination* observed as the laws of Mendel, reviewed in Chapter 18. Maturation of the gametes completes the processes of spermatogenesis or oögenesis following each meiotic division.

The genetic shuffling referred to in step 1 of the life cycle occurs (a) in the random separation of homologous chromosomes at meiosis, and (b) in the random recombination of chromosome sets when sperm and egg nuclei fuse in fertilization. Any one of the millions of sperms produced has as good a chance as any other to fertilize a particular egg. A third source of random mixing is the process of "*crossing over*" between parts of each chromosome pair during the first meiotic division. This follows the alignment of the two chromatids of each chromosome *plus* the two from the homologous chromosome when these chromosomes pair up during the first meiotic division. The four chromatids form a four-thread *tetrad* at this stage. Crossing over consists of reattachment of corresponding parts of chromatids between the paired chromosomes. This results in a reciprocal (balanced) exchange between corresponding portions of members of a tetrad. Hence, after meiosis, a gamete could possess an array of genes in each chromosome quite different from that found in the original gamete mother cell. Yet the entire genetic complement is *balanced* (one complete set). This continual source of variation by genetic recombination is the raw material on which natural selection can act during the course of evolution. There are, of course, variations and complications—fascinating and biologically significant material for your course in genetics.

■ **What, then, is the *biological significance* of separate sexes?**

Throughout this exercise, keep in mind the role of cells or cellular processes in relation to the total life cycle of the organism. Know the definitions of the following terms and their relationships to one another: *meiosis, fertilization, differentiation, mitosis, embryology, growth,* and *maturation.* Know when and where each occurs during an animal's lifetime.

Terms employed in describing the stages of mitosis and meiosis define convenient (though arbitrary) stages of a continuous process: *interphase, prophase, metaphase, anaphase,* and *telophase.*

■ **Diagram each stage for both meiosis and mitosis.**
■ **Distinguish between the two processes in terms of *essential* details, their overall differences, and the basic biological significance of these differences.**

2. Mitosis (Fig. 16.1)

I. Materials

Numerous organisms are useful to illustrate the mitotic stages. Large numbers of dividing cells can be seen in slides of onion root tip (Fig. 16.2), hyacinth root tip, salamander larva epithelium, *Ascaris* developmental stages, and whitefish embryo (Fig. 16.5). Note that these are all rapidly growing, hence rapidly dividing regions or organisms.

II. Instructions

Examine slides showing various stages of division. Remember that the stages of mitosis are employed only to describe the *continuing* division process and to help us study the successive changes undergone by chromosomes, nucleus, and cytoplasm. Chromatin material will be stained black by a nuclear stain, usually *hematoxylin*; the cytoplasm will probably appear gray (or reddish if counterstained with *eosin*, a cytoplasmic dye). Laboratory diagrams (Fig. 16.1) and models should be studied to help you visualize mitosis as a three-dimensional process.

1. **Interphase** (Figs. 16.1–16.4) During this stage the chromosomes are metabolically most active. They are *uncoiled*, giving the nucleus an irregular or netlike appearance once thought to indicate that the chromosomes literally broke apart during interphase. Recent studies have proven that the individual chromosomes, though extremely long when uncoiled during interphase, retain their individual integrity at all times. This is the period of exposure of the genetic material (its messages being transmitted to *messenger RNA*) and of the proliferation of RNA and proteins in the cytoplasm. It is also the time of *chromosomal replication* (formation of identical *chromatids*[1]) for later separation. Find the *cell membrane, nuclear membrane,* and *centrosome* (a small sphere in animal cells, near the nucleus, with a dark central granule, the *centriole*).

2. **Prophase** (Figs. 16.1, 16.2, and 16.6) This stage marks the onset of nuclear reorganization prior to division. It is characterized by *centrosome division* in animal tissues and the gradual thickening of individual chromosomes due to *coiling up* of the strands. Actually, each chromosome at this stage consists of two identical units, the *chromatids*, which separate later during the subsequent meiotic phases. The two chromatids of each chromosome are attached by one *centromere*. The centrioles divide and move to opposite ends of the cell to form *poles* toward which the chromatids will soon move.

3. **Metaphase** (Figs. 16.1, 16.2, 16.5, 16.7, and 16.8) This stage is marked by the alignment of individual chromosomes (each a chromatid pair) along a plane at the center of the cell (*equatorial plane*). This stage is also marked by dissolution of the nuclear membrane and appearance of astral rays or fibers around each centriole.[2] *Note that in mitosis the homologous chromosomes do not pair, but align themselves separately on the equatorial plane.* Look for *spindle fibers* between each chromosome and the centrioles. The point of spindle fiber attachment on each chromosome is the centromere.

4. **Anaphase** (Figs. 16.1–16.5, 16.9, and 16.10) The *centromeres split* and the chromatids separate; they then begin a rapid migration toward opposite poles, moving along an array of fine fibers (*spindle fibers*) that lie between the two centrioles. Movement of the separate chromatids may be caused by contraction of the spindle fibers, causing each centromere (and in turn the attached chromatid) to be drawn toward the centrioles. Each chromatid assumes a characteristic shape (V, J, or I) as it moves toward the pole since it is at least in part "dragged" through the cytoplasm by the centromere. It may be due also to mutual repulsion of the division products of each centromere. The result, however explained, is a cluster of chromatids assembled into two groups, one around each centriole. Note that the chromatid division products of each chromosome always move to opposite poles, *assuring equal division of the nuclear material.* The chromatids now can be considered separate chromosomes (though they will not replicate until the next interphase). The two chromosome clusters are now future daughter nuclei, each with a set of *precisely similar genetic material.*

5. **Telophase** (Figs. 16.1, 16.4, and 16.5) Each daughter nucleus becomes enclosed in a new nuclear membrane. The chromosomes uncoil and "disappear." New centrosomes appear.

Cytoplasmic separation (cytokinesis), as distinct from *nuclear* division, occurs in plants by ingrowth of a distinct plate or cell wall. In animals, it results from an infolding of the cytoplasm between the nuclei and a gradual pinching apart of the daughter cells.

Find as many of the above stages as possible in both *polar* view (looking down on the metaphase plate so that chromosomes are viewed

[1] Do not confuse *chromatids*, the two identical units of each prophase chromosome, with the pair of homologous *chromosomes* each carrying corresponding genetic units. One homologous chromosome of each pair can be traced to the sperm, the other to the egg of the original zygote from which the organism now undergoing mitotic divisions was derived, i.e., two chromatids per chromosome and two chromosomes (each a tightly bound pair of *chromatids*) per homologous chromosome pair. The actual pairing of the latter occurs *only* during meiosis, as will be seen in Section 3.

[2] Absent in plant cells.

WHITEFISH BLASTULA
cell
(X 1,500)

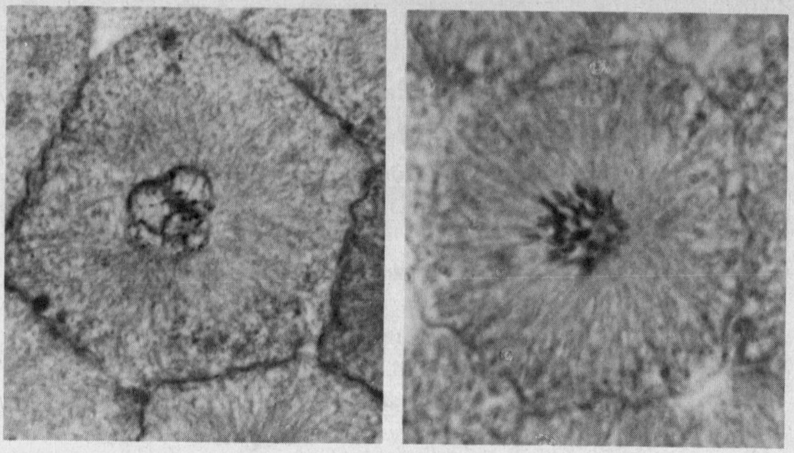

MITOSIS

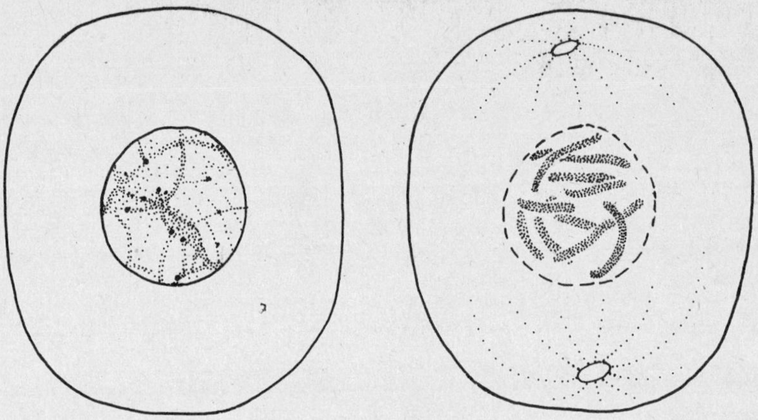

STRUCTURE	INTERPHASE	PROPHASE
NUCLEAR MEMBRANE	Intact.	Disintegrates.
NUCLEOLI	Visible; darkly stained.	Become dispersed in cytoplasm.
CHROMOSOMES and/or CHROMATIDS	Sometimes visible as randomly–coiled chromatin threads.	Condense; chromatids become visible, begin migration to equatorial plate.
CENTROMERES	Not visible.	Visible as dark dots on chromosomes.
CENTRIOLES	Visible proximal to nucleus.	Form spindles, migrate to poles.
SPINDLE FIBERS	Not formed, but astral rays may be present.	Being formed—head for centromeres.

Figure 16.1. Mitosis. Photos are of whitefish blastula cells, ×1500.

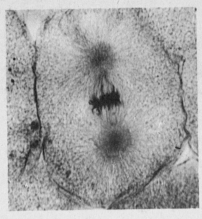 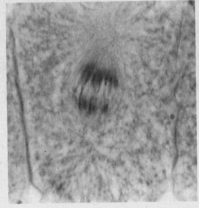 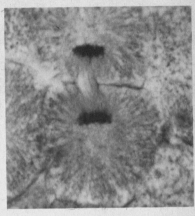

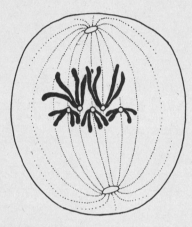 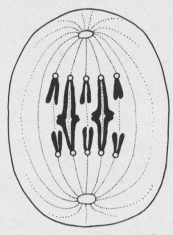 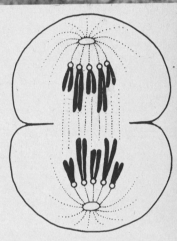

METAPHASE	ANAPHASE	TELOPHASE
Still disintegrated.	Still disintegrated.	Reforms in each daughter cell.
Invisible.	Invisible.	Reform in each daughter cell.
Line up on equatorial plate.	Split at centromeres; migrate toward poles.	Aggregate at poles; become indistinct.
Visible as dark dots on chromosomes.	Split: one centromere per chromatid.	Become invisible.
Remain at poles.	Still at poles; appear to pull chromatids toward them.	They've done their work; begin replication.
Attach to centromeres.	Visible between centrioles and between centrioles and centromeres.	Disappear.

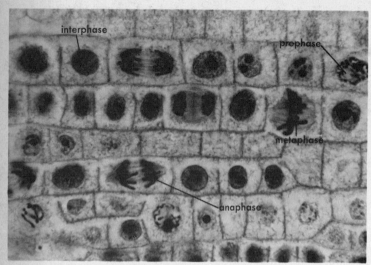

Figure 16.2. Various mitotic phases in the onion root tip. (Courtesy CCM: General Biological, Inc., Chicago.)

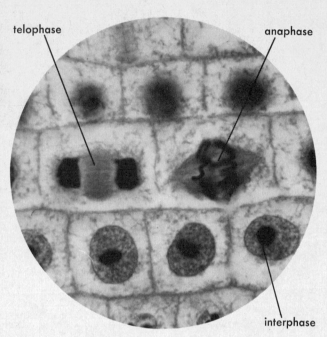

Figure 16.4. Onion root tip, interphase, anaphase, and telophase. (Courtesy CCM: General Biological, Inc., Chicago.)

arranged in a circle or in a central cluster) and in *lateral* view (showing both centrioles).

Draw each stage, including both an early and late prophase, selecting the view that shows each stage to best advantage. Label parts referred to in the above discussion, designate view shown, magnification, and other pertinent details.

■ **Where in the body does mitosis ordinarily occur?**

■ **What is meant by the statement: "*Unpaired* chromosomes appear separately on the equatorial plane, yet each is actually doubled"?**

■ **How is the exact equality of partition ensured?**

Figure 16.3. Onion root tip, early anaphase. (Courtesy CCM: General Biological, Inc., Chicago.)

Figure 16.5. Whitefish mitosis showing various mitotic figures. (Courtesy CCM: General Biological, Inc., Chicago.)

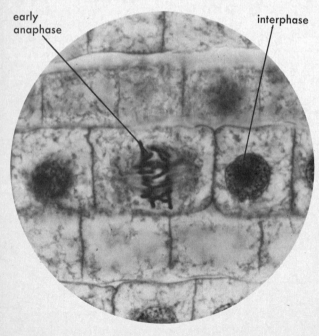

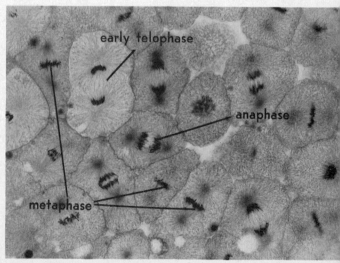

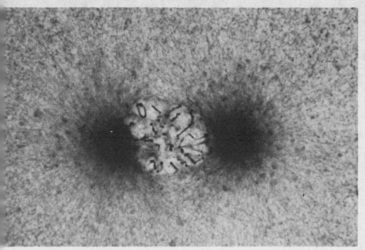

Figure 16.6. Whitefish, early prophase. (Courtesy CCM: General Biological, Inc., Chicago.)

3. Meiosis

I. Introduction

Meiosis is not only a means of producing variability, but also of producing *balanced* variability. Since each gamete ends up with a complete chromosome set (however recombined), it possesses a complete gene set as well. Nothing in nature is more precise—billions of gametes, each with one representative of each of untold thousands of genetic factors common to that species.

Figure 16.7. Whitefish metaphase. (Courtesy CCM: General Biological, Inc., Chicago.)

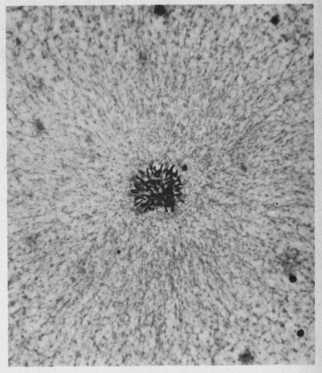

Figure 16.8. Whitefish mitosis, polar view of metaphase. (Courtesy CCM: General Biological, Inc., Chicago.)

Thoroughly familiarize yourself with diagrams of meiosis by which germ cells are produced in ovaries (oögenesis) and testes (spermatogenesis).

Note that in oögenesis (egg formation) the meiotic process results in *one egg* and *three*

Figure 16.9. Whitefish mitosis, anaphase. (Courtesy CCM: General Biological, Inc., Chicago.)

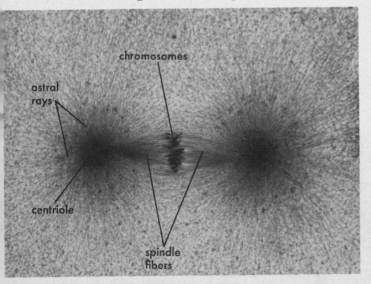

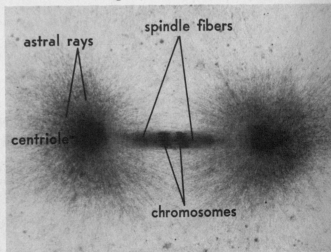

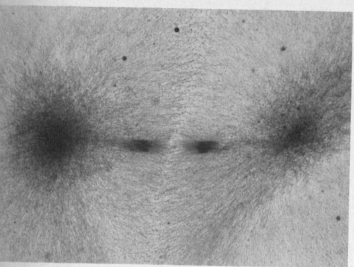

Figure 16.10.   Whitefish mitosis, late anaphase. (Courtesy CCM: General Biological, Inc., Chicago.)

polar bodies (usually only two are found since the first polar body often does not divide); the latter are essentially nuclei with very little cytoplasm (see Fig. 16.11). The single egg carries most of the cytoplasm of the other three cells, an apparent adaptation to the nutritive needs of the egg and early zygote.

Sperms, however, function as metabolically

Figure 16.11.   Sea star polar body formation. (Courtesy CCM: General Biological, Inc., Chicago.)

egg   polar bodies

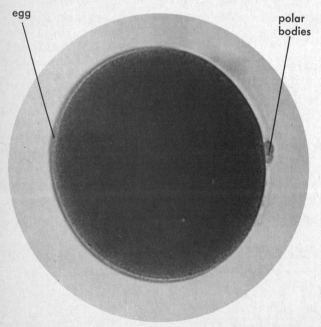

complex, tailed, haploid nuclei with very little cytoplasm. All four cells derived from the two meiotic divisions of a sperm mother cell become mature spermatozoa. Hence, comparing the meiotic process in sperms and eggs, note that four times as many sperm result from the same number of mother cells. Yet the basic process of meiosis is identical in each case and should be studied from the standpoint of chromosomal changes resulting in the formation of haploid gametes.

■ **What is the biological function of meiosis?**

■ **How is meiosis related to reproduction in general, and to sexual reproduction in particular?**

■ **How is meiosis related to fertilization? to mitosis?**

■ **What are the differences between oögenesis and spermatogenesis?**

II. Procedure

1. **Spermatogenesis**   The material showing spermatogenesis probably will be taken from the testis of a salamander, frog, or grasshopper.

Each lobe of the frog testis usually contains material at an approximately equal stage of development. Look for mature gametes near the opening of the sperm duct and for developing ones at the opposite end. In the grasshopper, the testes are slender *lobes,* each with its own duct. Lobes are subdivided into *lobules,* which have progressively more mature gametes as one goes from the tip of the lobule to the sperm duct (where mature sperm are found).

2. **Oögenesis**   For oögenesis, developing eggs of *Ascaris* are particularly favorable demonstration material. This organism has very few chromosomes (only two pairs in the species generally used in these studies, *Ascaris megalocephala,* a parasite of horses).

For orientation, examine under low power a prepared slide of testicular material showing spermatogenesis, and of ovarian material illustrating oögenesis. In each, attempt to find the characteristic stages of meiotic divisions summarized below.

Subsequently, examine the stages under high-dry or oil immersion for maximum detail.

Draw each stage, using selected examples from your slide. Be prepared to explain the *genetic significance of each stage* you have drawn.

3. **First meiotic division** (*reductional* division)

a. *Prophase*. This is the longest portion of the full division sequence. It is divided into a number of substages. Basically, this is the time when chromosomes coil and shorten and when the two chromosomes of each *homologous pair* come together. This is termed *synapsis* or *pairing*. The homologous chromosome pair is called a *bivalent*. As in mitosis, *each* chromosome has one *centromere* and two *chromatids*. The synaptic pair, or bivalent, thus forms a group of four chromatids (*tetrad*), usually arranged in the form of a doubled cross, with the two centromeres in the center, the extended "arms" being corresponding parts of matched-up chromatids, one from each homologous chromosome (see illustration in your text, laboratory models, or charts). The important process of *crossing over* occurs at this stage of meiosis. As in mitosis, the *centriole* divides, moves to opposite ends of the cell, and marks the poles of the dividing nucleus.

b. *Metaphase*. The nuclear membrane has generally disappeared by this time and pronounced *asters* form around the centrioles. The paired chromosomes are originally still arranged as tetrads at the metaphase plate, but during this stage the process of chromosome separation begins, possibly caused by repulsion of the paired centromeres. The centromeres do *not* divide, but simply *separate*, each conveying its own chromosome (from each homologous pair) to opposite poles (centriole plus astral rays).

c. *Anaphase*. Separation of homologous chromosomes continues, moving along the *spindle fibers* between the two centriole poles. Each chromosome, still consisting of its two *chromatids*, is led by its intact centromere and moves toward one of the two centrioles. One chromosome from each homologous pair goes to an opposite pole.

d. *Telophase*. End of division 1 of meiosis; one half the original chromosome number is in each daughter cell, though one representative of each homologous pair is present, each chromosome consisting of its two chromatids, tied together by a single centromere. This completes the reductional division, phase 1 of meiosis.

4. **Second meiotic division** (*equatorial* division) This division occurs after a very brief interphase. It is a direct continuation of the first division and part of the complete meiotic process. Each daughter cell from division 1 undergoes a rapid division with essentially no prophase required.

a. *Metaphase*. The chromosomes line up, each consisting, as noted above, of two chromatids, connected by a single centromere. The centriole splits and the products move apart, each surrounded by astral rays to form the two poles.

b. *Anaphase*. Centromeres now *split* and the chromatids pull apart rapidly along the spindle fibers, forming two groups, one at each pole (centrole). A new nuclear membrane forms around each cluster.

c. *Telophase*. This is the final division sequence, cytoplasmic separation. Each haploid nucleus now contains *one chromatid from each tetrad* (original pair of homologous chromosomes).

These stages are followed by completion of the gametogenic process, the production of mature spermatozoa or eggs. In the male, four haploid *spermatids* are the products of meiosis of a diploid mother *spermatogonium*. Each spermatid then elongates, develops a tail, and becomes the rapidly motile, fully developed *spermatozoan* (sperm).

In the female, the corresponding stages produce four haploid *oötids*, meiotic products of one diploid *oögonium*. However, only one viable egg results. The *three polar bodies*, seen on the egg surface, are the nuclei of three oötids, whose cytoplasm was retained by the fourth oötid through unequal cytoplasmic division. The fourth oötid becomes the mature egg.

■ **What is the adaptive importance of this divisional pattern in the production of male and female gametes?**

Fertilization is the consummation of the sexual process, in which two mature haploid nuclei fuse and form a single diploid nucleus. The product becomes a diploid *zygote*, and the process of development of a new, diploid individual begins with the first *cleavage division*.

■ **What type of cell division follows zygote formation?**

In some organisms, fertilization may occur before completion of oögenesis. In these cases sperm entry stimulates rapid completion of the meiotic process, and nuclear fusion follows. In any case, the formation of the fusion (diploid) nucleus stimulates an entire sequence of complex but wonderfully controlled physiological events that result in *development, growth, maturation,* and, ultimately, a *new adult*.

■ **What would be the result in the next generation if meiosis failed and *diploid* gametes were formed?**
■ **Does this actually occur in nature?**
■ **How is meiosis related to genetic variability? to evolution?**

Suggested References

Balinsky, B. I., 1970. *An Introduction to Embryology*, 3rd ed. Philadelphia, Pa.: W. B. Saunders.
Berrill, N.J., 1953. *Sex and the Nature of Things*. New York: Dodd Mead and Co.
Mazia, D., 1953. "Cell Division." *Sci. American*, 189: Aug.
——, 1961. "How cells divide." *Sci. American*, 205: Sept.
——, 1964. "Cell division," BSCS *Pamphlet* 14. Boston, Mass.: D. C. Heath.
Swanson, D. P., 1964. *The Cell*, 2nd ed. Englewood Cliffs, N.J.: Prentice-Hall.
Taylor, J. H., 1958. "The duplication of chromosomes." *Sci. American*, vol. 198.
Wilson, G. B., 1966. *Cell Division*. New York: Reinhold.

CHAPTER 17   EMBRYOLOGY

1. Introduction

I. General Considerations

Differentiation, development, and growth, summarized in the term *embryology*, embody a wondrous sequence of change. Our objectives during this laboratory period are not only to see some of the earliest stages of development, but also to become familiar with the evolutionary significance of the patterns disclosed. Both living and stained material will be used to observe cleavage stages and cell and tissue differentiation. Advanced embryological stages, best seen in stained whole mounts or in tissue serial sections, will be incorporated later in a complete embryology course. If time and facilities permit, however, you might enjoy observing developmental sequences in frog larvae or chick eggs, both of which are easily maintained in the laboratory throughout their full embryological period.

Fertilization, cleavage, blastula, and gastrula formation can be observed in a number of marine invertebrates. Echinoderm embryos are especially good owing to a lack of yolk, shells, and membranes, which renders them convenient to handle and easy to study. Their early pattern of

development, especially cleavage and gastrulation, is both dramatic and diagrammatically simple. For inland laboratories, frog eggs or those of any other amphibian collected during the breeding season are suitable. These vertebrates demonstrate embryological processes through metamorphosis and development of the adult organism. Chick embryos are also excellent under the specialized conditions of a yolk-laden egg enclosed in special membranes within a calcified shell. Fixed preparations of mammals may also be used. These are often the 6-, 12-, and 20-mm stages of pig embryos or even various stages of calf development. A variety of material helps to illustrate how comparable are the broadly similar patterns of development in widely differing vertebrates.

II. Early Developmental Stages

Descriptions of the stages listed below are based on *Pisaster ochraceus*, the common Pacific sea star.

1. **Unfertilized egg** (Fig. 17.1) The female gamete is recognizable by its large nucleus (at this stage termed a *pronucleus* or *germinal vesicle*) with a prominent *nucleolus*, and by the surrounding *cell membrane*. The *an-*

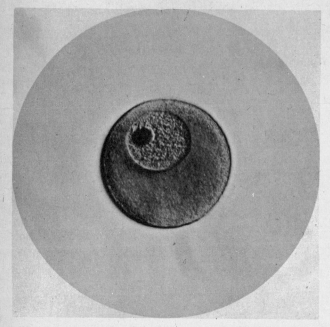

Figure 17.1. An unfertilized sea star ovum showing the nucleus and nucleolus. (Courtesy CCM: General Biological, Inc., Chicago.)

Figure 17.2. Two-cell stage of the sea star. The egg has divided by mitosis into two cells, with the cells remaining in contact with each other. (Courtesy CCM: General Biological, Inc., Chicago.)

imal hemisphere is the portion with least yolk; the *vegetal hemisphere* (not vegetable, please) has the greatest amount of yolk.

2. **Fertilized egg (zygote)** Immediately after sperm entry, a remarkable series of physical and physiological changes begin. Sudden alteration of the egg surface prevents entry of other sperm; then follows rapid formation of a *fertilization membrane* extruded under an outer jelly layer. The sperm head penetrates the egg cytoplasm, and its nuclear material forms a *male pronucleus* that fuses with the *female pronucleus* to form the fertilization nucleus.

3. **Cleavage** The early mitoses following zygote formation, *before any increase in size occurs*, constitute the *cleavage stage*. Furrows or cleavage planes can be seen in the cytoplasm. The resulting cells are *blastomeres*. During this initial phase of differentiation, cleavage divides the zygote into smaller units, establishes the basic symmetry, and blocks out future zones of specialized growth. Certain materials responsible for later differentiation appear to be passed to specific blastomeres during cleavage.

a. *2-cell stage* (Fig. 17.2). The first cleavage is the slowest; it divides the zygote into equal

meridional halves with the furrow passing through the animal and vegetal poles. This division establishes the primary *bilateral symmetry*.

b. *4-cell stage* (Fig. 17.3). The second cleavage plane, also meridional but at right angles to the first, produces four equal blastomeres, like quarters of an orange.

c. *8-cell stage* (Fig. 17.3). The third cleavage is equatorial, passing slightly above the equator and perpendicular to the other two cleavages. At this stage the yolk apparently slows division in the vegetal hemisphere and tends to permit more rapid cleavage in the animal hemisphere. The smaller animal cells, *micromeres*, possess little yolk. Vegetal cells, laden with yolk, are called *macromeres*.

d. *16-cell stage* (Fig. 17.4). No definite cell arrangement can be seen, only a solid ball of cells, the *morula*.

e. *32-cell and subsequent stages* (Fig. 17.5). Subsequent mitoses result in a single-layered, hollow sphere of ciliated cells. This stage initiates the larval phase, when the organism leaves the fertilization membrane and can swim about freely.

4. **Blastula** (Fig. 17.6) This single-layered sphere of about 1000 ciliated cells is arranged

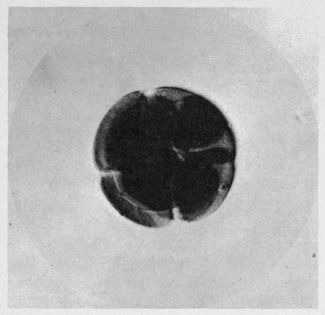

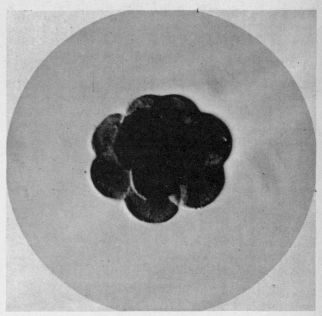

Figure 17.3. The 8-cell stage of the sea star results in two tiers of cells, one on top of the other. However, the division is not quite equal, with the result that the upper cells are smaller than those in the lower tier. (Courtesy CCM: General Biological, Inc., Chicago.)

Figure 17.4. The 16-cell stage of the sea star results in a solid ball of cells. (Courtesy CCM: General Biological, Inc., Chicago.)

around a central *segmentation cavity* or *blastocoel*. Cleavage is now complete and *blastulation* has occurred. The blastula is still approximately the size of the original egg, hence the blastomeres necessarily have become progressively smaller as cleavage continued.

■ **How does the genetic constitution of the blastomeres compare with that of the zygote? of the original egg?**

5. **Gastrula** (Figs. 17.7–17.9) This embryonic form results from differential growth and movement of cells in *gastrulation*. The several-layered embryo forms after the vegetal hemisphere of the blastula flattens, and the flattened cells gradually move into the blastocoel (*invagination*). Rapid cell division causes rolling in of cells (*epiboly*) at the margin of the *dorsal lip*. Together, these changes produce the *gastrula*. Though it varies widely in different groups, gastrulation is the first step in the formation of the *primary germ layers* (*ectoderm, mesoderm, endoderm*). In most animals, gastrulation is

rather strikingly modified from the simple echinoderm pattern, chiefly because of the large amount of yolk. In a hen's egg, for example, the gastrula is a tiny island of tissue

Figure 17.5. Sea star zygote, later cleavage stage. (Courtesy CCM: General Biological, Inc., Chicago.)

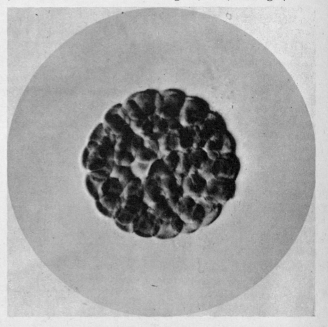

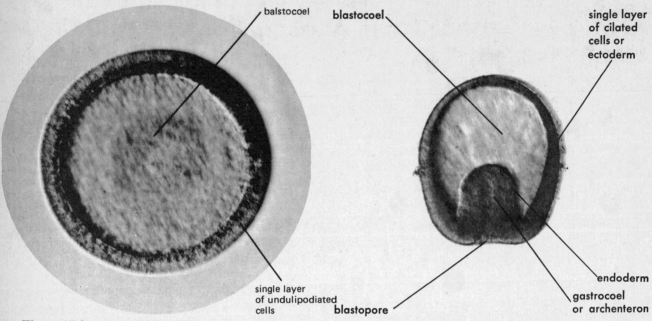

balstocoel

blastocoel

single layer
of cilated
cells or
ectoderm

single layer
of undulipodiated
cells

endoderm

gastrocoel
or archenteron

blastopore

Figure 17.6. Sea star blastula. (Courtesy CCM: General Biological, Inc., Chicago.)

Figure 17.8. Sea star gastrula showing blastopore and beginning of gastrocoel cavity. (Courtesy CCM: General Biological, Inc., Chicago.)

floating on a huge dense yolk. But the results are similar, whether cells sink in, roll in at the blastocoel margin, or simply move independently into the cavity and re-form as an inner layer.

Figure 17.7. Early gastrula stage of the sea star. (Courtesy CCM: General Biological, Inc., Chicago.)

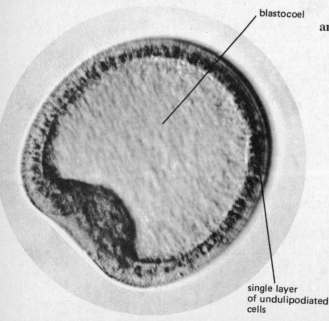

blastocoel

single layer
of undulipodiated
cells

As the inner layer forms, it gradually eliminates the old blastocoel and forms a new cavity, the *gastrocoel* (or *archenteron,* meaning "primitive gut"). The external opening of the blastocoel is the *blastopore*—the posterior end of the larva, and the future *anus* in echinoderms.

■ **What does the blastopore become in annelids? mollusks?**

The new embryo now changes rapidly. It elongates along the blastopore axis and begins to develop the important middle layer of the body—the *mesoderm.* The external cell layer, or *ectoderm,* forms the body covering and most of the nervous system. The *endoderm,* or inner layer, formed from cells that have invaginated into the blastocoel, eventually becomes both the gut lining and parts of many organs. In echinoderms, hemichordates, and primitive chordates such as amphioxus, three pairs of *enterocoels* (literally, "gut pockets") bulge from the roof of the gut. These form *coelomic pouches* that gradually fill the space between ectoderm and endoderm. These pouches then separate from the gut and grow to form mesoderm

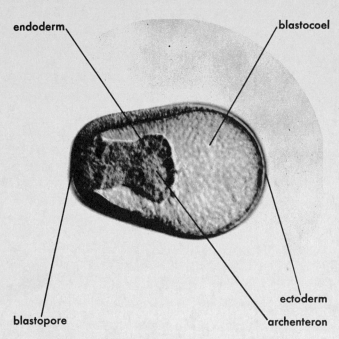

endoderm   blastocoel

ectoderm

blastopore   archenteron

Figure 17.9.   Late gastrula stage of the sea star. (Courtesy CCM: General Biological, Inc., Chicago.)

tissue sheets that in turn form the *coelom*. Where the sheets join above and below the gut along the dorsal and ventral axes of the embryo, they form a double-walled *mesentery*. The bulk of body substance in eucoelomates develops from mesoderm and is formed in close association with the mesodermal linings and coelom.

III. Evolutionary Considerations

The coelom has been discussed as a format for complexity and a highly efficient mode of animal organization. Coelomic construction, in fact, helps define "higher animals." All bilaterally symmetrical animals (Grade BILATERIA) are grouped partially on the basis of their coelom, on whether it is absent, partial, or complete (that is, the levels ACOELOMATA, PSEUDOCOELOMATA, and EUCOELOMATA; review Chapter 6, Section 1, for discussion of these categories).

The majority of animals are eucoelomates. Embryological study of the coelom, however, shows that it develops by two distinct methods. This difference, as well as the *manner of cleavage of the zygote* and possession of a particular *type* of *larva*, divides the level EUCOELOMATA into two large groups of phyla, the superphylum **ENTEROCOELA** (sometimes termed the "echinoderm–chordate

line"), and the superphylum **SCHIZOCOELA** (the "annelid–arthropod line").

The enterocoel type of mesoderm and coelom formation is considered to be of sufficient evolutionary and structural significance to indicate a fundamental relationship among animals developing in this manner. The superphylum **ENTEROCOELA**, therefore, was erected to include the phyla **POGONOPHORA, CHAETOGNATHA, ECHINODERMATA, HEMICHORDATA,** and **CHORDATA** (in spite of great modifications in higher chordates). *Radial* (as opposed to *spiral*) cleavage of the embryo and common possession of a *dipleurula* swimming larva also link some of these groups. The *bipinnaria* larvae of certain echinoderms (asteroids) and *tornaria* larva of acorn worms (hemichordates) are similar enough to have been mistaken for one another—yet they develop into adults so different that they are placed in distinct phyla. Note the variety of larvae named in Table 17.1.

It would be a worthwhile exercise to review the literature and study the morphology of these larvae, and then add to your chart enough information to compare patterns of undulipodia, special sense organs, length of larval life, and food for each kind of larva.

Another chart would be still more instructive, a comparison of the patterns of metamorphosis you were able to find. Highly complex embryological sequences have evolved in the course of evolution. One must remember, however, that all of these variations are derived from either the radial or spiral pattern of early cleavage—the fundamental clue to relationship.

■ **What does this indicate about the relative usefulness of characteristics expressed *early* in development as indicators of relationship among major groups?**

A distinctive pattern of mesoderm and coelom formation characterizes the other group of higher organisms, the annelid–mollusk–arthropod complex forming the superphylum **SCHIZOCOELA.** In these animals, mesoderm forms as *paired bands* growing from specific cells in the blastopore area. Zygote cleavage is *spiral*, not radial; and a *trochophore* larva is formed. The mesodermal bands later split to form the coelom. This process is quite different from the enterocoel pattern. The substances that later form mesoderm cells become differentiated far sooner than in enterocoels. In certain annelids and mollusks, future mesoderm

Table 17.1

Swimming Larvae of Various Phyla

NAME OF LARVA	TAXONOMIC GROUP
	CNIDARIA
Planula	HYDROZOA
Actinula	TRACHYMEDUSAE
Cornaria	CHONDROPHORA (*Velella*)
Ephyra or ephyrula	SEMEASTOMEAE
Cydippid	**CTENOPHORA**
Scypha	**PORIFERA**
Infusoriform	**MESOZOA**
	PLATYHELMINTHES
Müller's	POLYCLADIDA
Pilidium	**RHYNCHOCOELA**
Modified trochophore	**ENTOPROCTA**
Trochophore	**MOLLUSCA**
Veliger	GASTROPODA
Trochophore	**SIPUNCULOIDEA**
Trochophore	**ECHIUROIDEA**
Trochophore	**BRACHIOPODA**
Trochophore	**ANNELIDA**
	ARTHROPODA
Nauplius	CRUSTACEA
Metanauplius	CRUSTACEA
Protozoaea	MALACOSTRACA
Zoaea	MALACOSTRACA
Mysis	MALACOSTRACA
Cyphonautus	**BRYOZOA**
Actinotroph	**PHORONIDA**
Dipleurula (hypothetical ancestral type)	**ECHINODERMATA**
Ophiopluteus	OPHIUROIDEA
Echinopluteus	ECHINOIDEA
Bipinnaria	ASTEROIDEA
Brachiolaria	ASTEROIDEA
Auricularia	HOLOTHUROIDEA
Tornaria	**HEMICHORDATA**

cytoplasm can even be located within the *egg*. Coelom and mesoderm formation, combined with the distinctly different type cleavage and larval form, mark the two groups as genetically far separated. This, in turn, implies the origin of the embryological patterns *after* separation of the two lines from a still earlier common ancestor.

Though it appears much the same as a final product, the coelom of enterocoels and of schizocoels is formed by such disparate embryological steps that the similarity is thought to be an evolutionary *convergence* (similarity of two structures or animals derived from different evolutionary antecedents).

■ **Can you think of any other example of convergence?**

Embryological patterns are useful in defining relationships so ancient that traces of them are no longer apparent in the adult organism. The *adult* animal generally demonstrates more recent or specialized structures acquired later in evolution.

■ *Explain* or *rephrase* each of the following embryological generalities so that their meanings are clear to you:

i. Between *distantly* related animals (having separated from a common ancestor far back in time), developmental patterns show similarities chiefly in the *earliest* stages. Between *closely* related animals the pattern is similar throughout, diverging only late in development. (*Compare*: embryological sequence or *ontogeny* of man and chicken with that of man and ape.)

ii. An eye-catching statement, "Ontogeny recapitulates phylogeny" (development of an individual repeats the evolution of its species), was once considered a biological maxim. A more accurate, modern version would be, "Ontogeny repeats *embryological stages* in phylogeny."

2. Laboratory Instructions

I. Echinoderms

1. *Patiria*, the bat sea star, is sexually ripe at any time of the year. Laid out on moistened paper toweling, it will readily extrude eggs (orange) or sperm (whitish) Dissect out the gonads to assure an abundance of viable eggs and sperms. Wash eggs several times in fresh sea water in *clean* finger bowls. To a fairly large suspension of such eggs, add a drop or two of *diluted* testicular material (which still contains many millions of sperm). *Adding too heavy a sperm suspension will inhibit proper development by causing polyspermy (fertilization by more than one sperm per egg) and arrested or abnormal cleavage.*

At 15-min intervals, pipette out a few eggs (with a *clean* pipette—the eggs are sensitive to minute amounts of impurities) and examine microscopically. Keep accurate notes of observations, including a time schedule of important embryological changes.

Find and *draw* good examples of the following stages from your own or other students' living material or from stained slides:

a. *Unfertilized egg*
b. *Fertilized egg* (*zygote*). Show jelly layer, fertilization membrane, nuclear changes.

c. *2-, 4-, 8-, 16-cell stages*. At normal room temperature these should appear in 3 or 4 hours.
d. *Blastula. Draw* an optical section (cross section visible when you focus on the center of the organism). Show *blastocoel, micromeres,* and *macromeres.*
e. *Gastrula.* Find early stages of invagination. *Draw* an optical section.
f. *Advanced gastrula. Draw* an optical section showing outpockets of coelomic pouches.
g. *Bipinnaria and brachiolaria larvae* (Fig. 17.10). Examine these interesting larvae; they are thought to be derived from the hypothetical *dipleurula* ancestral larva.

2. Sea urchins from the U.S. Pacific coast (such as *Strongylocentrotus franciscanus*, *S. purpuratus*, and *Lytechinus*, and from the Atlantic coast (such as *Arbacia punctulata* and *S. dröbrachiensis*) will be used in this exercise. Eggs (and sperm) may be obtained either by injecting 0.5 molar KCL into the mouth, or by passing 10 volts of alternating current through the urchins in sea water (electrodes may be placed at any two points on the urchin test). The effect of the electrically induced discharge of gametes will be the appearance of reddish or orange eggs or

Figure 17.10. The swimming bipinnaria larva of the sea star. (Courtesy CCM: General Biological, Inc., Chicago.)

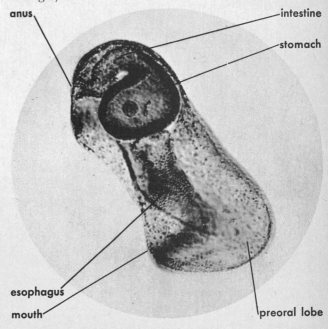

whitish sperm. As with the sea star experiment, add a *small* amount of sperm to the dish containing eggs (*reference*: Harvey, 1956, p. 49; Costello *et al.*, 1957, p. 184). Pipette a small sample onto a slide, cover, and observe microscopically. Cleavage should begin in approximately 1 hour. Once it does begin, pipette and examine small samples every 15 min, recording your observations.

Find and *draw* good examples of the following stages from your own or other students' living material or from stained slides:

a. *Unfertilized egg*
b. *Zygote.* Show jelly layer, fertilization membrane, and nuclear changes. Can you find evidence of stages of mitosis?
c. *2-, 4-, 8-, 16-cell stages.* At normal room temperature the fertilized ovum divides in approximately 1 hour with subsequent divisions occurring every 40 min thereafter, eventually forming the blastula.
d. *Blastula.* This should occur in about 7–8 hours from fertilization. *Draw* an optical section.
e. *Gastrula.* Try to find early stages of invagination. The gastrula should appear in about 12–15 hours from fertilization. *Draw* an optical section.
f. *Echinopluteus (pluteus) larvae.* The larvae appear in about 1 day from fertilization and will last about 3–4 days. Can they move? Do they have special locomotor structures? How do they resemble and differ from the sea star larvae?
g. *Special project.* The sea urchin eggs used in the preceding section have some interesting properties that you wish to observe. Since sea urchins exude massive numbers of eggs and sperm into the water with no assurance of fertilization, one might wonder whether the eggs of such sexually reproducing organisms could give rise to a new individual without fertilization. In some organisms unfertilized eggs regularly develop into new individuals. In the honey bee, for example, fertilized eggs become females, and the unfertilized eggs males. The males are therefore *haploid adults*, clearly a highly specialized adaptation.

■ Can you recall any other example of parthogenesis?

■ What chromosomal adaptations are required to make this possible?

■ Why is it highly unlikely that *sperm* could successfully develop into adults?

Obtain some sea urchin eggs (making sure that they are not contaminated with sperm). Traumatize a portion of them (by shaking, adding salt for a brief period, electric stimulus, sudden heat change, etc.), leaving the remainder as undisturbed *controls*.

■ Do any of the traumatized eggs develop? What happened to the untraumatized (control) eggs? If any eggs do develop, is their development normal in body form and timing of development?

■ In nature, what do you think could induce parthenogenesis in sea urchin eggs?

■ Does parthenogenesis give rise to *both* sexes or to only one? If only one, which? How might you account for this?

■ If parthenogenesis were to occur spontaneously in humans, a most unlikely hypotheses in any higher animal (although it can be done experimentally in frogs and even in rabbits), what would be the sex of the offspring? Could such an individual produce a child by normal sexual reproduction?

■ What would be required for this to be possible?

II. Chordates

1. Frog or salamander.

a. *Procedure.* Spring or early summer are the best times to collect breeding frogs. Ovulation can be artificially induced at other times of the year by injecting into the female frog's body cavity *hypophysis* material, containing the hormones stimulating sexual development. (The hypophysis is a small endocrine gland that you may recall struggling to dissect from the ventral surface of the brain.) After ovulation in the frog, which generally occurs 24–48 hours after hormone injection, eggs may be squeezed from the ripe female and covered with a diluted suspension of sperm from the ground-up testis of a male frog.

Fertilized eggs should be kept in a shallow dish, well lighted, and in *fresh, regularly changed, pond water.* Feed the developing tadpoles very small bits of bread crumbs or yeast. Normal feeding habits can be observed by placing tadpoles in a freshwater aquarium where they can feed on algae. Record *temperature* and *time* to reach successive stages of development. Vary the temperature with different groups of tadpoles and note results. Similarly, note the effect on development from varying the *diet*. If time permits, follow the entire metamorphosis and turn in a careful report on this special project.

b. *Stages to study and draw* (from stained or living material)

    i. *Zygote*

    ii. *2-, 4-, 8-, and 12- or 16-cell stage.* Show cleavage furrows, location of resulting cells, and disposition of pigment. Be sure to notice the relative size of cells in advanced stages of cleavage, and the relationship between *cell size* and *cleavage rate* (cell number) at the animal and vegetal hemispheres.

    iii. *Early blastula* (32 cells). Note presence of a blastocoel and gradual incorporation of yolk into vegetal cells (Fig. 17.11). If stained sections are available, study these for a better idea of relationships between blastocoel, embryonic hemispheres, and individual cells.

Fixed, unmounted specimens can be cut in half under a dissecting microscope or hand lens and examined.

    iv. *Gastrula.* Locate a crescent-shaped slit between the pigmented animal and nonpigmented vegetal cells. Pigmented animal cells cover more than half of the sphere, due to their more rapid division rate (yolk appears to slow down division of cells near the vegetal pole). Continued rapid division and downward migration cause these pigmented cells to cover most of the larva. A *germ ring*, originally an equatorial belt of cells, marks the advancing margin of animal hemisphere cells. The vegetal hemisphere cells move into the blastocoel, and the *overgrowth* of animal hemisphere cells continues. Eventually, the germ ring *encircles the remaining external yolk cells*. This ring marks the *blastopore* (Fig. 17.11). Continued growth of pigmented cells reduces the blastopore to a *yolk plug*, then to a minute opening. As gastrulation is completed, the embryo rotates 90° in its jelly envelopes, so that the dorsal surface is uppermost.

Draw a vegetal view of a frog gastrula, showing the blastopore; then draw a later stage in which the pore has become a yolk plug.

Figure 17.11.   Diagrammatic cross section of an amphibian blastula (*left*), and gastrula (*right*).

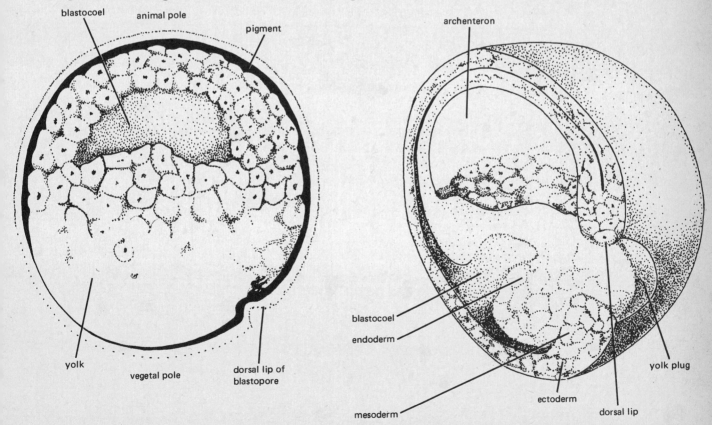

The *dorsal margin* or *dorsal lip* of the germ ring (the crescentic slit mentioned earlier) is of particular interest. This, the *dorsal lip of the blastopore*, is a key center of organization of the embryo. It is here that the *primary organizer* is produced. The latter is a diffusible chemical responsible for the onset of a chain or series of chemically induced differentiations, each successive change inducing the next. The first structure to be differentiated from invagination of cells near the dorsal lip is the *archenteron roof* (*chordamesoderm*). The archenteron (or primitive gut) in turn induces the differentiation of a *notochord*, which marks the middorsal region of the embryo, with the blastopore being posterior (the primitive anus). The notochord in turn becomes an area of intense activity and rapid differentation. Gastrulation in the frog is a distinct modification from the enterocoelic pouches that are formed in echinoderms and amphioxus, being considerably modified by the presence of yolk cells. It is fundamentally (evolutionarily) similar, however, as mesoderm is formed from particular archenteron roof cells, which then separate to form a double tissue sheet (as in echinoderms), which becomes the lining of the coelom.

Each step in differentiation acts as a stimulus, inducing another stage in this extraordinary domino chain of sequential changes. The total process is especially remarkable when we realize that throughout this rapid, highly integrated development sequence, the embryo must respire and metabolize as an independent functioning organism enclosed within its protective membranes. The echinoderm bipinnaria and the annelid trochophore are actually free-swimming organisms at this stage.

■ Can you guess what the *chick* embryo would look like at the corresponding stage? the human embryo? (Check in your text or one of the embryology references at the end of this chapter.)

v. *Later stages* of tadpole development.
a. Stained section of a *late gastrula*. Locate the blastopore, the reduced blastocoel, gastrocoel (archenteron), remaining yolk cells, and primary germ layers. The *mesoderm* is starting to form notochord from the roof of the archenteron at this stage.
b. Neural fold stage (Fig. 17.12). Find the mesoderm, notochord, and *neural plate* (somewhat later this becomes the neural *fold*, which may be the stage of development of your specimen).

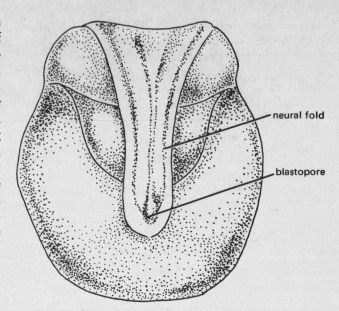

Figure 17.12. Diagrammatic amphibian neural fold stage.

■ Which are the *anterior, posterior, dorsal, ventral* axes?
■ Where is the anus? mouth? brain?

c. Neural tube stage. Note the dorsal fusion of the neural folds.

■ What do they form?
■ Find the *mesodermal somites.*
■ What will they produce?
■ What major changes are yet to occur in the developing tadpole?

Further observations can be made through successive stages of tadpole growth and metamorphosis observed in an aquarium, such as leg and mouth formation in the young adult, resorption of the tail, disappearance of the operculum. Tadpoles should be maintained and watched carefully in their aquaria over a long period of time if the entire growth cycle is to be observed. Examination of selected individuals at specific time intervals will demonstrate these changes and provide material for preparation of specimens for special study.

2. **Gallus gallus** The domestic chicken produces eggs with a large amount of yolk. As a result, cleavage is incomplete and restricted to an area on the yolk called the *blastodisk* (Fig. 17.13). The early mitotic divisions during cleavage form a flattened

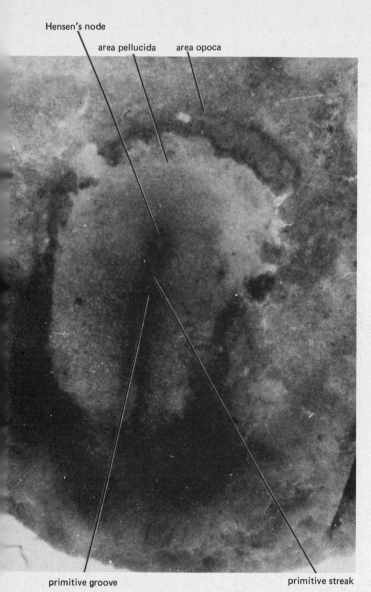

Hensen's node

area pellucida   area opoca

primitive groove   primitive streak

Figure 17.13. Dorsal view of 13-hour chick (*Gallus gallus*) blastodisc.

blastula consisting of several cell layers, with a shallow space (blastocoel) between the upper layer (*epiblast*) and lower layer (*hypoblast*). The hypoblast does not extend to the outer limits of the epiblast. The peripheral area (where the epiblast is in direct contact with the yolk) is called the *area opaca*, where breakdown of yolk occurs. This region does

not become part of the embryo. The central blastodisk (*area pellucida*) actually forms the embryo.

As the yolk is metabolized by the embryo, the relative weights of these two entities change. The yolk gradually lightens and the embryo becomes heavier. You can discover the rate of yolk metabolism by weighing the yolk and the embryo separately at regular intervals, and determining the *ratio* of these weights. Weigh your samples of embryo and yolk separately in a dish or bowl, subtracting the weight of the container. (Don't fail to use a container . . . a yolky weighing tray we don't need!) The embryos studied in Sections b–f should then be weighed for the above embryo–yolk ratio change determinations.

The movement of cells in gastrulation is complex, again owing to the interfering effect of the yolk. It cannot be a simple movement of sheets of cells but rather a directed series of inwanderings or migrations of many individual cells.

a. In the *pregastrulation stage* of the blastodisk, cells can be seen to migrate toward the center of the posterior blastodisk to form a *primitive streak*. In the center of the primitive streak a depression develops, the *primitive groove*, which widens anteriorly to form *Henson's node* (Fig. 17.13). Look at slides of a pre-24-hour chick embryo. *Draw* the embryo, labeling the parts just described.

b. *Gastrulation.* Cells "invaginate" during this stage, actually wandering or migrating into the blastocoel along the primitive streak, where they come in contact with the hypoblast. Beneath the epiblast, cells move just anterior to Henson's node and form there the *notochordal process.* Other mesoderm cells that invade the blastocoel then become the *somites, lateral plates,* and other embryonic structures. Cells from the epiblast, and some from the hypoblast as well, begin to form the *endoderm.* As the cells of the primitive streak (homologous to the blastopore) are used up, the streak starts to recede posteriorly, leaving in its wake a *neural groove.* Examine slides of a 24-hour chick embryo (whole mounts and serial cross sections). *Draw* what you see.

c. Examine a *fertilized egg.* Remove the shell carefully so as not to break any other membranes. Note the *air space* at the blunt end of the egg. Look at a piece of the shell under the microscope. (*Is it smooth or does it seem to have pores?*) The *albumen* (egg white) is of two types (one is more viscous than the other) and includes the *chalaza* (twisted masses at either end of the yolk), made up of the more viscous albumen).

The shell and other extraembryonic membranes are added to the massed yolk with its minute zygote passenger as they move down the oviduct of the hen. As the layers are being added, development proceeds so that an egg is usually released at the *gastrula stage* (24 hours). Locate the blastodisk island on the yellow yolk. *Excise* this small structure and examine it under a microscope. The most obvious structures should be the *somites*. (*How many are there?*) Between the somites locate the *notochord*, which extends anteriorly under the *neural groove*. Laterally are the *neural folds*. At the posterior extent of the notochord, locate the *primitive streak* (Fig. 17.14).

■ What is the function of the shell? of the other extraembryonic membranes? What materials can pass through the shell?

■ What is the function of the *yolk*? What might account for its color? How is it taken in by the embryo?

d.   Examine appropriate slides of a *33-hour chick embryo. How many somites are there?* Note the variation in number from counts of other students. The number of somites is actually a better indication of age than is the number of hours, since the hen often retains the egg until a particular time of day. Note that the neural fold is restricted to the area near the primitive streak. Follow the neural tube anteriorly. At its anterior limit, the neural tube is continuous with three lobes: the *hindbrain*, the *midbrain*, and the large *forebrain*. Part of the neural tube may be obscured by the *vitelline vein*, which is connected to the somewhat lateral *heart*.

e.   Observe a *living 48-hour chick* (Fig. 17.15). Describe any movement you see. The moving-pumping organ is the heart! *Extraembryonic mesoderm* and *endoderm* form a *yolk sac* that eventually will completely surround the yolk. They also will form conspicuous *blood vessels* to carry nutrients to the embryo.

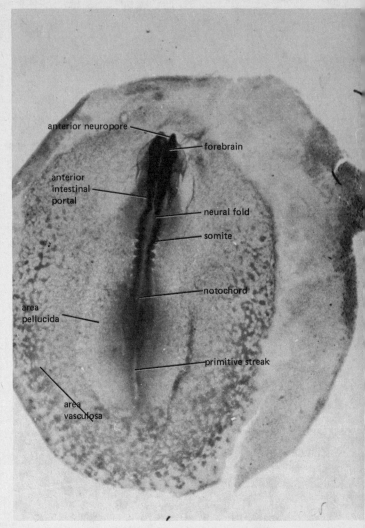

Figure 17.14.   Dorsal view of 24-hour chick blastodisc.

■ How many somites are there now?

Counting somites is difficult because the anterior portion of the embryo is probably lying on its side. Try to determine how much of the embryo has twisted (count the somites and compare with counts of the 72-hour chick). Follow the neural tube anteriorly into the hindbrain. The midbrain is the most anterior portion of the brain, and the forebrain is lateral. (*How do you account for this?*) The *disk* over the forebrain will become the eye. The ear area may be seen as an enlargement on the *anterior cardinal vein*, located near the somites opposite the heart. The larger neighboring vessel that loops anteriorly and enters the heart is the *aorta*.

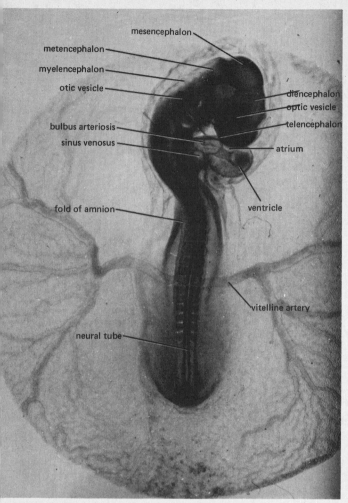

Figure 17.15.   Dorsal view of 48-hour chick blastodisk.

f. The *72-hour chick* may surprise you (Fig. 17.16). It is now completely covered with membranes. The mesoderm and ectoderm have joined and produced two new extraembryonic membranes: the *amnion* near the embryo, enclosing the fluid-filled *amnionic cavity*; and the *chorion*, which is continuous with the upper edges of the amnion and spreads outward to cover the embryo, the amnionic cavity, and the entire yolk sac.

■ **How many somites can you count now? Has the region of torsion moved?**

The heart now empties into three branches (*aortic arches*). Between these you should be able to see *gill slits*.

Figure 17.16.   Dorsal view of 72-hour embryo.

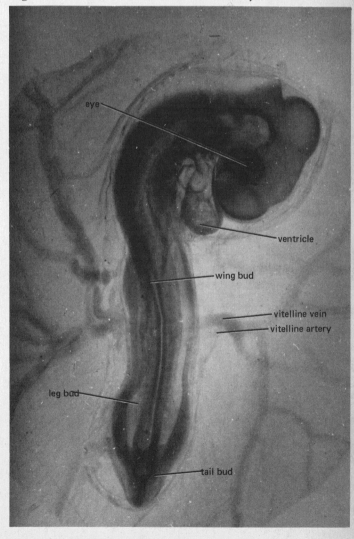

Locate and *draw* the structures described above, based on your live chick embryo and prepared slides. Study also the serial sections and try to develop a three-dimensional sense while reviewing the successive sections (watch how tubes and structures appear, run their course, and disappear in subsequent sections).

■ **What is the function of the yolk sac?**
■ **How might you account for the twisting (torsion) of the embryo? How is this related to the *chalaza*?**
■ **Is the 48-hour chick significantly larger than the 24-hour chick? What does this suggest about the developmental pattern?**
■ **Which region of the embryo seems to be developing the fastest?**

■ **How can you account for gill slits in a developing bird?**

Near the heart is the first of five regions of the brain (Fig. 17.17): the *telecephalon, diencephalon,* large, lateral *mesencephalon, metencephalon,* and the creased *myelencephalon.* You should also be able to see *limb buds,* slight enlargements of the embryo anterior and posterior to the *vitelline veins.*

■ **What is the primary function of the amnion? the chorion?**

Figure 17.17. Dorsal view of brain regions of the 72-hour chick.

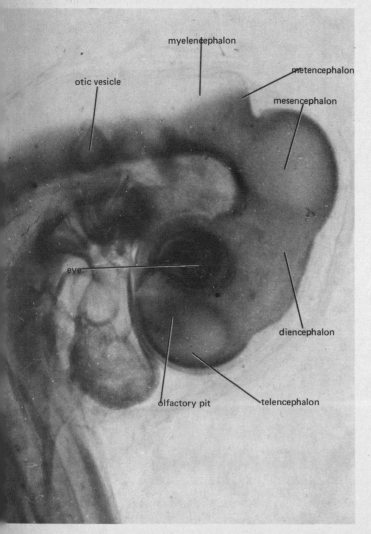

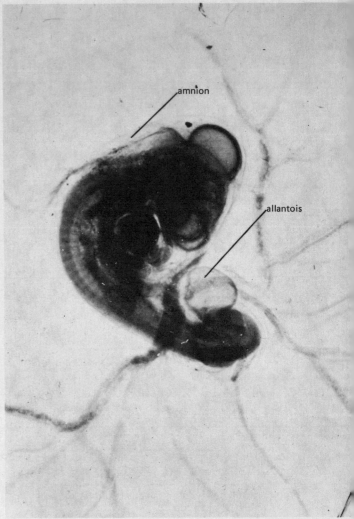

Figure 17.18. Dorsal view of the 96-hour chick embryo.

g. The *90- or 96-hour embryo* possesses still another membrane: the *allantois* (Fig. 17.18). It arises from the *gut tube* near the hind limb bud. It is a fluid-filled, bulbous bag composed of endoderm on the inside and mesoderm on the outside. It is located between the amnion and the chorion, and comes in direct contact with the chorion. Amnion and chorion fuse and the resulting *chorio-allantoic* structure develops blood vessels. These vessels connect with the embryo and serve in respiration. Note the present position of previously studied structures. Can you dissect out the allantois? If you can, weigh it (and its contents) separately from weighing the embryo and the yolk.

■ What are the functions of the allantois?

■ Can you trace out the development of *endoderm* in the chick embryo?

■ Excluding the exchange of gases (through the shell), what portion of the yolk was metabolized and transformed into embryonic tissues?

■ What portion became waste products? (Refer to your allantois weight.)

■ How does chick embryo development differ fundamentally from that of the other organisms we have reviewed in this chapter? How does this relate to their respective habitats and survival requirements? to their yolk concentrations?

Suggested References

Allen, R. D., 1959. "The moment of fertilization." *Sci. American*, 201: July.

Arey, L., 1965. *Developmental Anatomy of Man*, 7th ed. Philadelphia, Pa.: W. B. Saunders.

Balinsky, B. I., 1960. *An Introduction to Embryology*. Philadelphia, Pa.: W. B. Saunders.

Costello, D. P., M. E. Davidson, A. Eggers, M. H. Fox, and C. Henley, 1957. *Methods for Obtaining and Handling Marine Eggs and Embryos*. Lancaster, Pa.: Lancaster Press.

Fischberg, M., and W. Blackler, 1961. "How cells specialize." *Sci. American*, 205: Sept.

Gray, G. W., 1957. "The organizer." *Sci. American*, 197: Nov.

Hamilton, H. L., 1952. *Lillie's Development of the Chick*. New York: Holt, Rinehart and Winston.

Hamilton, W. J., J. D. Boyd, and H. W. Mossman, 1962. *Human Embryology*, 3rd ed. Baltimore, Md.: Williams & Wilkins.

Harvey, E. B., 1956. *The American* Arbacia *and Other Sea Urchins*. Princeton, N.J.: Princeton University Press.

Monroy, A., 1950. "Fertilization of the egg." *Sci. American*, 193: Dec.

——, 1963. *Foundations of Embryology*, 2nd ed. New York: McGraw-Hill.

——, 1962. *Experimental Embryology, Techniques and Procedures*, 3rd ed. Minneapolis, Minn.: Burgess Publishing.

Patten, B. M., 1951. *Early Development of the Chick*, 4th ed. Philadelphia, Pa.: Blakiston & Sons.

Rugh, R., 1951. *The Frog: Its Reproduction and Development*. Philadelphia, Pa.: Blakiston & Sons.

Spemann, H., 1938. *Embryonic Development and Induction*. New Haven, Conn.: Yale University Press.

Spratt, N. T., Jr., 1964. *Introduction to Cell Differentiation*. New York: Reinhold.

CHAPTER 18

GENETICS

1. Introduction

In the preceding sections we reviewed the series of changes in the life cycle of an organism, from fertilized egg of one generation to production of a fertilized egg in the next. *Cell division, differentiation, growth, gametogenesis,* and *fertilization* are all parts of this process, studied by the cytologist, the embryologist, and a number of other specialists. Now we turn our attention to *genetic recombination* and *variation,* the mechanics of *inheritance, chromosome structure,* and gene action. In recent years the double helix of DNA (deoxyribonucleic acid) and its precision mechanism for translating genetic messages into cell structure and functions has dominated biological research and produced some of the most spectacular breakthroughs in modern science, such as the incredibly successful efforts of Watson and Crick to determine the *structure of the gene* (an assemblage of DNA units); breaking the *genetic "code"*; understanding DNA–RNA (ribonucleic acid) relationships, and the *mechanics of gene action* (transport of the gene "message" and control of protein synthesis). These studies fall into the realm of *molecular genetics.* A valuable and fascinating account of the spectacular development of this field can be found

in readings from Scientific American (see references by Calvin and Pryor and Srb *et al.* at the end of this chapter).

Our survey of this field best follows the historical sequence, as this focuses on the principles that underlie *all* genetics: the Mendelian discoveries of particulate or nonmixing inheritance. These principles remain the central foundation for all subsequent knowledge of genetics and inheritance.

Genetics rapidly has become one of the theoretical bases of all biological knowledge. It includes a great range of fundamental biological research, from classical studies of the "laws" of inheritance, to the study of control and feedback mechanisms in development (*developmental genetics*), as well as the basic studies already referred to as molecular genetics. Also included is the hereditary control of the physiology of the whole organism (*physiological genetics*). All cellular activities, including the functioning of *adult* cells, tissues, and organs, reflect the continuing influence of *gene action.* Another, and equally important, aspect of genetics is the development of special statistical techniques for the quantative study of evolutionary processes in species formation (*population genetics*).

The laboratory exercises included here will demonstrate these essential principles of the *chro-*

mosomal theory of inheritance, based upon a clear recognition of the quantitative nature of the transmission of inherited traits.

The hereditary transmission of genetic factors through successive generations without "blending" or dilution is the first postulate of Mendel: the segregation or nonmixing of hereditary factors or genes (termed alleles by early workers). A century after its discovery, this principle is still the basis of modern genetics.

Many organisms have been utilized in the study of genetic principles: viruses, bacteria, molds, protists, worms, fruit flies, wasps, corn, peas, rats, mice, humans. The advantages of studying smaller, simpler organisms are (1) ease of handling, (2) rapid life cycles, (3) small numbers of obvious characteristics by which individuals or populations differ, and (4) availability of large populations for statistical evaluation.

Drosophila melanogaster, the common fruit fly, is one of the most widely used tools in genetics. During the explosive development of genetics following the "rediscovery" of Mendel's 1866 paper in 1900 (see references), Thomas Hunt Morgan and his students and associates developed the hereditary study of Drosophila practically into a science of its own. Most of the modern demonstrations of Mendelian laws, as well as newer principles of the chromosome role, sex determination, mutation, and gene recombination are based on the work of these Drosophila geneticists in their famous "fly room" at Columbia University. Subsequently, the use of viruses and bacteria, for the reasons already stated, has largely replaced Drosophila. However, many identifiable traits or "marker genes" can be used for study with Drosophila and make fascinating and readily available classroom demonstration material.

Marker genes are actually mutations—permanent, inherited differences from the so-called "normal" expression of some trait. These mutations usually result from exposure of fruit flies of earlier generations to x-rays (in nature the same mutations occur, but at a slower rate). All life cycle stages of the fly can be x-rayed, and any change produced is permanent (hereditary) if these changes happen to involve appropriate reproductive rather than somatic cells.

Mutant stocks have been carefully bred, catalogued, studied, and compared with "wild type" flies over the course of 70 years. Crosses between normal and mutant flies are followed in terms of number and proportions of progeny that show mutant or normal traits. Large numbers of offspring

must be raised and experimental crosses repeated many times in order to give statistical validity to the conclusions reached. Essential for these experiments are the procedures originated in the brilliantly lucid studies of Mendel during the last century, which are as valid today as then:

1. Maintenance of pure (homozygous) parent stocks for both normal and mutant expression of each particular character.
2. Controlled crossings between these parental stocks and their progeny.
3. Accurate counts of all progeny, statistical evaluation of these counts, and determination of ratios of classes of offspring showing the mutation.

Results expected under the particular theory being tested can then be compared statistically with results actually obtained. New theories are proposed or views developed as to the probable manner of transmission of the trait, based on interpretation of the proportions of progeny found expressing the marker. From such experiments, and many others considerably more complex, the current concepts of hereditary transmission of specific chromosomes and genes have been developed. This imposing body of knowledge has been verified and advanced by innumerable experiments. The history of its development is one of the most brilliant and exciting chapters of modern biology.

A special project section is also included to demonstrate some concepts in population genetics, in addition to the following experiments to elucidate the Mendelian laws of inheritance.

2. Laboratory Instructions

I. Introduction

The original members of an experimental cross are called the P_1 generation (parental). Their offspring are the F_1 (first filial) generation; and their offspring, the F_2 (second filial) generation.

Review Mendel's original experiments as described in his 1866 paper. (A reprinted edition is listed in the reference section; abstracts and interpretive summaries are also found in the genetics texts listed.) Note how he tested hereditary concepts using parents bred and tested as "pure" for the expression of the selected individual traits. Then observe how he simplified the results by formulating *ratios*, deriving from these a hypothesis. Using this hypothesis, Mendel predicted the outcome of a different cross, then tested it by making the cross and comparing ratios actually found with those he had predicted. Ultimately, he derived his famous two postulates, (1) *segregation of alleles* (which established the concept of discrete hereditary units or genes) and (2) *random reassortment of alleles* (which later were seen to result from the statistically random separation of chromosomes at meiosis and their recombination at fertilization).

■ **What would you say were the most important contributions and truly novel aspects of Mendel's approach?**

■ **How does his work illustrate the scientific method?**

II. Procedures in *Drosophila* Genetics

Stock cultures of various mutant strains as well as normal wild-type *Drosophila* are maintained in cotton-plugged milk bottles or smaller shell vials. These contain agar medium with an appropriate nutrient base and added yeast. Larval flies, or maggots, feed on bacteria growing in the yeast. Later the larvae pupate and emerge as adult flies. The total life cycle takes about two weeks, depending upon temperature.

Crosses are made by placing a male and a previously unmated female in a freshly prepared culture bottle, which is labeled and stored until eggs are laid. The parents are removed before the larvae hatch; the F_1 adults that develop from the larvae (in about a week) are etherized and examined individually under a dissecting microscope. Results are tallied and F_1 parents are selected for the next cross if another is needed. The resulting F_2 generation is handled similarly. Females for the F_1 cross must be *freshly emerged from their pupal cases to ensure virginity* in order to be assured of properly controlled experimental mating of these females. An alternative method is to use a "virginizer" which exposes the females to a brief period of high temperature to kill any sperm from an earlier insemination.

In order to undertake these experiments, a few simple but essential techniques must be carefully learned. Your teaching assistant will demonstrate and explain the following procedures:

1. Opening and stoppering the culture bottles; "shaking the flies down."
2. Transfer of flies to an empty bottle and etherization (not too much ether if you expect to use the flies again).
3. Handling of flies with a fine brush; arranging them for ratio counts and microscopic examination on a glass plate.

Besides the mechanics of labeling bottles, handling, transferring, etherizing, and counting the flies, you must know the following thoroughly:

1. The life cycle; appearance of larvae at various growth stages; difference between pupae and empty pupal cases; sex of adults (distinguished at a glance).
2. Difference between newly emerged and older adults for selection of virgin females.
3. Appearance of normal and mutant traits.
4. Methods of collecting and listing data, statistical procedures, determination of experimental results, and calculation of predicted results.

III. Study of Flies

1. Study an example of each stage of the *Drosophila* life cycle under low power.

a. *Egg.* Note size, shape, two posterior projections.

b. *Larva.* Note segmentation, means of locomotion, black chitinous jaws on head, and respiratory spiracles posteriorly.

c. *Pupa.* Observe pigmented *puparium* (pupal case from the preceding larval molt) containing metamorphosing last-stage larva; locate eyes, folded wings, legs, and anterior, hornlike spiracles; observe empty pupal case.

d. *Adult. Draw* a dorsal view of each sex, showing major external structures.

2. Observe culture bottles containing flies at different stages of development. Progressive changes in the medium and markings on the glass made by the crawling larvae help indicate phases of the life cycle present.
3. Obtain a fly culture for practice in etherizing and handling flies.
4. Examine microscopically as many examples of mutant stocks as are available.

■ **How many progeny can a single pair of normal *Drosophila* produce?**

■ **Where is *Drosophila* found in nature?**

■ **Are mutant types commonly found in nature? If not, why not?**

IV. Study of Giant Chromosomes in the Salivary Glands of *Drosophila* (Fig. 18.1).

This exercise affords the opportunity to prepare a temporary slide of chromosomes and to examine them microscopically for a better visual understanding of the topography of the giant salivary glands (roughly 100 times larger than are comparable chromosomes from ordinary cell nuclei.)

1. From a stock culture of *Drosophila* carefully select the largest larvae.
2. Place the larvae in a depression slide filled with a physiological saline solution.
3. Cut off the head of the larva and with a teasing needle or fine, pointed forceps remove the paired salivary glands connected to the visceral mass.
4. Quickly separate the glands from any visceral material and place them in a depression slide filled with aceto-orcein or aceto-carmine stain. Let them stand for at least 15 min.

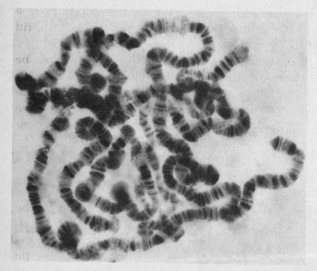

Figure 18.1. Giant salivary gland chromosomes of *Drosophila melanogaster.*

5. Remove the glands from the stain and blot off the excess surface fluid. Put them on a clean microscope slide. Place a coverslip over the gland and apply sufficient pressure to produce a uniform smear.
6. Examine your preparation microscopically.

■ **How do you account for the dark *bands* or rings of varying thickness over most of the length of the chromosome?**

■ **How many chromosomes can you count?**

■ **How do you account for the enormous size of the salivary chromosomes in comparison with those in other cells?**

■ **Of what use in genetics research are these "superchromosomes"?**

V. Genetics Experiments

Monohybrid or one-factor cross (*illustrating which law of Mendel?*)

1. **Autosomal mutation** You will be given a culture of flies with an *autosomal* mutation (mutation on a chromosome *other* than X or Y, the sex chromosomes) and another bottle of flies with wild-type or normal expression of this trait.

a. Study examples of your flies to identify the mutation clearly, to distinguish it from the normal expression of this gene.
b. In a culture bottle with fresh medium, place a virgin mutant female and wild-type male.
c. In a second bottle make the *reciprocal* cross (virgin wild ♀ X mutant ♂).
d. Label each bottle with your name and the date and show the proper symbols for the cross and generation.
e. Remove (and destroy) P_1 flies from each bottle after 5–7 days to prevent their mating with offspring.
f. When adults start to emerge in each bottle, select a freshly emerged female and a male for your F_1 cross. As you are running two experiments concurrently, keep the data for each separate.

■ **Must a virgin female F_1 parent be used?**
■ **Why can you select parents at random in both of your experiments?**

Prepare your F_1 crosses as you did the P_1, using a properly labeled new culture bottle or tube for each. Remove and destroy F_1 parents after 5–7 days.

g. Collect the remaining adult F_1 progeny as they appear over the course of several more days. Etherize and examine them, separate them into appropriate *classes* (male and female; mutant and normal within each sex). Count the members of each class.

Record your results on a data sheet in tabular form, showing all classes of progeny and the *observed* and *predicted* ratios of these classes. The predicted ratio is to be calculated or determined from the hypothesis of inheritance that you are testing.

h. Repeat step g for the progeny of your F_1 crosses (F_2). Keep results of the original reciprocal crosses separate so they can be compared later.

Large samples are important for these counts, so collect as many F_2 as time allows (but do not allow time for any F_3 to emerge).

■ **Why are these large samples necessary?**

i. Calculate results and summarize data for each experiment. Compare observed with predicted results.

■ · **Are the differences statistically valid?** (See Appendix III, pages 351–352.)
■ **What differences can you detect between the reciprocal crosses?**
■ **What conclusion can you draw from this?**

j. Prepare a report summarizing both of your experiments—including *objectives, methods, data* and *observations, results,* and *conclusions.* Include the proper notation to represent the crosses and their products.

Discuss briefly the genetic significance of the difference between mutations arising in *autosomal* and *sex chromosomes.*

2. **Sex-linked monohybrid cross** Repeat the procedures described above with a *sex-linked* mutation (one located on the X or sex chromosome). *White-eye* in *Drosophila* is a widely used example. It was historically important as the first *Drosophila* mutant to be discovered and explored and helped to estab-

lish some of the basic premises of the gene theory.

Make reciprocal crosses just as you did in the autosomal experiment and prepare a similarly organized report. Include a statement of the hypothesis you are testing and a statistical comparison between ratios expected and ratios found. Divide your report, as you did previously, into *objectives, method, data* (tabulated counts and necessary calculations), *observations, results,* and *conclusions.*

In your conclusions, include answers to the questions below and try to account for any experimental discrepancies from predicted results. It may be of interest to add to your data those of other student teams to increase your statistical sample.

■ **Will enlargement of your sample enhance the validity of your results?**
■ **How do results of your reciprocal sex-linked crosses compare?**
■ **How do the sex-linked and autosomal results compare?**
■ **How do you explain the differences?**
■ **What general principle is illustrated by these crosses?**
■ **What principle is demonstrated by a cross involving two distinct mutations (a dihybrid cross)?**
■ **Why would Mendel's postulates be difficult to discern if three different mutations (a *trihybrid* cross) were used?** Observe how this illustrates the significance of Mendel's use of the *simplest possible cases,* the one- and two-factor crosses.
■ **How would you explain the results of a tri- and a multiple-factor cross?**

Similarly illuminating experiments can be performed using other material such as corn seedlings, bacterial cultures, bacterial viruses, or mice.

The relevance and importance of genetic principles are confirmed by their universality of application.

VI. Study of Populations

This exercise affords the opportunity to observe the growth of a *population,* rather than a few individual progeny, and possibly *changes of allele frequencies* (evolution) in a *gene pool* (the full genetic component of *all* members of an inter-breeding population).

1. Select a male and female fruit fly (one homozygous wild-type and one with a visible marker trait, a homozygous recessive). Place them in a 250-ml flask with a mixture of agar, yeast, and fruit (such as banana). Stopper with a piece of cotton. (*Why should cotton be used?*)

2. Periodically (every second or third day) lightly etherize the adult flies, record the date and any special conditions, and count the number and allelic type of the dead flies (which should be saved). Be sure not to overetherize the flies while counting.

■ What special information can you derive from keeping track of the dead flies?

3. Continue to sustain and record data from the fruit fly colony for at least *six weeks*.

4. Plot your data on the three graphs A-C provided at the end of this chapter (Fig. 18.2).

■ How do you account for the shape of the curve in Graph A?

■ On which days was the population increasing most rapidly?

■ On what day did the death rate overtake the rate of reproduction?

■ Did the population size continue to increase, stabilize at some equilibrium number, or decrease?

■ How would you account for this final portion of the curve (that is, what factors do you consider responsible for the result)?

■ How would you test your hypothesis from the previous question?

■ Describe two ways of generating the data used in Graph B. (See Section VII, Special Project 2 on this page.)

■ Is Graph B congruent with Graph C? If not, how do they differ?

■ If they are congruent, what does this suggest about the relative *adaptive advantage* of the wild-type and mutant (marker) traits?

■ If they are not congruent, what has happened to the population gene pool? What remarkable phenomenon might have you witnessed?

■ Is there a statistically valid difference[1]

---

[1] See Appendix III for statistical treatment of data.

between the allelic frequencies at the beginning and end of this experimental study?

■ By comparing your results with those of other students, what can you say about the "relative fitness" of the tested traits?

■ Small islands are often inhabited by wingless insects. How do you account for this fact?

We now know from this experiment that (1) unchecked populations grow exponentially, and (2) in nature the population size remains approximately stable, from which we can deduce that more individuals are produced than can survive. We further know that individuals vary, that some of these variations are heritable, and that certain variations better enable an individual to survive, that is, differential survival among naturally occurring variants is the general rule.

■ What conclusions can be drawn from these facts?

■ What general mechanism appears best able to account for these observations?

■ What is the end result of the entire process?

VII. Special Projects

1. **Human population growth (1750–1975)** Look up figures for human population from 1750 until now. Plot them on a graph with *numbers* on the X axis and *time* in years on the Y axis. These figures may be found in *The Limits to Growth* (see Meadows *et al.*, 1972, in the reference section of this chapter) or in any standard encyclopedia.

■ Which part of the fruit fly curve (Graph 18.2A) does the human population curve resemble?

■ What do you predict that the rest of the curve will look like?

■ What factors could modify this curve? How does the slope of this curve relate to the *most basic problem facing mankind today—overpopulation?*

2. **Hardy–Weinberg law** (calculation of gene frequency in a population)

a. *Gene frequency in a* Drosophila *population without selection* (data from Section VI.1).   In exercise VI calculations of allelic frequencies

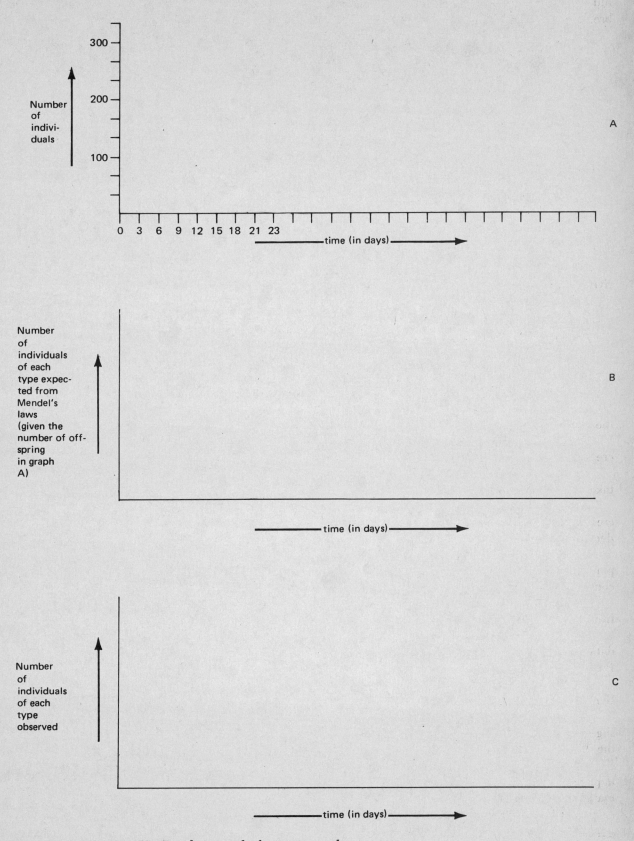

Figure 18.2.   Exercise VI. Population and phenotype graphs.

(relative abundance of a wild-type and mutant gene in a population) could be greatly facilitated by use of the Hardy–Weinberg formula. In this formula (using a single pair of alleles, such as A and a), let p represent the proportion of A alleles, and q represent the proportion of a alleles. Therefore, $p + q = 1$. If we know one side of the proportion, the other is easily calculated: $q = 1 - p$. If we expand $p + q$ to the second power, we have $(p + q)^2 = p^2 + 2pq + q^2$, which represents all genotypes of this allelic pair. The proportion of AA is represented by p^2, the proportion of Aa by $2pq$ ($Aa + aA$), and the proportion of aa by what? Therefore, frequency of the A allele in the gene pool is

$$A = p^2 + \tfrac{1}{2}\,(2pq)$$
$$= p^2 + pq = p\,(p + q) = p$$

■ **What is the allelic frequency of the gene a?**

Keep in mind that we are dealing with *gene* frequencies, not the frequency of *paired* genes for each trait found in the living organism. The gene frequency, therefore, is a *theoretical* measurement of the frequencies of individual genes as if they were plucked out of their "host" and put into separate groups according to type—hence the term "gene pool." This eliminates all gene + environment interactions that result in the actual *expression* of a gene (the *phenotype*). Use of this procedure greatly simplifies studies on randomness and assortative distribution of genes and makes possible the use of highly sophisticated and very illuminating statistical procedures—the basis for the development of *population genetics*.

Random mating will always result in constant values for p and q, assuming a *large* population is present and no variables interfere with the true randomness of the gene distribution.

■ **What external or internal influences might *interfere* with random mating and hence with the application of the Hardy–Weinberg principle?**

If in exercise VI there *was* random mating, and if the population size were not too small, one could easily and confidently determine the expected *genotypes* and *phenotypes* of any gene pair (with a mutant expression in one of the two) in the following way (by simple binomial expansion):

$$p = \text{frequency of } A = 50\% \text{ or } 0.5$$
$$q = \text{frequency of } a = 50\% \text{ or } 0.5$$

Then

$$p^2 = 0.25 \text{ homozygous wild-type } AA$$
$$2pg = 0.50 \text{ heterozygous wild-type } Aa$$
$$q^2 = 0.25 \text{ homozygous recessive } aa$$

Thus in the case of a simple recessive, under ideal, unchanging conditions, we would *always* expect ¼ of the fruit flies to show the marked trait and ¾ to appear as wild types. If the actual result varied significantly from these values (and, in fact, we *were* dealing with a simple recessive gene trait), we must conclude that either (1) errors were introduced by faulty lab techniques, or (2) the population was not large enough to be relatively free of sampling errors, or (3) individuals did not mate randomly (which is equivalent to saying that *natural selection* was operating).

Since the Hardy–Weinberg formula is a simple mathematical expression of Mendel's basic postulate—the nonmixing of alleles—the question of accuracy of the formula does not apply. Given the simple gene expression hypothesized (dominant wild type; recessive mutant) and assuming nonmixing (nonblending) of the expression of these genes, the 25%–50%–25% *genotypic* frequency or 75%–25% *phenotypic* ratio must result.

We have also omitted the possibility of *gene flow* (since the operation is presumed to occur in a closed system). Other modifying factors (except for natural selection) such as meiotic drive and mutation pressure usually have negligible influence on allelic frequencies and are therefore also disregarded here.

This experiment illustrates the slow rate of change of a gene in a population once it is established there by mutation frequency and natural selection. Even if natural selection is subsequently relaxed, the gene tends to remain in the population at about the same level. An example is the presence of sickle-cell trait in black Americans, despite the absence of favorable selection due to malaria resistance (which no longer applies) and despite that fact that homozygous recessives are almost entirely lost to the breeding population (hence their *genes* are lost).

LABORATORY INSTRUCTIONS 317

The application and usefulness of the Hardy–Weinberg formula for such studies as this is clearly evident. Much of our understanding of the retention of genes in populations is based on the Hardy–Weinberg principle.

■ Would you be able to calculate it with *more* than two gene pairs? (The complications are endless!) What would be the result if you selected a series of *phenotypic* expressions of the same gene locus for your study?

■ Could the same approach be used to calculate the gene frequency of *two* mutant pairs (loci)?

b. *Gene frequency in a Drosophila population with selection*

i. Select a male and female fruit fly (one homozygous wild-type and one with a visible marker trait, a homozygous recessive). Place them in a flask with appropriate nutrients.

ii. Periodically (by generations) record the day and count the number of individual flies and the number of each *type* of individual. Then *remove any homozygous recessives of each generation* (before they can breed).

iii. Replenish food supplies as needed and maintain the colony and the periodic counts for at least six generations.

iv. Compute the *expected* gene allelic frequencies. (Remember, *you have removed homozygous recessives* of each generation.) Compute also the expected *phenotic* frequency (ratio) and compare these results with your *observed* ratios.

■ Were the allele frequencies statistically altered?

■ Why were the homozygous recessives removed before they could mate? What would have been the resulting frequencies had they *not* been removed?

■ If 9% of a population were homozygous recessives, what then would be the allelic frequency of a recessive trait in the gene pool?

■ How many generations would it take, using the artificial selection technique above, to *reduce* the allele frequency of the recessive trait by 50%? by 90%?

c. *Nonbiological demonstration of frequency of two "alleles."* Consider the allele *H* to be the "heads" side of a coin and *h* to be the tails side. Flipping the coin, fill in the table provided at the end of this chapter (Fig. 18.3). For trial 1, box 1 of the table, flip the coin twice. If it lands heads up both times, mark the box 100; if it lands heads up and tails up, 50; and if it lands tails up both times, 0. Complete the table through trial 5, or through 10 if your results are "strange" (worn coin? too brief a flip?), or you are convinced you can beat the system. Prepare a graph plotting the number of flips against the percent of *H* for each trial.

■ Why would it be undesirable to plot the total number of flips and the average percent for all 10 trials?

■ Does the percent of *H* vary more when fewer flips (that is, greater variability) are involved?

■ What trends can you detect in your table?

■ If the number of flips were the number of individuals, and the percent of *H* the allelic frequency of *H*, what could we say about the effect of *small populations* on allelic frequencies? of large populations?

■ A large population of Martians has

Figure 18.3. Exercise VII, special project 2c. Table for results of coin flipping.

Trial Number	2 Flips % of H	4 Flips % of H	6 Flips % of H	10 Flips % of H	50 Flips % of H
1					
2					
3					
4					
5					
6					
7					
8					
9					
10					

either long or short chins regulated by the alleles *L* (for long) and *l* (for short). With urbanization came disease that wiped out all but four individuals (two females and two males). After many generations, the population had only short chins. Give a possible explanation for this phenomenon (discovered and mathematically analyzed by Sewell Wright).

■ Of what possible evolutionary importance might this phenomenon be?

Suggested References

Calvin, M., and W. A. Pryor, 1973. *The Organic Chemistry of Life: Readings from Scientific American.* San Francisco, Calif.: W. H. Freeman.

Carlson, E., 1966. *The Gene: A Critical History.* Philadelphia, Pa.: W. B. Saunders.

Carson, H. L., 1963. *Hereditary and Human Life.* New York: Columbia University Press.

Darwin, C., 1859. *The Origin of Species.* Many reprint editions.

Demerec, M., and B. P. Kaufmann, 1945. *Drosophila Guide.* Washington, D.C.: Carnegie Institute of Washington. (Directions for culturing and handling *Drosophila* in the laboratory.)

Dobzhansky, T., 1962. *Mankind Evolving. The Evolution of the Human Species.* New Haven, Conn.: Yale University Press.

Haskell, G., 1961. *Practical Heredity with Drosophila.* London: Oliver and Boyd.

Huxley, J., A. C. Hardy, and E. B. Ford (eds.), 1958. *Evolution as a Process*, 2nd ed. London: G. Allen and Unwin.

Kempthorne, O., 1957. *An Introduction to Genetics Statistics.* New York: J. Wiley & Sons.

Li, C. C., 1955. *Population Genetics.* Chicago, Ill.: University of Chicago Press.

Malthus, T. R., 1798. *An Essay on the Principle of Population.* Many reprint editions.

Mayr, E., 1966. *Animal Species and Evolution.* Cambridge, Mass.: Harvard University Press.

Meadows, D. H., D. L. Meadows, J. Randers, and W. W. Behrens III, 1972. *The Limits to Growth.* New York: Potomac Associates.

Mendel, G., 1967 (reprint). *Experiments in Plant Hybridization.* Cambridge, Mass.: Harvard University Press.

Peters, J. A. (ed.), 1959. *Classic Papers in Genetics.* Englewood Cliffs, N.J.: Prentice-Hall.

Scheinfeld, A., 1965. *Your Heredity and Environment.* Philadelphia, Pa.: J. B. Lippincott.

Simpson, G. G., 1950. *The Meaning of Evolution.* New Haven, Conn.: Yale University Press.

Smith, L., 1947. "The acetocarmine smear technique." *Stain Tech.*, 22:17.

Sonneborn, T. M. (ed.), 1965. *The Control of Human Heredity and Evolution.* New York: Macmillan.

Speiss, E. B., 1962. *Paper on Animal Population Genetics.* Boston, Mass.: Little Brown & Company.

Srb, A. M., R. D. Owen, and R. S. Edgar, 1965. *General Genetics.* San Francisco, Calif.: W. H. Freeman.

——, 1970. *Facets of Genetics: Readings from Scientific American.* San Francisco, Calif.: W. H. Freeman.

Stebbins, G. L., 1966. *Processes of Organic Evolution.* Englewood Cliffs, N.J.: Prentice-Hall.

Strickberger, M. W., 1968. *Genetics.* New York: Macmillan.

Sutton, H. E., 1965. *An Introduction to Human Genetics.* New York, Holt, Rinehart and Winston.

Wallace, B., and A. M. Srb, 1964. *Adaptation*, 2nd ed. Englewood Cliffs, N.J.: Prentice-Hall.

Watson, J. D., 1971. *Molecular Biology of the Gene*, 2nd ed. New York: W. A. Benjamin.

CHAPTER 19

ANIMAL BEHAVIOR

1. Introduction

An astonishing range of behavioral patterns has been recorded in recent years as a result of active interest in the relatively new discipline, *ethology*, which may be considered an effort to determine the "psychology" or methods by which nonhuman animals relate, communicate, and interact with one another. It is now well established that ants communicate by using chemical signals (pheromones); that some fishes can detect changes in electric currents and pressure in their medium; that bats find food by echolocation; that their moth insect prey can hear the bats' high-frequency sounds and respond with avoidance maneuvers. Many other ethological data are documented and being collected. All of these examples fall under the category of sensory phenomena and are related to highly specialized neurological and biochemical mechanisms (see reading list at end of chapter). The challenge for us is to select from such facts the common features of behavior that contribute to our understanding of biological principles.

Students of zoology should learn to differentiate among the various kinds of behavior and to distinguish their grades of complexity from relatively simple *kineses, taxes,* and *reflexes* up to com-

plex *inborn behavior patterns* and to *learned behavior.*

■ **What are some generalizations about the behavior of animals that these studies suggest and support?**

1. Behavioral responses are triggered by stimuli. These are often remarkably specific, as in the case of *releasers*. These stimuli in turn are related to the animal's environmental setting, which influences the organism's internal state, often in a feedback fashion. There is a close relationship between physiological determinants, such as hormones, and the behavior expressed.

2. Behavior alters with development, undergoing maturational changes, and may be modified by the environment. Some long-term environmentally induced and behavioral changes may occur through *conditioning*.

3. Behavior usually is *adaptive*. It is an indispensable function of animal survival—whether for orientation, courtship, nest building, feeding, defense, communication, herding or schooling, or any of a remarkable array of social interactions. Among subhuman primates this even includes *speech*

(or ability to discriminate among and reproduce a variety of symbolic signs or cues), as recent work with chimpanzees indicates.

4. Behavior is often *periodic*. Activity cycles vary from a few hours duration through circadian, seasonal, and annual to multiyear cycles.

5. Some behavior appears to be *hereditary*. It is controlled by the action of genes, whose products in turn are affected and modified by factors in the environment, the interaction being subjected to *natural selection*, ultimately resulting in a genetically controlled behavior pattern.

6. Behavior has *evolved*, and it can in turn affect the evolution of a species. Such activities as distinctive courtship behavior and new mating habits may separate a population reproductively, a basis for more complete reproductive isolation and, ultimately, the evolution of a distinct species.

7. Behavior is important at the *population* level. Many kinds of animals form intra- and interspecies associations, which vary from loose aggregations to herds, flocks, and schools of varying internal cohesion and organization, up to highly integrated and interdependent societies. Division of labor, morphological specialization into *castes* (as in ants and termites), communication between individuals, and complex social relationships are all reflections of the behavioral pattern between or within species.

8. Behavior often determines *dominance relationships* and intragroup rank (hierarchy or "pecking order") and *territoriality*.

The study of animal behavior is a young and rapidly growing discipline. The recent award of the Nobel Prize to two pioneer ethologists, Konrad Lorenz and Niko Tinbergen, attests to the growing importance and respect won by this field. Yet, much remains to be discovered, much considered factual now may prove to be incomplete, misconstrued, or simply wrong in the near future. Most of ethology is yet to be learned.

2. The Laboratory

Despite the fact that many behavior patterns are seasonal, nocturnal, or restricted to the animal's normal undisturbed environment, a number of behavioral traits can be studied by students in the laboratory.

Laboratory studies on behavior should provide opportunity for students to observe the work in several areas. Projects should emphasize observation and analysis of one pattern in considerable detail. This is far more significant—and interesting—than a casual survey of many patterns. Precise description of the behavior of an animal and (where possible) *quantitative* study is not easy. It demands perception, patience, and persistence.

Suggested problems for laboratory study are the following.

I. Behavior of Day-Old Chicks

Day-old chicks are available from local hatcheries and can be maintained in heated brooders. Identify the "pleasure call" and the "distress call." Determine the circumstances under which each call is given, for example, with respect to water, food, cold, sight and sound of other chicks. Tape-recorded observations will be very useful and can be made available to other students (and to stimulate responses in isolated test chicks). Color preference of chicks can be demonstrated and quantified by using colored plaster balls and by counting the number of pecks on each ball per unit time. *Design your own tests* to determine the effect of various stimuli on the *timing, onset, frequency,* and *duration* of the two basic call notes of your chick. *Record* your observations carefully (perhaps on tape as well as notebook) and *tabulate* your results.

■ **Under which conditions do the chicks emit these "distress calls"? their "pleasure call"?**

■ **What variables, other than *fear, hunger,* and *presence of other chicks*, appear to elicit these calls.**

■ **How do you account for the differential response to various colors?**

The effect of *hormones* on chick behavior can be shown by implanting subcutaneously a pellet of testosterone and contrasting the behavior of implanted and control (sham-operated) chicks one week later.

Procedure. Carefully shave about 0.5 cm² skin on the back of each chick, in a location such that the chicks cannot reach the area with their beaks. Wipe the shaved area with an antiseptic solution. Holding the chick somewhat firmly in one hand, make a short shallow incision through the skin with a sterile scalpel or razor blade in the shaved area.

Little bleeding should occur. Lift one side of the cut with blunted dissection forceps and have your lab partner place the testosterone pellet well under the skin using fine forceps. For the control group, do everything the same, including the lifting of the skin, but place no pellet (or insert an *empty* pellet) inside. Seal over the incision with a small piece of first-aid tape. Allow a day or so for acclimation, then set up tests as before to study the chicks' call response to various stimuli (for example, color, feed, presence of other chicks, tape-recorded calls, intense or dim light, temperature changes). Note *differences* between hormone-treated (experimental) and nontreated (control) chicks.

■ **Which stimuli cause behavioral (call-note) responses that are *most* affected by the testosterone? Which stimuli + response seem *least* affected by the testosterone?**

II. Conditioning of Planaria

Planaria exhibit moderate cephalization and possess a nervous system that runs nearly the full length of the body (Fig. 19.1). At the anterior end there are two eye spots, which are photosensitive, yet cannot discriminate shapes, as they lack lenses.

For this experiment we will *condition* planaria to react to bright illumination in a non-characteristic manner. First, isolate individual planaria and observe their reaction(s) to sudden bright illumination. Keep them in the dark or subdued light; then turn on a 75-watt light placed directly over them (but not close enough to modify the temperature). Repeat this procedure several times to observe a typical response.

Follow the same general procedure for the following conditioning experiment. With the onset of illumination, touch the animals with a prodding stick (a Q-tip will do). Repeat this for a series of trials. You will note that the reaction to touch is an immediate "shrinking" by the planarian. Turn off the light after each trial. Do not disturb the subjects in any way at any other time. After a number of these "training trials" (at least 20), turn on the light and do *not* stimulate with the prod. Do this at least 20 times. Record and tabulate your results.

Figure 19.1.   The nervous system of a planarian.

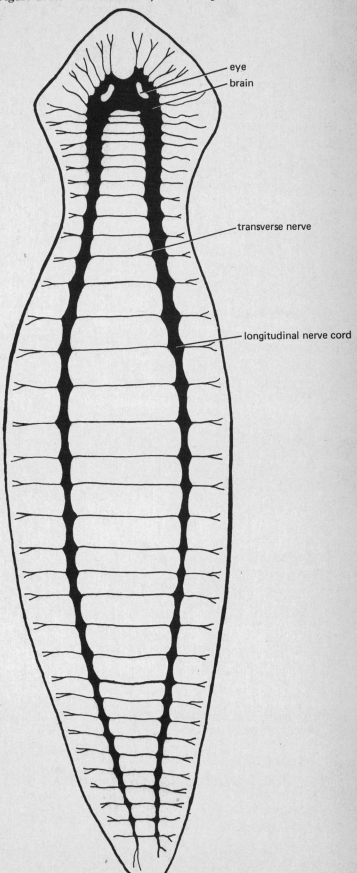

eye
brain
transverse nerve
longitudinal nerve cord

■ How does the reaction of the subjects to light *after* conditioning differ from their reaction *before* training?

■ After a number of posttraining trials, does the behavior become *extinguished*?

■ Describe the behavioral decay.

■ Was it rapid? Did the subjects stop responding for a period and then start up again (as in a sudden "recall" response)?

III. Chemical Transfer of Information in Planaria

A fascinating set of experiments was done by Dr. James V. McConnell at the University of Michigan on the chemical transmission of conditioned responses in planaria. We will attempt an approximate duplication of his "far-out" experiment.

Retrain the planaria used in the preceding experiment, but this time allow at least 100 training trials. Do not allow the retrained behavior pattern to become extinguished. Immediately cut and macerate these subjects into very small pieces. Retain the aqueous medium. Set up two groups of planaria. Place the *experimental group* into the medium containing the cut-up pretrained planaria. Place the *control group* in a similar medium but with pieces of *non*trained planaria. Allow the groups to feed on their respective broken-up planaria for one or two days.

After the waiting period, "train" each planarian by pairing the prod with the light a single time. See if onset of illumination a second time will then elicit the response. Continue this train + test procedure, noting carefully the number of trials required to retrain both the experimental and control groups.

■ Did you find a difference in the number of trials required to train the experimental and control groups? What might account for this?

IV. Light Responses of *Daphnia*

The freshwater cladoceran *Daphnia* responds differentially to light, varying with intensity. It moves *upward* toward dim light, *downward* away from bright light. *Daphnia* also responds differentially to different wavelengths, moving *toward* a yellow light source of relatively long wavelength, and *away* from a blue light source of relatively short wavelength.

Place a *Daphnia* at the top of a water-filled 50-ml graduated cylinder; immediately place a lit 25-watt light bulb directly above the top of the cylinder. Record the time required for the *Daphnia* to descend to the bottom of the cylinder or otherwise respond to the light stimulus. Repeat these observations several times with a series of individuals, keeping track of each response period (latency).

With the same individual *Daphnia*, tested one at a time, repeat the above procedure with a 50-watt, 75-watt, 100-watt, and 200-watt bulb. Vary the order in which you present the light stimuli to obviate any "order effect." Keep separate records for each level of illumination.

Compute the mean latency period for each level of illumination, combining the data for all individuals.

Prepare a graph of the response pattern by plotting illumination intensity on the X axis and time of descent (or to reach equilibrium) on the Y axis.

■ What conclusions can you draw from the graph? Is the relationship between successive levels of illumination a linear one?

■ What variation (variability) was found in the responses of the different individual *Daphnia*? Did the times vary more during bright or dim illumination? How can you account for these differences?

V. Reaction of Terrestrial Isopods to Humidity

Sow bugs (or pill bugs) respond to decreasing humidity levels with an increase in their rate of movement, called *orthokinesis*. This can be demonstrated by placing equal numbers of isopods in two vessels, one moist (not wet) and the other dry.

■ How might you set up a humidity chamber for this test? (Try a petri dish with moistened filter paper in the lid.)

Record the *rate* of movement or the *number* of isopods that are observed moving in each vessel after given time intervals under moist and dry conditions (at constant temperature and light). Several recording techniques can be used. For example, the isopods can be placed on a cut-out piece of smoked kymograph paper[1] in matched

---

[1] See Appendix IV, page 353, for a description of the kymograph.

petri dishes (one with the moist-chamber lid as suggested above). Keep them in a darkened place at constant temperature. Record the number of marks on the paper counted after 1 hour.

■ **How else might you measure response of the sow bugs? What other variables should be controlled?**

■ **Of what adaptive use might be the differences you observed? (Where do sow bugs normally reside?)**

■ **Which other organisms would you suggest have patterns of behavior at least in part controlled by humidity? (What do these others have in common with sow bugs?)**

VI. Proboscis Extension in the Blowfly

Upon detection of a sugar solution, the blowfly *Phormia* extends its proboscis and feeds until satiated. There is an interesting series of experiments that can be performed to analyze this behavior by demonstrating:

1. **Locations of the receptors** Etherize a group of *Phormia* by placing a piece of cloth containing a few drops of ether into a jar. Add the test flies and cover the jar. As soon as the blowflies have stopped flying about, open the jar and shake them out. (If left in too long, they will die.) *Do not etherize near any electrical appliances or near flames, lit cigarettes, etc.; ether is explosively inflammable.*

Touch a wooden applicator stick to liquid melted wax from a candle. Before the wax hardens, touch it to a *Phormia* fly to attach it on the stick. By this means attach a series of flies by various parts of their anatomy to applicator sticks. This will permit the blowflies to be exposed to testing at various points on their bodies. It can be done by the simple procedure of dipping exposed body parts of the wax-held blowflies into various sugar solutions.

Prepare a saturated sugar solution for these tests. Place several grams of sugar in a petri dish and add sufficient water to bring the sugar into solution.

Proboscis extension is a reflexive action. After the flies have fully recovered from the effects of the ether, touch various parts of the fly to the sugar solution, watching the proboscis for its extension. Be careful that only *one* portion of the anatomy touches at a time. For example, touch the antennae alone, then the wings, and any other exposed portions.

When the fly's *glucoreceptors* (sugar-sensitive receptors) contact sugar, the proboscis extends by a reflex action. Locate these glucoreceptors from your tests and prepare a sketch of the fly, designating the *receptor sites*.

2. **Sensitivity of the receptors to dilute sugar solutions** Once you have identified the site(s) of glucoreceptors, establish an approximate threshold level for sugar detection by testing flies against a series of sugar concentrations. Pipette a constant amount of water into a series of petri dishes, then add carefully weighed aliquots of sugar to make up a graded series of solutions (for example, 0.25 g, 0.50 g, 0.75 g, 1.00 g, 1.25 g, etc.) varying with the water volume used.

Using an individual fly and previously identified glucoreceptor areas, find the solution of *lowest* concentration still able to induce a proboscis extension response. Do this for a series of flies, using the same solutions each time.

■ **How much variation did you find among your test subjects?**

■ **How might you have altered the paradigm (experimental method) to reduce the amount of variation caused by external factors (experimental error)?**

3. **Ability of glucoreceptors to differentiate between sugars and other substances** Prepare the following solutions in petri dishes at room temperature. (For additional studies, prepare test solutions of your own selection.)

   a. A glucose solution above the threshold level of response.
   b. A comparable solution of *sucrose* (or other sugars that might be available).
   c. A solution of an artificial sweetener.
   d. A solution of 10% NaCl.
   e. Pure water.
   f. A solution of unsweetened coffee or tea.

■ **To which solutions did the fly show a positive response?**

■ **Having established the *presence, location, threshold of response,* and *specificity of re-***

sponse, what further tests of these receptors do you feel would now be indicated?

4. Adaptation of the receptors Neurological *adaptation* is a reduced response or an *habituation* induced by receptors that are under frequent or continuous stimulation. Touch the glucoreceptors of a previously untested fly to a sugar solution repeatedly until the proboscis fails to respond. Attempt this for several flies and record your observations.

■ What factor *other* than neural adaptation might have caused the negative response in this series of experiments? (A primary responsibility of every investigator is to avoid being deceived by possible artifacts or a misinterpretation of observed results caused by unrelated or unrecognized external factors. "Wishful thinking" can introduce an unintentional but highly deceptive bias in a series of observations where certain results are desired.)

VII. Feeding Behavior of Hydra—A Study in Chemoreception

Hydra normally feeds on a variety of small crustaceans. Some of the factors that elicit the feeding response can be tested by the following laboratory experiments. Report your conclusions in the following chart after recording fuller experimental details with procedures, sketches, and observations in your notebook.

EXPERIMENT	FOOD OFFERED	RESULT
1	Crushed *Artemia* (undiluted)	
2	*Artemia* juice (crushed *Artemia* in 2 ml water)	
3	*Artemia* juice aged for 24 hours or oxidized by hydrogen peroxide	
4	Inert material (filter paper)	
5	Filter paper dipped in *Artemia* juice	

■ How is the hydra's reflex feeding response normally elicited?
■ What is the sequence of events that you observed?

VIII. Chemical Communication in Ants

Ants communicate with one another by means of *pheromones*, chemical substances secreted by various animals that affect the behavior of *other* animals of their species (interorganismic hormones). Trail-following in ants is dependent upon such a pheromone (see references by Wilson at end of chapter). Bring an ant colony (commercially available) into the laboratory, or make your study a *field* effort.

To determine which part of the ant body is responsible for the pheromone production, divide the insect into its three body parts: head, thorax, and abdomen. Using a *separate* rod and microscope depression slide for each, crush each of the three body parts.

At the other boundary of the ant colony, draw a 15-cm trail with the rod dipped in the extract from the ant head.

■ Do the ants follow the trail?

Follow the same procedure for extracts from thorax and abdomen. Be sure to use a clean rod for each.

■ Which part of the body is responsible for pheromone production?
■ Does this imply that the pheromone is produced in this area of the body?

Make similar extracts of anatomical parts (for example, antennae, legs, caudal portion of abdomen) and repeat the testing procedure.

■ Could you further localize the site of pheromone concentration?
■ If *more* than one part appears to contain pheromone, how might this be explained? How might you test this hypothesis?

IX. Mating Behavior in the American Cockroach; *Periplaneta*

Isolate a pair of sexually mature *Periplaneta* and place them in a 10 cm by 10 cm container. Under dim light, preferably at night (*why?*), observe and record the sequence of events during courtship, pairing, and copulation.

■ What specific stimulus initiates male *wing-raising*?

■ What is the female response to male wing-raising?

■ How might you determine if a pheromone is involved in this mating behavior?

X. Aggressive Behavior of Siamese Fighting Fish under Various Pharmacological Conditions[2]

Beta splendens, the Siamese fighting fish, is normally unaggressive in its natural freshwater tropical habitat—so long as it remains within its own territory and has a sexual partner during its mating period. When crowded or stimulated by the proximity of a sexual rival, however, these fish exhibit characteristic aggressive responses.

For this experiment, each group of students will need:

a. Two male and two female *B. splendens*.
b. A 2-gallon rectangular fish tank.
c. A mirror slightly smaller than the tank's smaller cross section.
d. A fish net.
e. A stopwatch.
f. Four beakers for holding the fish.
g. Four 500-ml beakers for the drug solutions (fill each with 480 ml water).
h. Two 1000-ml beakers.

1. Place one male fish in the main tank and the others in separate beakers. Allow a period of acclimation. *Record: coloration, fin positioning and movements, gill characteristics*, and the *general behavioral pattern* (activity, swimming pattern, location in the beaker or tank).

a. Now place the mirror in the main tank, start the stopwatch, and *record changes in the male's behavior as a function of time*. Include information on *taxes* (directional movements), *color changes, rapidity of movement, flaring of tail and fins, specific behavior*. Remove the mirror.
b. Place the other male in the tank and observe the behavior of the two fish together. Note similarities and differences from the previous mirror experiment. Remove the males to separate beakers before any physical damage is done. (In Thailand, fish fights to the death are

staged and elicit as much betting and excitement as do cock fights.)
c. Repeat the procedures outlined above for the *females*, making a similar set of observations.
d. Next, place a *male* and a *female* together in the tank and observe as before. Again, separate before damage occurs.

2. Ask your teaching assistant to dispense into your 500-ml beakers the drugs he has prepared. You should have two dextroamphetamine and two Librium solutions (or other comparable test drugs).

a. Place two of the fish (your choice as to sex) in separate dextroamphetamine beakers, and allow the drug to act for 15 min. *Observe* the progressive changes in the fish's appearance and behavior over a period of time.
b. Pour both solutions into a 1000-ml beaker. *Observe and record:* behavior and color changes, again noting the time span. Remove the fish before they are injured. Place them in fresh water and return them to your teaching assistant.
c. Repeat the preceding experiment, but this time with the Librium solutions used on the remaining two fish.
d. Describe the behavior of male and female fish in their *drugged* and *nondrugged* states when they are isolated, and, later, when they are combined. Make your observations and remove, wash, and return the fish as before.

■ What is their behavior, color, etc., when the fish are combined?

■ What might account for the behavioral sex differences?

3. Preparatory Instructions for Teaching Assistant

Prepare sufficient solution of 2 mg/ml dextroamphetamine and 1 mg/ml solution of Librium in H_2O so that 10 ml of each can be dispensed into each dextroamphetamine or Librium beaker. These solutions must be freshly prepared prior to each laboratory period and not used for more than one day, since deterioration of the drugs soon occurs. Each *Beta* exposed to a drug must be allowed a 24-hour rest period in fresh water for drug metabolism to be completed before it can be retested.

---

[2] Instructions for the teaching assistant's preparation of the drug solutions will be found in Section 3 on this page.

Suggested References

Armitage, K. B., 1960. "The use of *Daphnia* to demonstrate biological phenomena." *Turtox News,* 38(4):118–121.

Bermant, G., 1963. "Intensity and role of distress calling in chicks as a function of social contact." *Anim. Behav.,* 11:514–517.

Best, J. B., 1963. "Protopsychology." *Sci. American,* 208: Feb.

BSCS, 1965. *Biological Science, Interaction of Experiments and Ideas.* Englewood Cliffs, N.J.: Prentice-Hall.

Dethier, V. G., 1962. *To Know a Fly.* San Francisco, Calif.: Holden-Day.

——, and E. Stellar, 1964. *Animal Behavior,* 2nd ed. Englewood Cliffs, N.J.: Prentice-Hall.

Edney, E. B., 1957. *The Water Relations of Terrestrial Arthropods.* New York: Cambridge University Press.

Fisher, A. E., 1964. "Chemical stimulation of the brain." *Sci. American,* June.

Fraenkel, G. S., and D. L. Gunn, 1961. *The Orientation of Animals.* New York: Dover Publications.

Greenough, W. T., 1973. *The Nature and Nurture of Behavior: Readings from Scientific American.* San Francisco, Calif.: W. H. Freeman.

Hess, E. H., 1956. "Natural preferences of chicks and ducklings for objects of different colors." *Psychol. Rep.,* 2:447–483.

Jacobson, M., 1965. *Insect Sex Attractants.* New York: J. Wiley & Sons, p. 154.

Kaufman, I. C., and R. A. Hinde, 1961. "Factors influencing distress calling in chicks with special reference to temperature changes and social isolation." *Anim. Behav.,* 9:197–204.

Lenhoff, J. M., 1961. "Activation of the feeding reflex in *Hydra littoralis,*" in *The Biology of Hydra,* H. M. Lenhoff and W. F. Loomis (eds.). Miami, Fla.: University of Miami Press.

Loomis, W. F., 1955. "Glutathione control of specific feeding reactions of *Hydra.*" *Ann. N.Y. Acad. Sci.,* 62: 209–228.

Lorenz, K. Z., 1952. *King Solomon's Ring.* New York: Crowell.

——, 1966. *On Aggression.* New York: Harcourt Brace and Jovanovich.

McConnell, J. V., 1967. *A Manual of Psychological Experimentation on Planarians.* Mental Health Research Institute, University of Michigan, Ann Arbor.

McGough, J. L., N. M. Weinberger, and R. E. Wilson, 1967. *Psychology, The Biological Basis of Behavior: Readings from Scientific American.* San Francisco, Calif.: W. H. Freeman.

Marler, P. R., and W. J. Hamilton, III, 1966. *Mechanisms of Animal Behavior.* New York: J. Wiley & Sons.

Smith, F. E., and E. R. Baylor, 1953. "Color responses in the *Cladocera* and their ecological significance." *Amer. Natural.,* 87:97–101.

Tinbergen, N., 1962. "The curious behavior of the stickleback." *Sci. American,* Dec.

Van der Kloot, W. G., 1968. *Behavior.* New York: Holt, Rinehart and Winston.

Wilson, E. O., 1963. "Pheromones." *Sci. American,* May, pp. 100–114.

——, 1960. "Source and possible nature of the odor trail of fire ants." *Science,* vol. 129, no. 3349, pp. 643–644.

——, 1965. "Chemical communication in the social insects." *Science,* vol. 149, no. 3688, pp. 1064–1071.

——, 1971. *The Insect Societies.* Cambridge, Mass.: Harvard University Press.

CHAPTER 20   ECOLOGY

1. Introduction

Ecology is the study of the myriad interrelationships of organisms and their environment. It is concerned with both *abiotic* factors, the nonliving components of the environment such as light, water, and temperature, and *biotic* factors, the living components, the plants, animals, microbes, and all of the organisms occupying various niches in the environment, each with a specific *role* in the complex of behavior patterns that make up a specific biological *community* (predator, prey, parasite, competitor, etc.). The total amalgam of communities in specific areas (*biotopes*) present a major interacting complex called an *ecosystem*.

Ecology is an interdisciplinary study. The ecologist must utilize a variety of specialized sources of information. Chemists, geologists, meteorologists, engineers, physicists, behaviorists, taxonomists, geneticists, and many other biological and physical specialists contribute data that enable the ecologists to quantify, codify, and interpret the great range of information from the natural world. From them he attempts to determine the general principles underlying the communal interaction of plants, animals, microbes, and their physical surroundings. The pioneer biological study of natural history is

here put into a scientific and extremely relevant context in the discipline of ecology.

A number of methods and techniques are now available that enable us to measure, describe, and analyze the influence of various abiotic environmental factors on plant and animal populations and communities. Some of these techniques will be used in this laboratory to demonstrate a few ecological principles amenable to such tests. However, a *laboratory* approach very much limits our view of ecological relationships, as we are manipulating conditions only in an artificial microenvironment. The most rewarding ecological inquiries invariably must be found in the *field*, the "living laboratory" of the real world. Several field exercises will be suggested to help you ask meaningful questions about animal populations and communities based upon your observations in nature.

As in all new disciplines, one first must become acquainted with the basic linguistic tools—the key terms ecologists use to suggest some of what has already been learned of ecological concepts. The student is urged to browse through the books listed at the end of this chapter, and to select one, such as Komandy's *Concepts of Ecology* or Odum's *Ecology*, for detailed study. You should quickly familiarize yourself with such conceptional terms as *ecosystem, community, habitat, biotope, niche,* and *role,* whose meanings have already been introduced to you. Further reading should soon build up your recognition of the endless ramifications and application of these and related terms.

Ecosystem and community concepts are also relevant to *human* societies which, as we are belatedly beginning to learn, are just as much a part of the natural interdependent world as any of the organisms we mistakenly believe we control—or even "own." In the final sections of this chapter we will turn our attention to human ecology.

An understanding of how human individuals, societies, and cultures interact with one another, with other organisms, and with the physical environment has become critical, political, and economic as well as a scientific and esthetic problem of our time.

■ What effects on our environment are still reversible? Which are irreversible?

■ How do human *numbers* (population) affect the answers and the eventual outcome?

■ What do you consider to be the most important problem facing us today? facing your children tomorrow? and the generation after that?

2. Ecology

I. Biological Succession in a Hay Infusion

Hay, properly treated, yields nutrients that are highly suitable for the development of many protistan populations. Boil 2 gm of hay in a small amount of distilled water, and then add water to bring the total to 500 ml. Pour this into a jar in which you have placed a thin layer (2–3 mm) of rich soil. Cover the jar and let the mixture stand overnight. (For further information on culturing techniques of protists, see the Special Projects section of Chapter 3.) Examine the infusion at regular intervals by taking samples from the surface, bottom, and center. (A cover glass may be floated for a surface sample.)

Record the *types* and *numbers* of protists found in each sample. A protozoology guide, such as *How to Know the Protozoa* by T. and F. F. Jahn, will help you to identify the organisms in your culture. Try to maintain the culture for up to four weeks in order to determine the series of successively dominant forms (*succession*) leading to a stable balance (*climax association*).

■ Which organisms are predominant at first?

■ When is this form succeeded by another? When do predators appear in the culture?

Make a composite graph in which you include the population growth (and decline) of each of the predominant species in your samples. Plot the *number of individuals* of each species to be included on the Y axis, and *time* on the X axis. Use different colors or types of points to separate the curves of the different species.

■ What kind of overlap do you find in your graph? Which major ecological groups or types of organisms tend to succeed one another?

■ At any point in time are there two or more similar species of similar habits with *equally large* numbers of individuals? What does this suggest? (*Competitive exclusion principle* is the title sometimes used to describe this phenomenon.)

■ How does the succession observed in your infusion culture compare with those of other students?

■ How do the results differ if different hay sources and soil types are used?

Prepare several hay infusion cultures, keeping each at a different constant temperature.

■ How does temperature affect the succession of organisms in the sample?

■ What other parameters might you manipulate in this laboratory system to test and determine some of the relationships between organisms and their physical environment?

II. Competition among Carpet Beetles

In their natural environment, four species of carpet or flour beetles are known to compete with one another for food and space. Much of our knowledge of this system comes from the work of Thomas Park and his students at the University of Chicago. We will study aspects of this model of interspecific competition using small jars (½ or 1 pint), with competing species of flour beetles of the genus *Tribolium*. Members of this genus often infest flour sacks and stored grain (Fig. 20.1). The entire environment (or "universe" in a statistical sense) under study consists of a half-pint jar of flour with its population of 50–100 beetles, representing one or two species in a rigorously controlled

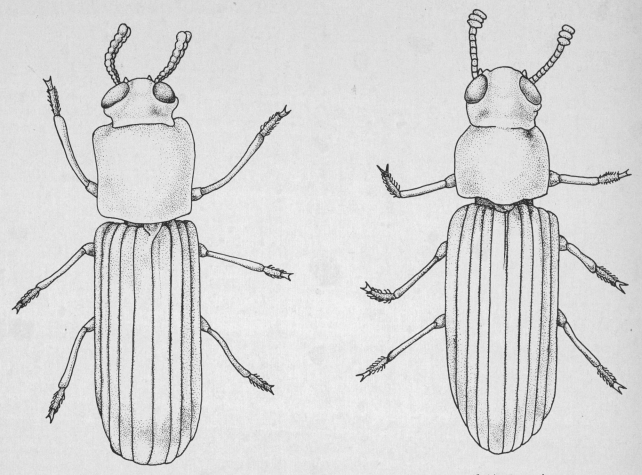

Figure 20.1. Two species of the flour beetle, *Tribolium*: (*right*) *T. castaneum*; (*left*) *T. confusum*.

system. However simple it appears, the study requires several weeks to determine the results. Consequently this project is best set up as a continuing one throughout this course.

Set up the initial culture with a standard number of beetles, such as 100 individuals of one species or of two interacting species. The number should be enough to provide intense interaction but not cause a toxic reaction or some other immediate effect of overpopulation.

Using two competing species, *T. confusum* and *T. castaneum*, in a mixed culture make weekly population counts of the adult beetles by pouring them out into an enamel tray and counting *rapidly*, aided by a small brush to keep them grouped. Run separate cultures of the individual species as controls.

Test the *population curve* of each species when mixed or one-species cultures are provided with (a) one type of food, or (b) two or more different kinds of food (for example, combinations of flour, different whole grains, dog biscuits, dry cereal). (See Pimentel, 1965, in reference section.)

Measure and record temperature and humidity at regular intervals. Record and graph population data and keep careful observations of counts and other observable effects, such as distribution of the beetles in the container, number of larvae and pupae (reproductive rate), number of dead beetles, etc.

■ How do the two species (separately and mixed) appear to allocate the resources available when *one* food type is present? when *two* food types are present?

■ How do the populations of the two species (separately and mixed) fare under the particular environmental conditions you have set up?

■ Can you observe any fighting or competitive behavior between individuals of *different* species (*inter*specific competition)? of the *same* species (*intra*specific competition)?

Vary the population ratios in your system, as suggested below:

Trial 1. Equal numbers of species A and B.
Trial 2. More individuals of species A than B.
Trial 3. More of B than A.
Trial 4. Beetles added after varying periods in small groups until the required starting population of A or B is reached, compared with a control where all of the beetles are placed in the flour at the same time.

■ **Does an initial advantage in numbers confer a competitive advantage to either of the two species?**

■ **What other parameters can you vary to determine survival ability?**

■ **What is the result of overpopulation of A or B separately or A and B together?**

Any number of competition–interaction experiments suggest themselves, and it is possible to develop an entire ecological study of species interactions based on this model (as Park did). Other pairs of species can be used to study the same phenomena. *Drosophila* and *Paramecium* are also excellent models, very easy to use, and with more rapid life cycles and easier reading of results than with *Tribolium*.

III. Intertidal Ecology

The seashore is really a variety of habitats, each with a markedly distinctive, highly localized set of environmental and conditions (Figs. 20.2–20.4). The intertidal region, the area between high and low water marks, is characterized by a number of *zones*, each having a different set of conditions (for example, different temperature variations, direct sunlight exposure, wave action, water level, desiccation, etc.). This zonation in environmental conditions (habitats) is reflected in a corresponding zonation of animal and plant species in each subregion of the intertidal habitat zone.

Before beginning your own study of intertidal zonation, you may want to refer to a natural history study of your local coastal area, such as, Flattely and Walton, 1922; MacGinite and MacGinite, 1949; Ricketts and Calvin, 1952; Smith et al., 1954; Stephenson and Stephenson, 1972; Yonge, 1949 (see reference section).

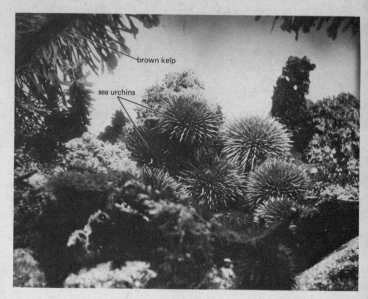

Figure 20.2. A small tidepool community of sea urchins (*Strongylocentrotus purpuratus*). Sea urchins use brown kelp (*upper left*) for food.

We will study zonation by setting up a *horizontal transect* in the intertidal region. First drive a large pole into the substrate at the point of recent high tide. Do the same at the point of low tide. Run a taut string between these two poles. From this line suspend a plumb bob (a string with a heavy weight that will reach ground level) that can slide along the transect line. Starting at one end of the transect, move the plumb bob 25 cm from the pole. Search the substrate immediately below the new plumb bob position and try to *count* and *identify* all the animals immediately in view, without overly disturbing them. *Be sure to return any rocks moved back to their former position.* Move the plumb bob another 25 cm and record the animal population and kinds as before. Continue this until you reach the other pole.

In the same manner record the *plant* species (chiefly algae) that you can identify that occur on the line. Select a *sample* of each kind to remove for study and identification if necessary. It is helpful to divide the class into groups, one concentrating on *animals*, another on *plants*, and a third on *environmental conditions* (physical factors such as water temperature, salinity, kind of substrate, etc.). After each group has made its detailed measurements in their special area of concentration, the groups should meet to relate their findings and combine their data.

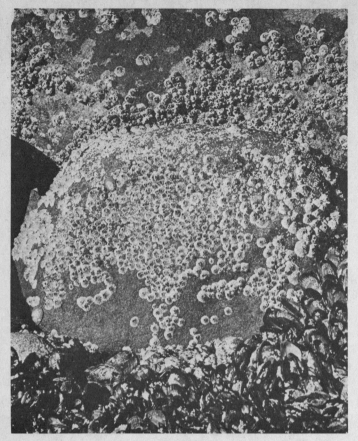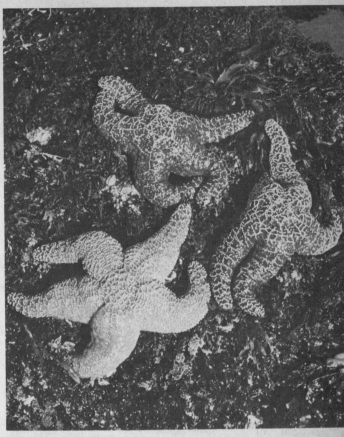

Figure 20.3. The demarcation between zones in sessile animals is often dramatic as this boundary between barnacles and mussels (*left*) clearly shows. The sea stars (*Pisaster ochraceus*) (*right*) prey upon black mussels and are usually found in the same regions.

When these observations and measurements are complete, you will have a list of animal species occurring at each sampled spot along your line. Arrange your data as shown in Figure 20.5, to illustrate at which points along the transect animals of the same *phylum* live (or the same *order* in the larger phyla such as **ARTHROPODA**).

Prepare a similar table for the plants.

■ Which organisms are present throughout the transect?

■ Which organisms are most limited in their distribution?

■ Did you record any fast-moving organisms (crabs, for example) on your line? Discuss several of the problems or limitations in sampling with the "line-transect" method.

■ Are there any organisms found *exclusively* on rock? or sand? or tidepools?

■ Do your data show a *zonal distribution* of organisms? Is this related to the physical factors along the transect line?

■ Does this suggest the presence of *limiting factors* in animal distribution? Compare several such line transects in the area studied by your class.

You can study *succession* in the intertidal zone by using another valuable sampling technique —the *square quadrant* method. Construct a grid by building a square wooden frame (1 m by 1 m in its inner margins). Drill small holes, 10 cm apart, along all sides of the frame. Thread the holes on one side of the frame with nylon string (fish line will work) and extend the string to the opposite side, threading the corresponding holes. The nylon line should be pulled taut when it is tied into place on the frame. Thread the adjacent sides of the frame in the same manner until all paired opposite holes are connected. Your wooden square is now

black m—
barnacle—
brown k—

Figure 20.4. Intertidal vertical zonation. Mussel, barnacle, and kelp zones are clearly distinguishable here.

divided into 100 smaller squares, each 100 cm² in area.

Lay the grid down on an area within the intertidal region. (You may wish to choose several sites within different intertidal zones.)

Carefully record the organisms occurring in each square. Counting the individuals of each species in your square plot may not give an accurate picture of the entire population. (*How would the existence of colonial organisms affect this count?*) But the method can still be very illuminating. Using your grid, estimate the total *area* inhabited by each species. An alternate method would be to count the number of squares in which each species occurs in your grid.

Now remove the grid, marking the four corners of the sampling plot (with stakes) so that

it can be relocated. Completely scrape all living material from the sampled area and return to study the reinvasion and repopulation by nearby species. Make weekly measurements (using your grid) of the area occupied by each of the invading species. Use your weekly measurements to graph the population growth (or decline) of each species (X axis = *area* inhabited by the species; Y axis = *time*).

■ Which organisms are first to invade the sampling plot?

■ In the earliest stages of succession (just after clearing of the area), which organisms increased in the area most rapidly?

■ Discuss the relative importance of *migration* and *reproduction* in this study.

■ Are the successional patterns in popu-

Figure 20.5. Data table showing distribution of animals in selected phyla.

DISTANCE ALONG LINE TRANSECT (in cm)	PORIFERA	CNIDARIA	PLATYHELMINTHES	ASCHELMINTHES	ANNELIDA	ARTHROPODA	MOLLUSCA	ECHINODERMATA	CHORDATA
0									
25									
50									
75									
100									

lation growth for the intertidal region similar to those studied earlier in the hay infusion? Can you draw any generalizations from this?

■ What percentage of the *original* species (those living in the sample plot before clearing) occur in the plot at the end of week one? week two? (and so on).

IV. Small Mammal Sampling

Using the capture–recapture trapping method, you will be able to estimate population densities of several species of small mammals in an area. A book on the local fauna will be helpful in identifying the small mammals you trap in your study (such as Booth, 1961; see reference section).

Divide the class into groups, each group sampling a different area. (There areas should be well apart so that there is no possible overlap in populations.) Groups should make careful observations of the vegetation in the area they are sampling. (Is it chaparral, desert, montane, marsh, or some other plant-community type?) Gather information on the climate (rainfall, temperature, humidity) for each area.

■ How would you determine and measure the microclimate?
■ How does the topography vary from one area to the next?

INSTRUCTIONS FOR EACH GROUP

Establish a square sampling area in which you set traps in a gridlike fashion. (The size of this area will depend on the number of traps available.) Use stakes to designate the four corners of your trapping area. Then set up parallel trapping lines, each 5 m apart. *Sherman live traps* work well in this experiment. Set out a series of traps along each line, 6 m between traps. Draw a map of your trapping area and number each trap site. A convenient method is to give vertical and horizontal components to each trap site as indicated in Figure 20.6.

Record your trap numbers and use this list in reporting the mammals captured.

Several kinds of bait can be used, depending on the animals sought. Peanut butter or oatmeal work well with most rodents; bacon will often attract very different animals. Set the traps in the morning and return to them the following morning at the same time (a full 24 hours later). An exception is desert or areas where nocturnal animals are trapped. These traps *must be emptied by sunrise* or the animals will die of heat shock. Use the *same* trapping method and bait if you are running a sequential study of population changes over time. Service one trapping line at a time. Empty each trap, carefully grabbing the animal inside. (We sug-

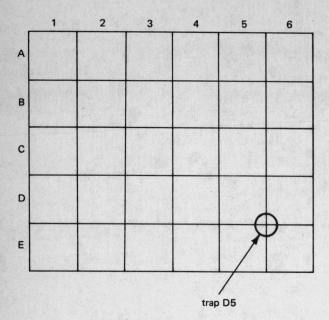

Figure 20.6. Trap map showing numbering system.

gest gloves!) Record trap number, date, species, sex (or degree of maturity if sex is uncertain), and any additional information you feel important to your study. (For instance, you may also wish to make weight or body-length measurements. Weigh the occupied trap and then the empty trap, the difference being the captive's weight.)

In order to identify recaptured animals, use a *marking* system. Toenail clipping is a good method as it has little effect on the animal. (Ear punching or ear tagging introduce the possibility of infection and thus alters the health of the captured animals.) Number the captured animals by clipping the first toenail on the right forelimb of animal 1, the second toenail on animal 2, and so on. (As you run out of toes on one limb, start with the next.)

Having recorded the necessary data and marked the captured individuals, release them. After a week or more, reset the traps, using the same sites, same bait, etc. Service the traps the following morning and (1) identify and record observations on newly captured individuals and (2) record data on *recaptured* marked individuals. Arrange your data to show the number of individuals of each species captured.

■ Which species seems to be *dominant* in this area?
■ In what other ways can you arrange the data?

A formula known as the *Lincoln index* makes it possible to calculate the total population of an area:

$$\text{total population} = \frac{\text{total marked individuals captured}}{\text{recaptured marked individuals}} \times \text{total captured}$$

Use this equation to determine (1) the total population of mammals in the area sampled and (2) the total population of each species.

In your own writeup of this experiment, include a brief discussion of some of the possible *sources of error* and limitations imposed by the method used.

COMBINING DATA FROM INDIVIDUAL GROUPS

Compare the data and results generated by each group.

■ How does the *population density* differ from one area to the next?

$$\text{population density} = \frac{\text{population sample in one area}}{\text{total population sampled}}$$

■ Can you relate differences in population densities to resource availability in the different environments (richness of the habitat, vegetation types, abundance of home sites, etc.)?
■ How can one determine or measure the availability of resources in a habitat? Discuss the meaning of this concept in terms of the animal's specific requirements and the peak populations the habitat can support.
■ In which area did you trap the largest number of species? in which the smallest number? How do you account for the differences?

Diagram several possible *food chains* derived from evidence in the sampled areas.

■ How would one determine the more complex *food web* from this information (or from additional studies you might suggest)?

V. Human Ecology

Of the many habitats on this planet, the one that is probably most familiar to you is the urban or suburban habitat. Here extremely dense enclaves of humans produce altered or even entirely reconstituted ecosystems. (Yet, are they *independently* able to sustain their energy and other life-support needs?) Closely packed structures in urban areas cover enormous areas that formerly were habitats for a vast array of animals and plants now largely displaced or reduced to remnants of their former distribution. Instead we have asphalt, cement, buildings, factories, motor vehicles, and houses. The "green space" is now limited to gardens, lawns, and yards that constrict living conditions so much that only highly adapted and/or very prolific species (animal and plant "weeds") can survive. Instead of large predators, humans must now contend with disease organisms and scavengers, domesticated and food animals, and "domiciliated" wild animals (rats, cockroaches, various ectoparasites, and internal parasites).

Urbanized humans live in structures where temperature is kept relatively constant via external energy sources, and they insulate themselves with a protective "skin" made from animal products, plant fibers, or synthetic mineral and organic products. Rain, wind, and sunlight are blocked or filtered out (though dust and toxic products may not be). Therefore human adaptations to deal with heat and cold have far less survival value than formerly. The acquisition of sweat glands by humans, together with their hair loss as a diurnal mammalian predator, and the function of skin color to regulate vitamin D production (correlated with the amount of sunlight) all have lost their former adaptative value.

Large human populations are no longer able to provide for their physical or biological needs where they reside, so food, water, and wastes must be transported considerable distances—all at the expenses of other areas and external energy sources. Whenever wastes accumulate faster than humans can dispose of them (or are willing or able to pay the price to do so), the environment becomes saturated; then it is seriously eroded. The multiple forms of *pollution* we see increasingly about us is the inevitable result. Unchecked population growth is another form of environmental oversaturation—the most pervasive and destructive of all. It produces crowding that arouses interpersonal pressures, violence, reduction of the quality of life, and compounded pollution problems. Various expressions of this tragic imbalance and threats of ecologic disaster face the future of all the "overdeveloped" nations, where energy and resources are recklessly consumed at an enormous rate. Such nations experience increasing dependence on constantly rising consumption in order to sustain an increasingly costly energy-demanding technology. The predicament of

millions in overcrowded impoverished areas (such as India and Bangladesh) and of other millions in ecologically disturbed lands (such as the drought-ridden Sudan grasslands across Africa south of the Sahara) is all too familiar a problem.

Yet the human population is still doubling at an unprecedented rate, though with considerable variation among different nations and groups, ranging from 1 to 3% *net increase* per year.

■ **What is the doubling rate at 3% per year?**

■ **What is the current *rate* of *increase* in the United States? India? Mexico? U.S.S.R.? worldwide? How does this compare with their respective energy consumptions?**

■ **How could population *growth* continue even if the growth *rate* were zero? (Think of *growth* as an expression of velocity, and *growth rate* in terms of acceleration.)**

All of these factors are part of *human ecology*, to be considered in the following exercises. This study can be conducted individually or in groups. It is designed to give you a biological perspective of your own habitat. Mark off areas in your own neighborhood (two blocks, for example). Observe and sample the physical and biological environments within these areas, following the lists given below of factors to *observe, measure, record,* and *interpret.* Collect as much information as possible and prepare an oral or written report on what you find. Compare data with those from other areas and look for similarities and differences in fauna that can be related to environmental observations.

A series of projects can be performed from those suggested here or you may wish to select one of your own design for classroom use and possibly as a longer term study to pursue on your own. The determination and measurement of environmental decay is a matter of increasing importance and of practical urgency. Measures to counter this reduction in the quality of our lives must be developed, accepted, and undertaken in time if the environmental impact is not, for all practical purposes, to become permanent and our lives, or those who follow us, to be blighted or even rendered impossible.

For some of the data needed you will have to enter homes. *Always* gain permission first, explain your purpose, describe what you are doing, and ask if you may talk to members of the household and inspect or be shown the premises. Try to determine the basic facts of food availability, quality, and amount; sanitary and health conditions; temperature control; clothing adequacy, that is, the meeting of basic physical and biological needs. Include an analysis of these data in your discussions. Then extend your observations and analysis to the yard or garden, streets or walks, and the immediate neighborhood. Include in your evaluation *group needs and facilities* such as schools, parks, recreation, and play areas, group gathering places (for children, youngsters, teen-agers, older, and retired persons).

■ **How do the streets serve these community needs? the schools?**

These elements form part of the *sociocultural* as well as *biological* and *physical* base of a community.

1. **Physical environment**

a. *Location of area.* General and specific; general description and orientation.
b. *Size of area*
c. *Climate.* Data from your own observations and from local meteorological information sources.
   i. Seasonal variation—rainfall, snowfall, dry, windy, hot or humid periods.
   ii. Seasonal and daily average temperatures (day and night).
   iii. Seasonal and daily average humidity (day and night).
d. *Ground cover*
   i. Percent of ground covered by cement, asphalt, or buildings.
   ii. Percent of ground covered by plants (lawns, gardens, etc.).
   iii. Percent of ground uncovered (dirt, empty lots, etc.).
e. *Noises.* Is there a continuous background noise? sounds of special intensity—buses, trucks, industry, aircraft? Do the residents appear bothered or concerned?
f. *Traffic density*
   i. Danger to residents, especially playing children.
   ii. Disturbance to residents—smoke, smog, noise, interference with normal community functioning.
g. *Dwellings*
   i. Average *size* (and range of sizes) of the houses in your study area; number of persons, families, or groups per dwelling.
   ii. *Characteristics* of dwellings—concen-

tration per unit area; type of buildings, presence of yards, gardens, open space, play areas.
   iii. Factors within the *households* observed:
   a. *Temperature*—control; type of heating or cooling.
   b. *Humidity*—level compared with exterternal; control method.
   c. Approximation of possessions per person—clothing, furniture, food.
   d. Approximation of possessions per household (or family).
   e. Major energy-using devices per household—appliances, TV, heating and cooling systems, automobiles.
h. *Other structures*
   i. *Stores*—number, size, type, products.
   ii. *Factories and office buildings*—light or heavy; type of product; effect on neighborhood; pollution (noise, effluents, aerosols); number of employees. (Local or from outside areas?) Their effect on traffic? on the neighborhood characteristics? Are there daytime or 24-hour shifts? (Community effect of the two patterns of employment?)
   iii. *Community structures*—prisons, museums, art galleries, cinemas, dance halls, religious buildings, restaurants, bars, etc. (Distribution? Impact on the local environment? on the wider urban environment? on the biological environment?)

■ **How do each of the above factors relate to economic stratification? community unity, or disunity? (Pride, sense of responsibility, and a sense of neighborliness, or, on the other hand, tension, isolation, distrust, fear, or violence in the neighborhood?)**

2. **Biotic environment** By use of methods learned here (transects, grids) or other appropriate means, count or measure and record the type, number, and distribution of the following groups of organisms you encounter in your test area. Determine also, whether these are *transient* or *permanent, dependent* or *independent* of human activities, and, if possible, their relationship to the human ecological system you are studying.

a. *Microorganisms* (disease, contaminants, etc.). Prepare agar culture media in a series of test petri dishes. (Check with your instructor for media and methods.) Test samples of house dust or sweepings, soil, and other samples for fungi and other organisms. Check with inhabitants about *incidence* and *prevalence* of infections (*definitions of these key epidemiological terms?*) and possible environmentally related diseases (dust, pollution products, such as industrial toxins and airborne contaminants, automobile, and other combustion products).

b. *Plants* (edible, decorative, etc.). Consider space occupied by plants and their importance and contributions to the neighborhood ("green space").

c. *Insect pests* (cockroaches, bed bugs, silverfish, mosquitoes, flies, etc.). Look in household, garden, street.

d. *Other invertebrates* (snails, spiders, sow bugs, centipedes). Where might you look for each of these?

e. *Vertebrates* (domestic, domiciliated). The latter are wild organisms adapted to human abodes, his presence, and his products.

f. *Humans.* Number, age stratification, primary occupations, ethnic composition.

g. *Wastes.* Amount and type of garbage and trash carried away, per day, week; efficiency of sewage disposal. (What happens in a heavy rainstorm when the storm drains overflow and backup? Will this overload the sewage plants?)

Try to work out and describe some of the major *food chains* in your test area.

■ **Why is it not possible to construct a valid food web for the area?**

■ **What is the approximate total food flow in and waste flow out of your area?**

Construct an *energy pyramid* for the total estimates you were able to make of energy sources and consumption.

Would it be possible for *speciation* to occur in your area (for example, insects that specialize in eating synthetic carpets, ants adapted for survival inside houses, household spiders, roaches, etc.)? Discuss the possibilities and limitations.

3. **Pollution** One aspect of human ecology that has become a major public issue and a continuing, in fact, increasing problem is *pollution.* All organisms produce waste prod-

ucts, but in the course of evolution, waste-consuming organisms have usually evolved that recycled these materials. Pollution results when the environment becomes over-saturated with a waste product that no longer can be eliminated, detoxified, or recycled. Unfortunately, the vast increase of human productivity through the development of technology, the population explosion, and intense crowding in urban concentrations (technological and industrial as well as human) has caused a concentration of waste products—solid, liquid, and gaseous—that we have been unable, or unwilling, to control. The result is the too familiar phenomenon of pollution. There are simple ways to detect and measure some kinds of pollution, a few of which are discussed below. The types of pollution you can detect depends upon the location and traffic density in your test area, the kinds and concentration of industry, and the population density. Pick experiments from the following suggestions that you feel are appropriate for your test area.

a. *Water pollution*
      i. *Heat.* Measure water temperature near a factory effluent point. Power-generating stations, including "clean" nuclear plants, are a particularly good place to test for heat pollution. (*Why?*) Measure water temperature in a series of other locations of the stream, river, lake, or ocean. Try to determine the *thermal gradients* and from them construct a *heat map.* (One could also use aerial infrared photographs to get this information.) Prepare several aquaria in the laboratory and heat them to the temperature extremes you found in your heat map. Try to collect examples of local aquatic organisms (copepods, snails, small fish, algae, and water plants). Test their differential survival and numbers after various periods in these aquaria and record your results.

■ **Do you find that one species dies out and is replaced by another? Predict the fauna you would expect to see in water at these temperatures. (If possible, sample some of the areas on your heat map and see if your predictions are correct.)**

■ **How, then, would you expect nuclear plants to affect the composition and quality of the indigenous flora and fauna? Which type of water body (lake, ocean, coastal, river) would be most affected? Why?**

As water is heated, its ability to hold oxygen is decreased, resulting in *deoxygenation*.

■ **What effect might this have on aquatic organisms?**

ii. *Deoxygenation*. There are many ways in which water can become deoxygenated. One of the most significant (ecologically) is the consumption of oxygen by bacteria that decompose organic matter. Clearly, the more organic the waste (detritus), the greater the oxygen loss. Other waste water with detergents containing phosphorus find their way into bodies of water, as these are not necessarily decomposed in water-processing plants (nonbiodegradable detergents). As algae is often limited in its growth rate by the amount of phosphorus (*a limiting factor*), a sudden influx of this element may produce enormous algal blooms, forming algal mats. These then die and sink to the bottom (or wash on shore) where they are decomposed by equally enormous populations of oxygen-using bacteria. The large numbers of oxygen-consuming bacteria deplete the limited aquatic oxygen supply, after which they metabolize *anaerobically*, producing methane and the familiar rotten-egg smell (SO_2 and other products). As noted above, heating the water has a similar oxygen-depleting effect.

Take a sample of fresh water and mud from the water body you have under study. Pour the water into a clean aquarium. Add diluted phosphorous-containing detergent.

■ **Do algal mats form after 1–2 weeks? After the algae sinks what happens?**

Add small goldfish or guppies.

■ **How long can they suvive?**

Set up comparable conditions in another tank, but bubble air through, using this as a control. Another control tank could be set up using bubbled *nitrogen* to remove the oxygen (a negative control test). Many variations of this experiment can be performed.

If you live near a body of water where algal mats have sunk and coat the bottom as a black scum and sludge (often evidenced by decaying algae on the shores), test the water using goldfish, gupplies, or other test organisms in an aquarium sample of the water. If you do not have such access, place some of the dead algae and decaying organic matter in an aquarium and test with goldfish (much as the quality of the air in a deep mine shaft is tested by bringing in a caged canary, which is highly sensitive to the presence of methane, carbon monoxide, or dangerous reduction of the oxygen level).

Organize your results and prepare a report.

iii. *Turbidity*. Take two samples (one of turbid water from a factory effluent and one of clear water) from a local body of water and place in two aquaria. Allow 1 or 2 days for the particles to settle out (if they are too small to do so, they may form a *colloidal suspension* with clay particles). Determine the water volume and amounts of debris and compare them.

■ **What is the concentration of particulates in the turbid water?**

Stir the water in both tanks to again raise the turbidity with the particulate matter and shine a light through each aquarium. Measure the amount of light that passes through the medium with a light meter.

■ **How else might you measure the turbidity?**

■ **How do they compare?**

■ **If the light in one tank is reduced, what effect might this have on photosynthesizing primary producers (algae and water plants)?**

■ **What effect would this loss have on the aquatic food *production*, on the food *chains*, and, ultimately, food *web* (that is, the *ecologic impact*)?**

iv. *Toxicity*. Although most synthetic detergents are disposed of (detoxified) in sewage disposal plants, some of the anionic types (producing anions) do enter natural waters, passing through the sewage processing system. It is known that detergents denature the cytoplasmic membrane and damage the protoplasm of certain microorganisms, and it is thought that they interfere with oxygen uptake and transport in fish.

To determine the toxicity of various household detergents on fish, select two or three different brands of detergent from a local market. Using small goldfish or guppies, place five or six fish in a 20-gallon aerated aquarium.

Run several tests on each detergent, putting different concentrations of the detergent into the aquarium to determine the toxic level (killing effect per unit time for a given proportion of the test animals; for example, LD_{50} (48 hours) equals the lethal dose in 48 hours for 50% of the tested animals. Run each experiment for two days. Check every 4 hours when high concentrations of detergent are used; every 12 hours when low concentrations are used. Examine the fish immediately upon death.

■ What is the condition of the gill surfaces?

■ What does the mucus buildup suggest?

■ Compare detergents advertised as *biodegradable* (meaning?) with nonbiodegradable.

Conduct toxicity tests immediately and also after the detergents have been in solution for periods of several days to a week or more.

■ How do the two types of detergents compare?

■ What is the ecological significance of your results?

b. *Air pollution*

i. Weigh a white filter paper disk and place it in an uncovered petri dish. Place these in sampling areas (in city streets or alleys, in homes, backyard, etc.) and allow them to stand for a period of time (at least a day). Retrieve your test papers and weigh them. Compare the weights of settlings or contaminants.

■ Can you relate these findings with possible environmental factors (nearby factories, high concentrations of automobiles, agricultural dust, prevailing winds, etc.)?

■ Might these findings be related to local differences in levels of asthma, emphysema, respiratory infection, or other pulmonary-related conditions?

■ How would you find out?

■ Does this suggest why it is so difficult to trace and attribute effects on health and well-being to specific environmental factors? (Review the difficulty in "proving" that smoking increases lung cancer, emphysema, circulatory, and health problems, since the evidence is statistical and necessarily indirect.)

ii. Collect rain water in suitable containers from different areas. Test these samples with pH paper. Are there regional differences? Rain falling through the atmosphere absorbs atmospheric carbon dioxide making the rainwater slightly acidic. High concentrations of CO_2 will increase the acidity of the rainwater. "Acid rain" is already having a severe effect on forest growth near industrial areas that is a serious and increasing hazard to their future potential or even survival. This is one of the most ominous and difficult forms of environmental decay to control or prevent without a major reorientation of our way of life. (How would you go about reducing the CO_2 output into the atmosphere?)

■ What are the primary sources of CO_2? What is the percent increment from human activities? Which might be reduced?

If you find regional differences in pH, try to relate these differences to regional CO_2 sources. Some workers consider the rising atmospheric level to be a potential threat to future climatic changes, possibly leading to a new *ice age*.

■ How might this be possible?

iii. *Automobile exhausts*. At this point you should be able to design a set of experiments to test or measure environmental effects or pursue your own hypothesis subject to such tests. Pose some question concerning automobile exhausts (for instance, What are some of the major biological effects? Can those be correlated with automobile concentrations? What are the important constituents of auto exhausts? How do these components separately act on test plants or other organisms?). Design an experiment to answer one of these questions subject to your ability and means to carry it out. Undertake the experiment and prepare a report with your hypothesis, data, discussion, and conclusions. Include an analysis of sources of error and limitations of your approach.

■ What other experiments can you suggest that are needed to determine the type and amount of auto exhaust and its effect on human health and that of the entire biological community?

■ Does your local government have regulations limiting auto emissions? Are they enforced?

■ If all automobiles in your area met the current standards set by the local government, what difference might this make on the composition of the air? What effect would such compliance have on human health? plant growth? the entire ecological system, local and regional?

c. *Solid waste pollution*
    i. Garbage collection method and fre-

quency? Litter left on streets or alleys? How is the garbage ultimately disposed of? long-range effect? any of it recycled? Analyze and criticize the local garbage disposal system.

    ii. Sewage disposal system—make a similar evaluation and critique.

Suggested References

Allee, W. C., A. E. Emerson, O. Park, T. Park, and K. P. Schmidt, 1949. *Principles of Animal Ecology*. Philadelphia, Pa.: W. B. Saunders.

Bates, M., 1960. *The Forest and the Sea*. New York: Random House.

Bernarde, M. A., 1970. *Our Precarious Habitat, An Integrated Approach to Understanding Man's Effect on His Environment*. New York: W. H. Norton.

Bonghey, A., 1968. *Ecology of Populations*. London: Macmillan.

Booth, E. G., 1961. *How to Know the Mammals*, 2nd ed. Dubuque, Iowa: William C. Brown.

Brown, M. E. (ed.), 1957. *Physiology of Fishes*, vol. 2. New York: Academic Press.

Carson, R., 1962. *Silent Spring*. Boston, Mass.: Houghton Mifflin.

——, 1964. *The Sea Around Us*. New York: Oxford University Press.

Coker, R. E., 1954. *Lakes, Streams, and Ponds*. Chapel Hill, N.C.: University of North Carolina Press.

Cox, G. W. (ed.), 1969. *Readings in Conservation Ecology*. New York: Appleton-Century-Crofts.

Deevey, E. S., Jr., 1958. "Life in the depths of a pond." *Sci. American*, 199: Aug.

Detwyler, T. R., 1971. *Man's Impact on the Environment*. New York: McGraw-Hill.

Ehrlich, J., and A. Ehrlich, 1970. *Population Resources Environment: Issues in Human Ecology*, 2nd ed. San Francisco, Calif.: W. H. Freeman.

——, 1973. *Human Ecology*. San Francisco, Calif.: W. H. Freeman.

Flattely, F. W., and C. L. Walton, 1922. *The Biology of the Seashore*. London: Macmillan.

Griswold, G. H., 1941. "Studies on the biology of our common carpet bettles." Cornell Univ. Agr. Exp. Sta. Mem. 240.

Hardon, G. (ed.), 1969. *Population, Evolution, and Birth Control*. San Francisco, Calif.: W. H. Freeman.

Henderson, C., Q. H. Pickering, and J. M. Cohen, 1959. "The toxicity of synthetic detergents and soaps to fish." *Sew. and Ind. Waste*, 31:295–306.

Hutner, S. N., and J. J. A. McLaughlin, 1958. "Poisonous tides." *Sci. American*, 199: Aug.

Jaeger, E. C., 1957. *The North American Deserts*. Palo Alto, Calif.: Stanford University Press.

Kennedy, D., 1974. *Cellular and Organismal Biology. Readings from Scientific American*. San Francisco, Calif.: W. H. Freeman.

Kormandy, E., 1969. *Concepts of Ecology*. Englewood Cliffs, N.J.: Prentice-Hall.

Leopold, A., 1949. *A Sand County Almanac*. New York: Oxford University Press.

MacGinite, G. E., and N. MacGinite, 1949. *Natural History of Marine Animals*. New York: McGraw-Hill.

Meadows, D. H., D. L. Meadows, J. Randers, and W. W. Beherens, III, 1972. *The Limits to Growth*. New York: Universe Books.

Moen, A. N., 1973. *Wildlife Ecology*. San Francisco, Calif.: W. H. Freeman.

Odum, E. P., 1963. *Ecology*. New York: Holt, Rinehart and Winston.

Peterson, A., 1953. *A Manual of Entomological Techniques*, 7th ed. Ann Arbor, Mich.: J. W. Edwards.

Pimentel, D., 1965. "Competition between carpet beetles," in *Research Problems in Biology*. Ser. 4, BSCS, pp. 142–143.

Richard, P. W., 1952. *The Tropical Rain Forest*. New York: Cambridge University Press.

Ricketts, E. F., and J. Calvin; rev. by J. Hedgepeth, 1952. *Between Pacific Tides*, 3rd ed. Palo Alto, Calif.: Stanford University Press.

Sigler, W. F., 1965. "The toxicity of synthetic detergents on fish, in *Research Problems in Biology*. Ser. 4, BSCS, pp. 178–181.

Slobokin, L. B., 1961. *Growth and Regulation of Animal Populations*. New York: Holt, Rinehart and Winston.

Smith, R. I., F. A. Pitelka, D. P. Abbott, and F. M. Weesner, 1954. *Intertidal Invertebrates of the Central California Coast—S.F. Light's "Laboratory and Field Test in Invertebrate Zoology."* Berkeley, Calif.: University of California Press.

Stephenson, T. A., and A. Stephenson, 1972. *Life Between Tidemarks on Rocky Shores*. San Francisco, Calif.: W. H. Freeman.

Van Dyne, G. (ed.), 1969. *The Ecosystem Concept in Natural Resource Management*. New York: Academic Press.

Wagner, R. H., 1971. *Environmental and Man*. New York: W. W. Norton.

Wilson, E. O. (ed.), 1974. *Ecology, Evolution, and Population Biology. Readings from Scientific American*. San Francisco, Calif.: W. H. Freeman.

Yonge, C. M., 1949. *The Sea Shore*. London: Collins.

APPENDIX I

MEASUREMENTS AND FORMULAS

I. Student Equipment

It is recommended that each student provide the following materials unless otherwise directed by the instructor:

Laboratory notebook (quadrille)
Drawing paper
Drawing pencil (4H)
Sandpaper pad for sharpening pencil
Eraser, soft rubber
Ruler, celluloid, millimeters–inches (6 in long)
Set of dissecting instruments:
    Scalpel

Scissors, straight, medium point
Scissors, heavy
Forceps, straight, milled tips (10–15 cm long)
Forceps, curved, milled tips (10–15 cm)
Dissecting needles, in handles (10–15 cm)
Probe
Medicine droppers (pipette)
Instrument case
Safety razor blades
Microscope slides, glass
Microscope coverglasses, no. 1 (22 mm), square or circular
Package of ordinary pins

II. Reference Tables

MULTIPLE OR FRACTION	NUMERICAL VALUE	PREFIX	SYMBOL	SCIENTIFIC NOTATION
one million	1,000,000	mega-	M	10^6
one thousand	1000	kilo-	k	10^3
one hundredth	0.01	centi-	c	10^{-2}
one thousandth	0.001	milli-	m	10^{-3}
one millionth	0.000001	micro-	μ(mu)	10^{-6}
one billionth	0.000000001	nano-	n	10^{-9}

In the following tables equivalent values are all shown on one horizontal line (for example, $0.305 \text{ m} = 304.8 \text{ mm} = 12 \text{ in} = 1 \text{ ft} = 0.333 \text{ yd}$). Standard abbreviations are shown in parentheses.

Linear Measure

METRIC						UNITED STATES			
Kilo-meters (km)	Meters (m)	Milli-meters (mm)	Micro-meters (μm)	Nano-meters (nm)	Angstroms (Å)	Inches (in)	Feet (ft)	Yards (yd)	Miles (mi)
1.0	1000.0						3280.8	1093.6	0.621
	1.0	1000.0				39.37	3.281	1.093	
	0.0001	1.0	1000.0			0.039			
			1.0	1000.0					
				1.0	10				
	0.025	25.4				1.0	0.082		
	0.305	304.8				12.0	1.0	0.333	
	0.914	914.4				36.0	3.0	1.0	
1.61	1609.3						5280.0	1760.0	1.0

Fluid Measure

METRIC		U.S. STANDARD			
Liters (l)	Milliliters or Cubic Centimeters (ml or cm³)	Fluid Drams (fl dr)	Fluid Ounces (fl oz)	Quarts (qt)	Gallons (gal)
1.0	1000.0	270.5	33.8	1.056	0.264
0.001	1.0	0.271	0.034		
0.004	3.69	1.0 (60 minims)	0.125		0.001
0.029	29.6	8.0	1.0	0.0625	0.008
0.946	946.3	256.0	32.0	1.0 (2 pt)	0.25
3.78	2785.3	1024.0	128.0	4.0	1.0 (231 in³)

Weight Measure

METRIC			AVOIRDUPOIS		
Kilograms (kg)	Grams (g)	Milligrams (mg)	Grains (gr)	Ounces (oz)	Pounds (lb)
1.0	1000.0		15,432.0	35.27	2.204
0.001	1.0	1000.0	15.43	0.35	
	0.001	1.0	0.015		
	0.065	65.0	1.0		
0.028	28.35		437.5	1[a]	0.062
0.453	453.6		7000.0	16.0	1.0

[a] Apothecary or troy ounce = 31.103 grams = 480 grains.

Time Measure

FUNDAMENTAL UNIT	SECOND	
0.001 sec $= 10^{-3}$ sec	millisecond	(msec)
0.000001 sec $= 10^{-6}$ sec	microsecond	(μsec)
0.000000001 sec $= 10^{-9}$ sec	nanosecond	(nsec)

Temperature

TEMPERATURE CONVERSION. To convert Celsius (centigrade) (°C) to Fahrenheit (°F), multiply °C by 1.8 and add 32; hence, for 10°C, $10 \times 1.8 + 32 = 50°F$. To convert °F to °C, substract 32 and multiply by 0.55.

III. Formulas[1]

Formulas are reduced to a final volume of 100 ml. All weights are given in grams and volumes in milliliters (ml = cm³).

1. *Acetic acid solution* (fixative)

Glacial acetic acid	10	ml
Distilled water	90	ml

2. *Acetocarmine solution* (stain)

Glacial acetic acid	45	ml
Distilled water	55	ml

Bring water to boil, adding powdered carmine until no more will dissolve; filter; use cold. Should be prepared under hood to avoid inhaling fumes.

3. *Acid fuchsin solution* (stain)

Acid fuchsin	0.5	g
Distilled water	100	ml

4. *Alcohol* (fixative)

Laboratory ethyl alcohol or ethanol (C_2H_5OH) is a 95 percent solution in water. Stock alcohol contains about 5 percent water. *Approximate* dilutions can be made by treating the alcohol as though it were 100 percent (absolute) alcohol and using as many milliliters of 95 percent alcohol as the final percentage strength of the solution requires and adding a sufficient volume of distilled water to make a total of 100 ml. For example, to prepare about a 45 percent solution, use 45 ml of (95 percent) alcohol and 55 ml of water, a suitable approximation for most purposes (actually 42.7 percent).

5. *Benedict's solution* (test for sugar)

Solution A:

Sodium citrate $Na_3C_6H_5O_7$, $11H_2O$	17.3	g
Sodium carbonate, $NaCO_3$, anhydrous	10.0	g
Distilled water, to make	85.0	ml

Chemicals should be dissolved in 50 ml of water heated to approximately 60°C; then cool; filter; and make up to volume (85 ml).

Solution B:

Copper sulfate, $CuSO_4$, $5H_2O$	1.73	g
Distilled water, to make	15.00	ml

Add A to B slowly, with constant stirring. This solution keeps well. It yields a green, yellow, or red precipitate (cuprous oxide) when heated with solutions containing reduced sugars. The color depends on the amount of sugar present.

6. *Bouin's fixative*

Picric acid, saturated aqueous solution (about 1 g will dissolve)	71.5	ml
Formalin (40 percent formaldehyde)[2]	38.0	ml
Glacial acetic acid	4.7	ml

After fixation, wash in ethyl alcohol, 45 percent or stronger, until the yellow color disappears.

7. *Congo red stain* (fat stain)

Congo red	0.5	g
Distilled water	100.0	ml

---

[1] In solutions where a precipitate forms, pour the solution through a piece of filter paper to remove the precipitate and impurities. In each case, precipitate or not, freshly made solution should be poured through filter paper to remove impurities.

[2] Commercial formalin is a saturated solution of formaldehyde gas in water, for example, 40 percent formaldehyde. *But this equals 100 percent formalin for the purpose of computing formalin dilutions.*

8. *Copper sulfate solution*

Copper sulfate, $CuSO_4$	5.0	g
Distilled water	100.0	ml

9. *Embalming fluid for mammals*

Formalin[2]	1.5	ml
Carbolic acid, melted crystals	2.5	g
Glycerin	10.0	ml
Water	86.0	ml

10. *Fehling's solution* (test for sugar)

Solution A:

Copper sulfate, $CuSO_4$	7.0	g
Distilled water	100.0	ml

Solution B:

Sodium hydroxide, NaOH ...	52.0	g
Potassium sodium tartrate, $KNaC_4H_4O_6$	34.6	g
Distilled water	100.0	ml

Shortly before using, mix equal volumes of solutions A and B. Heating this with an equal volume of a solution containing a reducing sugar yields a brick-red precipitate (cuprous oxide).

11. *10 percent formalin solution* (for preserving material)

Formalin[2]	10	ml
Water	90	ml

Use for fixing entire specimens. Solution will decalcify invertebrates with calcium carbonate skeletons, crustaceans, and skeletons of small vertebrates. Depending on size, specimens are fixed from a few days to several weeks in formalin. Be sure to have at least 10 times as much fluid as bulk of specimen; then wash out in running tap water for at least 1 day. Preserve specimen in 70 percent alcohol.

12. *Glassware cleaning fluid*

Potassium bichromate, $K_2Cr_2O_7$, technical grade	11	g
Water	50	ml
Sulfuric acid, H_2SO_4	50	ml

Heat water to dissolve biochromate, cool, then add acid cautiously.

13. *Glycerine-carmine* (fixing, cleaning, and staining nematodes)

Carmine	1	g
Ammonia, NH_4OH	1	ml
Sodium chloride, NaCl	1	g
Glycerin	50	ml
Distilled water	49	ml

Dissolve carmine in ammonia by adding a little water. Dissolve NaCl in glycerine, mix in above, and add water.

14. *India ink* (for *Paramecium* and other ciliated organisms or tissues)

Rub solid ink stick (used for photographic retouching) in a small volume of water to obtain a black suspension. Fluid India inks contain chemicals toxic to protists and therefore should not be used.

15. *Iodine-KI solution* (for starch test)

Potassium iodide, KI	0.7	g
Iodine, I_2, crystals	1	g
Distilled water	100	ml

16. *Iodine solution* (for staining flagella of spermatozoa, *Euglena*, and other organisms or cells)

Prepare a strong solution of iodine crystals in 50 percent ethyl alcohol. Each case must be tested separately, varying the strength as necessary.

17. *Locke's solution* (physiological salt solution for mammalian tissues)

Sodium chloride, NaCl	0.9	g
Potassium chloride, KCl	0.042	g
Calcium chloride, CaCl	0.025	g
Sodium bicarbonate, $NaHCO_3$	0.02	g
Distilled water	100	ml

Add 0.1 to 0.25 g glucose if tissue is to be kept for extended periods.

18. *Lugol's solution* (test for starch)

Potassium iodide, KI	0.02	g
Distilled water	10	ml
Iodine, I_2, crystals	1	g
Ethyl alcohol, 95 percent	90	ml

19. *Methyl cellulose solution* (for slowing down protists)

Methyl cellulose	1	g
Distilled water	90–100	ml

To 50 ml of boiling water, add methyl cellulose and allow it to soak and stand for 45 min; add remaining water and stir until smooth. For use, a ring of the solution on microslide should be made. Place a drop of culture in center, add cover slip. As protists swim outward, they are slowed by the increasing viscosity.

20. *Methyl green*   (counterstain)

Methyl green	5	g
Distilled water	100	ml
Glacial acetic acid	5	drops

21. *Methyl violet*   (indicator stain)

Methyl violet	0.05	g
Glacial acetic acid	0.2	ml
Distilled water	100	ml

22. *Methylene blue solution*   (indicator stain)

Methylene blue	5	g
Distilled water	100	ml

23. *Narcotizing agents*

No one agent can be used for all organisms. It is desirable to have a variety of solutions on hand. Prior to preservation one should narcotize the animal first in order to preserve structures fully expanded. Maintain organisms in minimal volume of water before adding any of the agents below.

a. *Carbon dioxide*. Either add charged water to the container with the organism, or bubble CO_2 gas directly into container.

b. *Chloral hydrate*. Concentrations from 3 to 10 percent are used, added drop by drop until organism is fully expanded.

c. *Eucaine*. A solution of B-eucaine hydrochlorate (1 g), alcohol 90 percent (10 cm^3), and distilled water (10 cm^3) is prepared and added drop by drop to cultures of microorganisms.

d. *Magnesium sulfate* (relaxing agent for marine organism). Prepare a saturated solution. Add drop by drop until organism is fully relaxed.

e. *Menthol crystals* (relaxing agent for mollusks). Add a few crystals to the water containing the organisms.

f. *Methanol*. Dilute 5–10 percent, add drop by drop to solution containing organisms.

24. *Nickel sulfate solution*   (1 percent)

Nickel sulfate	1	g
Distilled water	99	ml

This solution inactivates flagella and cilia.

25. *Potassium chloride solution*   (0.1 percent)

Potassium chloride	0.1	g
Distilled water	99.9	ml

Interferes with ciliary and muscular movements.

26. *Picroacetic acid*   (fixative and stain)

Glacial acetic acid	1	g
Distilled water	99	ml

To above mixture add an excess of crystalline picric acid in order to obtain a saturated solution.

27. *Picrocarmine*   (stain)

Ammonium hydroxide, NH_4OH	5	ml
Distilled water	50	ml
Carmine	1	g
Picric acid, saturated acqueous solution	50	ml

28. *Ringer's solution*   (isotonic for frog tissues)

Sodium chloride, NaCl	0.65	g
Potassium chloride, KCl	0.014	g
Calcium chloride, $CaCl_2$, anhydrous	0.012	g
Sodium bicarbonate, $NaHCO_3$	0.02	g
Distilled water, to make	100	ml

29. *Ringer's modified solution*   (for amphibian tissues, Holtfreter's solution)

Sodium chloride NaCl	0.35	g
Potassium chloride, KCl	0.005	g
Calcium chloride, CaCl	0.01	g
Sodium bicarbonate, $NaHCO_3$	0.02	g
Distilled water, to make	100	ml

30. *Safranin stain*

Safranin	1	g
Anilin water	90	ml
Ethyl alcohol, 95 percent	10	ml

Anilin water can be prepared by shaking 4 ml of anilin in 90 ml of distilled water and then filtering.

31. *Saline solution*   (isotonic for frog tissues)

Sodium chloride, NaCl	0.7	g
Distilled water	100	ml

32. *Saline solution*   (isotonic for mammalian tissues)

Sodium chloride, NaCl	0.9	g
Distilled water	100	ml

33. *Sudan IV (Scharlach R) solution*   (stain specific for fats)

Sudan IV	0.1	g
Ethyl alcohol, 95 percent	50	ml
Acetone	50	ml

APPENDIX II

SCALE
OF
SIZES

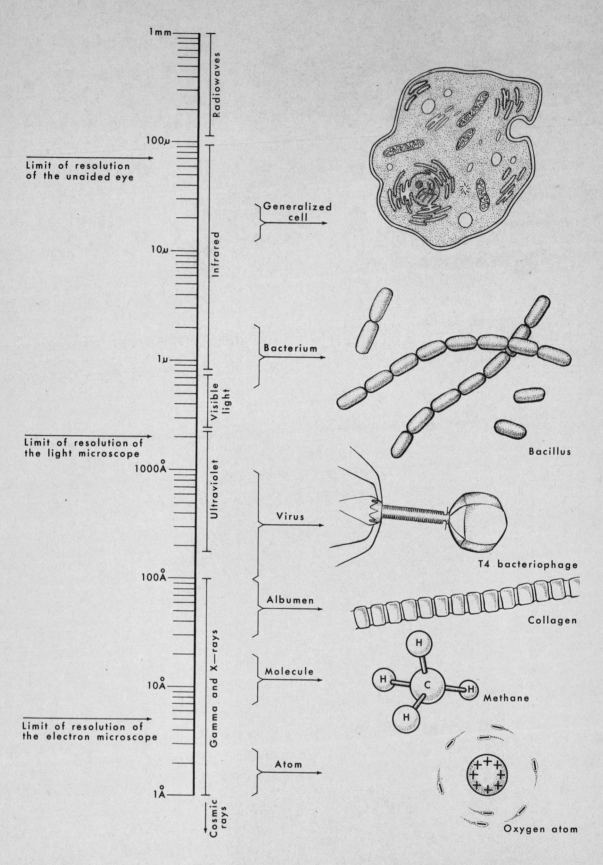

1mm

Radiowaves

100μ

Limit of resolution
of the unaided eye

Generalized
cell

Infrared

10μ

1μ

Bacterium

Visible
light

Limit of resolution of
the light microscope

1000Å

Ultraviolet

Bacillus

Virus

T4 bacteriophage

100Å

Albumen

Collagen

Gamma and X—rays

Molecule

10Å

Methane

Limit of resolution of
the electron microscope

Atom

1Å

Oxygen atom

Cosmic
rays

Modified after Stadhouders, A. M., 1967. De micro-structuur de cel (2) prepareer-technieken in de
elektronenmicroscopie. Uit *Natuur en Techniek*, 35e jrg. nr. 2, febr.

APPENDIX III

SIMPLE STATISTICAL TREATMENT FOR DATA IN CHAPTER 18

In Chapter 18 you mated a homozygous dominant with a homozygous recessive *Drosophila melanogaster* and observed the frequencies of homozygous recessives after several generations, (1) when the population was left alone throughout the experiment (except for removal of the dead), and (2) when homozygous recessives were selected against (removed) and each successive generation was separated from its parental generation.

In both cases, we wish to see if the resulting allele frequencies were significantly different from those predicted by the Hardy–Weinberg law. For both cases we can do this by generating the expected numbers of phenotypic recessives (necessarily homozygous) and the expected numbers of phenotypic dominants (all the rest, including homozygous dominants and heterozygotes) based on the Hardy–Weinberg law, and comparing these expected frequencies with the observed frequencies in the last generation.

To accomplish this there is a simple statistical test, the *chi square* (χ^2) *test of significance*. The χ^2 value obtained from the data is a measure of the variation of the observed from the expected values. The *critical values* of χ^2 are obtained from statistical tables and represent, in the present case, the χ^2 values above which it is 70% ($\chi^2_{0.30}$) and 95% ($\chi^2_{0.05}$) certain that the observed and expected

values are, in fact, different. This takes into account the differences that might be expected from random variation.

For the present experiment, the χ^2 critical values are

$$\chi^2_{0.30} = 1.07$$
$$\chi^2_{0.05} = 3.84$$

Any obtained χ^2 value larger than the critical value is statistically significant at the critical value given. Note that $\chi^2_{0.05} > \chi^2_{0.30}$. This is reasonable since to be 95% sure requires a greater variation from an expected value than to be only 70% sure.

The formula for computing the observed χ^2 for the data in this experiment is

$$\chi^2 = \Sigma \frac{(O-E)^2}{E}$$

where Σ indicates "the sum of," O represents the observed frequency, and E represents the expected frequency.

Here we are concerned with two groups, the phenotypic recessives and the phenotypic dominants. Thus, the obtained χ^2 value will be the sum of two values.

You already have the observed frequencies. What are the expected frequencies?

The Hardy–Weinberg law says that the

proportion of homozygous recessives is equal to q^2. Since in both experiments $p = q = 0.5$, $q^2 = 0.25$. Therefore, we would expect the number of homozygous recessives to be 0.25 N, where N is the total number of individuals in the last generation (or in all generations in the case of experiment A). Similarly, the number of phenotypic dominants ($p^2 + 2pq$) should be $1.00 - 0.25 = 0.75$, and the expected number is then 0.75 N.

For example, assume there are 100 individuals in the last generation. Then

$$\frac{E}{\text{recessives}} = 0.25 \times 100 = 25,$$

and

$$\frac{E}{\text{dominants}} = 0.75 \times 100 = 75$$

If we *observed* 40 phenotypic recessives and 60 phenotypic dominants, then

$$\chi^2 = \frac{(40-25)^2}{25} + \frac{(60-75)^2}{75} = 12$$

Since $12 > 3.84$, we can conclude that the obtained values are further from their expected values than would be expected by chance 95% of the time, or, in other words, there is a statistically significant difference.

Now try this simple statistical treatment on your own data.

APPENDIX IV   THE KYMOGRAPH

In the macroscopic study of the physiology of contractile tissue, one frequently uses the *kymograph* (Gr. *kyma*, wave, + *graph*). This is an instrument consisting of a rotating drum upon which is applied a piece of smoked paper. A *stylus* (pointed object) is brought into essentially frictionless contact with the paper, and as the drum rotates the stylus scrapes off some of the smoke, yielding a permanent record (in the form of a line or wave).

To prepare kymograph paper, first cut a piece of paper so that it will wrap around the drum with about 2–3 cm overlap. Hold the paper several centimeters above a kerosene lamp flame (high enough so that it does not catch fire), and move the paper over the flame to get an even coat of smoke. After the smoking process, try not to touch the smoked side.

Now place small pieces of tape on one end of the paper and carefully tape it to the drum. Wrap it around the drum and secure it with tape at the top and bottom but not in the middle of the paper.

Take the contractile tissue to be studied and secure it to a fixed point beside the drum. Attach the other end of the tissue at about midstylus (one end of the stylus should already be attached to the strand which contains the tissue suspensor). The point of the stylus should just touch the kymograph paper, approaching it at about an 80° angle. Start at the bottom of the paper and give the drum one complete rotation to be sure that the stylus scrapes the smoke off evenly all around the drum.

When contractile tissue such as muscle is attached to an immovable point and to the stylus, then when the tissue contracts, the stylus is moved (deflected). Rather accurate representations of contractile movements are possible with the kymograph; its speed of rotation can be set, allowing for a record of the rapidity of response, the number of responses possible until exhaustion, the varying magnitude of response, various latencies, and other parameters. For each successive measurement, reset the drum's position so that a fresh area is exposed. Label each tracing by using the head of a pin to scratch on the paper.

INDEX

cauda epididymis, 253
caudal artery, 250, 265
caudal fins, 211, 213, 216
Caudata, 24
cavernosum urethra, 254
cecum, 89, 181, 259
cell membrane, 285, 293
cell size, 301
cells, 272–277, 278; defined, 271–272; division, 283–292; meiosis, 284, 289–292; mitosis, 283, 284–289
cellular immunity, 249
cement body, 182
centipedes, 21
central nervous system, 88–89, 218, 276
central reservoir, 45
centriole, 291
Centroderes, 20
centromere, 285
centrosome, 285
cephalic arteries, 149
cephalic cartilages, 184
cephalic ganglia, 184
cephalic vein, 183
Cephalobus, 105
Cephalochordata, 18, 23, 208, 210–212
Cephalodiscus, 23
Cephalopoda, 23, 164, 178–186
cephalothorax, 132, 143
Ceratium, **30**, 35
Ceratomyxa, 36
cercariae, 89
cerebellum, 235, 255, 267
cerebral cortex, 267
cerebral ganglia, 115, 152, 159, 167, 177, 186
cerebral hemispheres, 235, 268
cerebropedal ganglia, 177
cerebropleural ganglia, 173, 177
cerebrospinal fluid, 269
cerebrum, 235, 255
Cerianthus, 19
cervical groove, 143
cervix, 262
Cestoda, 20, 86, 89, 90, 94–97
Cestodaria, 20
Cestum, 19
Cetacea, 25, 242
Chaetognatha, 18, 23, 208, 297
Chaetogordius, 20
Chaetonotus, 20
Chaetopleura, 165, **166**
Chaetopterus, 20
chalaza, 304
Chelicerata, 22, 130, 132–133
Chelifer, 22
chelipeds, 133, 135, 143
Chelonia, 24
chemical communication, 324
chemoreception, 324
chick embryo, 302–307, **303**, **304**, **305**, **306**
chickens, 24
chicks, day-old, 320–321
Chilomonas, 34, 52
Chilopoda, 21
Chimaera, 23, 209
Chiroptera, 25, 242
chitons, 22, 163, 165–167
Chlamydomonas, 35, 41
Chlonorchis, 90
chloragogue cells, 114
Chlorohydra, 69
Chloromonadida, 35
Choanichthyes, 23, 209, 219
choanocytes, 59, 60
Choanoflagellida, 35
Chondrichthyes, 23, 209, 215–219
chondrocranium, 218
Chondrophora, 19, 298
chordamesoderm, 302
Chordata, 15, 18, 23–26, 207–222, 297, 300–302; *See also* Amphibia; Mammalia
chorio-allantoic structure, 306
chorion, 305
choroid coat, 233
choroid plexus, 235, 255
chromatids, 285, 291
chromatophores, 181
chromosomal replication, 285
chromosomal theory of inheritance, 309–310
chromosomes, 312
chrondrocytes, 278
Chrysomonadida, 34
cicadas, 153
Cichlidae, 219
cilia, 89, 191
Ciliatea, 19
ciliated epithelium, 280, **280**
Ciliophora, 19, 36–37
Cimex, 22
Ciona, 23
circulatory system: Amphibia, 228–231; Annelida, 113; Arthropoda, 148, 149–150; Chordata, 217; Echinodermata, 196; Mammalia, 248–251, 265–267; Mollusca, 171, 178, 182–184
circumesophageal connectives, 152, 159
circumferential lamellae, 278
circumoral ring, 190
circumpharyngeal connective nerves, 115
Cirripedia, 21, 130
cirrus, 93, 95, 122, 200, 210–211
Cladocera, 20
clam shrimps, 20
clams, 23, 163–164
clamworms, 109, 119–122
clasper, 216
class, 15
classification, 2, 15–27
Clathrulina, 36
cleavage, 294, 301
Climatius, 23
climax association, 328
Cliona, 19
Clione, 23
clitellum, 110
clitoris, 254, 262
cloaca, 103, 226
cloacal pit, 213
Clonorchis sinensis, 93–94, **93**
clypeus, 154
Cnidaria, 17, 19, 67–86, 298
cnidoblasts, 68
CNS-directed motility, 88
coat-of-mail shells, 22
Coccidia, 19, 36
cockroach, 21, 130, 133, 152, 153–159, **154**, 324–325
Codosiga, 35
Coelenterata, 17, 19, 67–86
coeliac artery, 230, 250
coeliac axis, 265
coeliacomesenteric artery, 230
coelom, 88, 170, 199, 220, 226
coelomic pouches, 296–297
coelomic spaces, 212
Coelomata, 99, 196
coevolution, 131
Coleoptera, 22, 153
Colias, 22
collecting tubules, 228, 252
Collembola, 21
colon, 247, 259
Colpidium, 52
Columba, 24
Columbiformes, 24
columella, 174, 234
columella muscle, 174
columnar epithelium, 272, **273**
comb jellies, 19
commisures, 159, 255, 268
community, biological, 327–328
competition, 328–330
compound eyes, 150
compound microscope, 5, 6, **6**
concentric lamellae, 278
conch jellies, 68
concha, 243
Conchostraca, 20
conditioning, 319
conjugation, 48, **48**, 49
conjunctivae, 154, 232
connective tissue, 274–275, **274**
control group, 322
conus arteriorsus, 229
Convoluta, 19
Copepoda, 21, 130
corals, 19
cornaria, 298
cornea, 150, 181, 186, 232
Cornuspira, 36
coronary arteries, 250, 266
coronary sinus, 266
corpora cavernosa, 254
corpora quadrigemina, 255
corpus callosum, 255, 268
corpus epididymis, 253
Corrodentia, 22
Corynactis, 19
Corynorhinus, 25
Cotylosaurs, 24
Cowper's glands, 260
coxa, 155
coxopod, 145
crab spiders, 132
crabs, 21
cranial nerves, 219, 235, 254–255, 268, 276
Craniata, 23–26, 208
cranium, 267
Cranson, 21
Crassostrea, 167, 168
crayfishes, 21, 130, 133, 135, 143–152, **144**, **145**, **148**, **151**
cremaster muscle, 253
crickets, 21
Crinoidea, 23, 190–191, 200–201